MICROPHONES

3RD EDITION

MARTIN CLIFFORD

My thanks to the Coral Springs Branch of the Broward County Library System for supplying me with their resources and the support of the staff.

THIRD EDITION
SECOND PRINTING

Printed in the United States of America

Library of Congress Cataloging in Publication Data

Clifford, Martin, 1910-
Microphones.

Includes index.
1. Microphone. I. Title.
TK5986.C58 1986 621'.38'028'2 86-5849
ISBN 0-8306-0475-8
ISBN 0-8306-2675-1 (pbk.)

TAB BOOKS Inc. offers software for sale. For information and a catalog, please contact TAB Software Department, Blue Ridge Summit, PA 17294-0850.

Questions regarding the content of this book should be addressed to:

Reader Inquiry Branch
TAB BOOKS Inc.
Blue Ridge Summit, PA 17294-0214

Cover photograph courtesy of AKG Acoustics, Inc.

Contents

Introduction

Of all the components in a high-fidelity system the microphone is the least appreciated and the least understood. Practically an unwanted orphan in the high-fidelity sound hierarchy, the quality microphone for amateur recording use is a "Johnny come lately," a newcomer. This is not only an odd situation, but one that is inexplicable, for the microphone is the fountainhead, the source, the starting point of all recorded sound—without exception. Every bit of recorded tape, every phonograph record, every compact disc, every broadcast has a single, common ancestry—the *microphone*.

That this state of affairs is shocking is beyond question. It is only when we come to analyze the reasons behind it that we can first begin to understand why the microphone is the last arrival in a long progression of high-fidelity components.

Quality microphones have always been used in broadcasting stations and recording studios. Until just a few years ago, any musical group that wanted to "cut a record" had only one route and that was to use the services of a professional recording studio. It was in the studio that the would-be recordist received his first introduction to a complex maze of electronic instruments. Placed in a situation that was extremely confusing, the recordist had no choice but to be guided by studio directors and recording engineers. It was, and still is, an expensive bit of education, with no assurance of recording success. The recordist went in bewildered and came out clutching a precious master-tape—still bewildered.

All that is now in the process of change, fortunately. At one time, tape decks for in home use had severe limitations. And the microphones with which they were supplied came in the same category. But today a number of high-fidelity manufacturers are supplying tape decks for the amateur recordist that are professional in every sense of the word. They are electronically designed to produce recordings that are superb. And accompanying this upward movement in tape deck quality, the manufacturers of the software, the tapes, began to produce tapes with new formulations that extended frequency response and had much higher signal-to-noise ratios. Various noise reducing techniques also became part of the in-home recording scene.

The result of all this is that the recordist can now work in his own environment. Outboard mixers and equalizers and a full line of quality microphones and accessories now give him an opportunity to produce professional recordings. So what we have today is a growing hobby, one that has great appeal for the amateur and professional musician, or for the high-fidelity listener, or for anyone having an interest in recording sound. Further, there are now two types of tape decks that can be used for quality recording—the open reel and the cassette.

The manufacturers of the hardware, the various tape decks, and the manufacturers of the software, the various open reel and cassette tapes, didn't waste any time in supplying the general public with detailed information on how best to use their devices. There is no shortage of advice on how to set up a recording facility. But they omit one important factor, undoubtedly the most important one—the microphone.

Although a tape deck for the amateur recordist can cost only 1/10 that of a studio unit, it can produce comparable quality results. And the tape being used with such tape decks is every bit as good as the tape used in studios. The difference in amateur versus recording in the studio is the use of cheap, poor-quality microphones, a lack of understanding on how to position microphones, and a general failure to appreciate what a microphone is and what it can do. The microphone is the first link in the recording chain. The amplifiers and speakers that follow the microphone can alter the sound or modify it in some way, but they cannot improve it.

The purpose of this book, then, is to supply you with practical working information about microphones. This book will not show you how to engineer a microphone, how to design one, or how to repair one. This book does explain what microphones are, how they work, and the different types available, with considerable emphasis on practical use.

Numerous references are made throughout this book to manufacturers of microphones and associated equipment using specified model numbers. However, new components are constantly being produced, older models retired, and model numbers changed. It is almost impossible to keep up with it, and so manufacturers' model numbers that are supplied here should be considered as a starting point. In some instances changes are purely cosmetic; in others they represent an improvement, either physically, electrically or electronically, or possibly a modification of specs.

There are two abbreviations for microphone–mike and mic.Of the two, the author's preference is for mike, but industry wide mic is used more often and so mic is used throughout this book. Mic is pronounced mike, mics is pronounced mikes, and micing is said as though it were written as miking. You can add this information to your collection of things that just don't seem to make sense.

No book writes itself, and this book is no exception. It is the collective effort of a large number of individuals, each of whom in his own way made a substantial contribution. Ideas and recommendations were made by the executives and engineers of AKG Akustiche U. Kino-Gerate GES.M.B.H. Vienna, specifically Dr. Rudolf Gorike, co-founder and Chief Scientist; Ing. Werner Fidi, Technical Director; Ing. Konrad Wolf, Supervisor, Condenser Microphone Development; Ing. Karl Peschel, Special Projects Engineering; and Ing. Norbert Sobol, Consulting Engineer.

This book is the end product of international cooperation. A special acknowledgment is due to AKG Acoustics, Stamford, Conn., 06902. Help came in the form of discussions about microphones, technical literature, specification sheets, and encouragement. So a special vote of thanks must go to S. Richard Ravich, Vice President and Larry Klein.

While AKG Acoustics supplied considerable data used in the preparation of this book, help was algo given by other microphone manufacturers including Beyer Dynamic, Inc., Crown International, HM Electronics, and Shure Brothers, Inc. Useful material was also forthcoming from my friends, notably Eugene Pitts, Editor of *Audio Magazine*.

Because of their constant work in the design and practical use of microphones, manufacturers are a rich source of technical information and it was helpful to borrow from their considerable expertise. Many of them were extremely cooperative, talking to me, answering my questions and supplying me with data I could not obtain otherwise.

I would especially like to thank AKG, Astatic, Audio-Technica U.S. Inc., Beyer Dynamic Crown International, Inc., HM Electronics, Inc., Nakamichi, Revox, Sansui, Satt Electronik A.B., Sennheiser, Shure Bros., Inc., Sony, and Turner. In some instances these fine companies supplied drawings and/or photos. Electro-Voice was kind enough to let me excerpt some text material and drawings from their fine and highly informative publication, *The PA Bible*.

Finally, there is always one man who sparks a project of this kind, who reads manuscripts and supplies advice, suggestions, guides, and corrections, always with a view to supplying the reader with a readable and practical work. In this case it is George A. Garnes, formerly Director of Advertising, North American Philips Corporation, New York, N.Y., to whom cooperation was a word to be taken quite seriously.

To all of them, my thanks and appreciation.

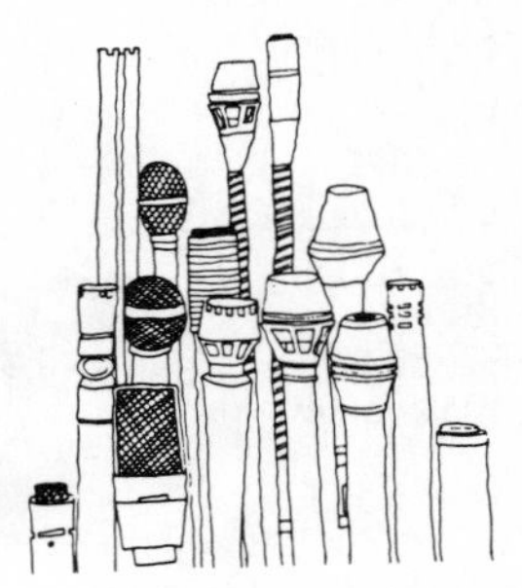

Chapter 1
The World of Sound

The input to every microphone (abbreviated as *mic* or *mike*), no matter how it is designed or constructed, is *sound*. Before any microphone can function, before it can do the necessary work of converting sound-energy input to an electrical output, sound must be present.

The difficulty with sound is it is so common. We are born hearing sounds and, except for those who have physiological defects, sound accompanies us throughout our lifetime. Sound is a necessary partner to one of our five senses—our sense of *hearing*. But because it is so much with us, we tend to take it for granted, perhaps not realizing it belongs to the energy family, a rather small group that includes heat energy, electrical energy, chemical energy, and mechanical energy.

We use sound for communications and for pleasure. We are perhaps not as critical when we use sound for communications, for the criterion is intelligibility, not authentic reproduction. With high-fidelity music, we are more concerned with an exact replica of the original sound source. That concern is a practical one, for reproduced sound that is distorted not only produces listening fatigue, but diminshes our enjoyment of music.

Learning more about sound is essential if we are to manipulate it, if we are to understand why microphones work the way they do, if we are to appreciate how we are to select particular types of microphones and position them effectively. The way we can understand this component is to know just what *sound* is, the relationships of sounds to each other and to microphones, and the behavior of sound under various conditions.

WHAT IS SOUND?

We all know what sound is and that's why it is so difficult to define.

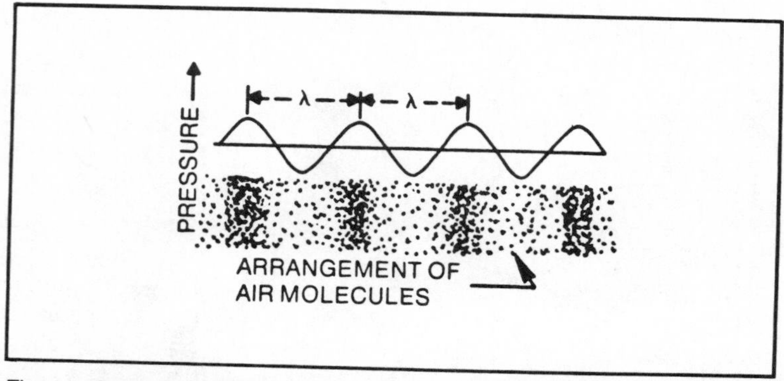

Fig. 1-1. Graphic representation of a sound wave. The distance from one peak to the next is a wavelength, represented by the Greek letter lambda (λ). The scattered dots below the sound wave represent the arrangement of air molecules.

Sound is generated when we displace the normal random motion of air molecules. If you clap your hands or bow a musical instrument, the result is an alternate rarefaction (expansion) and compression of air molecules, with large groups of molecules forming bunches and fewer molecules between these bunches. Sound travels as a wave (Fig. 1-1). Each wavelength is the distance from the peak of one sound wave to the peak of the next, represented by the Greek letter *lambda* (λ). An ocean wave, for example, contains a large amount of water rising to form some sort of peak, followed by a trough, then another peak and another trough. More elegantly, sound can be defined as consisting of longitudinal pressure waves in the air, with alternate regions of high and low pressure—high where we have larger numbers of air molecules, low where the number of air molecules is fewer.

SOUND PRESSURE

Sound is a pressure increase and decrease above and below a barometric mean. The air particles compress and become more dense and alternately disperse, but they usually remain in the same general area in which they were originally at rest. Sound energy is transmitted as variations in pressure.

If you were to throw a stone into the unruffled surface of a pond, the disturbance would result in the formation of waves. Some distance away, near the shore, a toy boat would be bounced up and down. It takes energy even to move a toy and in this case that energy was supplied by the falling stone. The energy was transmitted from the stone to the boat, but the actual waves themselves remained essentially in the same place. The boat on the waves moves only up and down. It doesn't move in a horizontal direction unless the wind or current pushes it independently of the vertical pressure movements.

BLACK SOUND AND WHITE SOUND

A sound, that has a pressure level so low that it is completely inaudible is called *black sound*. This does not mean there is no sound pressure, simply that it is not enough to produce the sensation of sound in a listener. The opposite of black sound is *white sound*, and, by definition, is any sound that is audible.

SOUND AND ITS TRAVELS

Sound can travel through liquid and solid bodies, through water, or steel, and other substances. "Sound" carries an implication that it is something we can hear, in short, that it is audible. However, sound exists above the threshold of our hearing. We call it *ultrasound*. And it also exists below our hearing range and is known as *infrasound*.

Sound travels much more rapidly through liquids and solids than it does through air, as indicated in Table 1-1. In air, sound travels at a speed of about 1130 feet per second. If an auditorium is large enough, people in that auditorium will hear the same sound at different times. The time separation, of course, is quite short. Sound moves through water about four times as fast as it does through air and through iron, its speed becomes about 14 times as rapid. Send sound through a rubber band and it moves along at a speed of about 131 feet per second. Sound, however, will not travel through a vacuum.

PRODUCTION OF SOUND

Any vibrating object can act as a sound source and produce a sound

Table 1-1. Velocity of Sound in Liquids and Solids.

	Sound Velocity	
Material	**Feet/second**	**Meters/second**
Alcohol	4724	1440
Aluminum	20,407	6220
Brass	14,530	4430
Copper	15,157	4620
Glass	17,716	5400
Lead	7,972	2430
Magnesium	17,487	5330
Mercury	4,790	1460
Nickel	18,372	5600
Polystyrene	8,760	2670
Quartz	18,865	5750
Steel	20,046	6110
Water	4,757	1450
Air	1,130	344

wave, and the greater the surface area the object presents to the air the more it can move. The object could be the vibrating string of a violin, a weak sound at best, but considerably reinforced by the vibrating wood body of the violin. It can be produced by a slap, or a dropped plate, or a bat striking a ball. The compressions and rarefactions of the air molecules are pressure variations that correspond to the vibrations of the sound source. Where the molecules of air are bunched together, the air pressure is above normal. In between the larger-than-normal groups of air molecules, we have lower air pressure simply because we have fewer molecules. This condition will continue to exist as long as the original sound source continues vibrating. When it stops, the air molecules distribute themselves more or less equally, the pressure differences between groups of molecules disappear, and we no longer hear sound.

When a sound is produced by a source, the air molecules around the source are disturbed; that is, they are moved out of their random scattered condition. Their normal pattern of approximately equal distribution is changed.

The disturbance of air molecules around a sound isn't restricted to a single source. You could have two sound sources more or less immediately adjacent, and the air molecules around each of these sources would be disturbed by each of them. In other words, the air can support a number of independent sound waves produced at the same time.

TYPES OF SOUND VIBRATION

Musical *tones* are produced by regular vibrations. Because of their orderliness, we find these tones pleasant. *Noise* produces irregular vibrations. The variations in air pressure are random and our sense of hearing interprets them as unpleasant.

A musical *note* consists of a fundamental wave and a number of overtones, called *harmonics*. The composite waveform comprising the fundamental and its harmonics can look quite irregular, with many sharp peaks and valleys, and yet the fundamental tone and each harmonic is made up of a very regularly shaped waveform. Our ears, apparently, are able to recognize the regularity of the individual waves in such a complex waveform and are both pleased and satisfied.

While we can describe sound as the result of a variation in air pressure, it does have three fundamental characteristics. These are *pitch, timbre,* and *loudness*. Every sound has a definite frequency, a recognizable character. We can identify a sound as being produced by a violin, the human voice, or a trumpet. Sound is also recognized by its strength.

PITCH

Pitch is the fundamental or basic tone of a sound and is determined by the *frequency* of that tone (Fig 1-2). The frequency of a wave is a measure of the number of complete waves per second. The greater the number

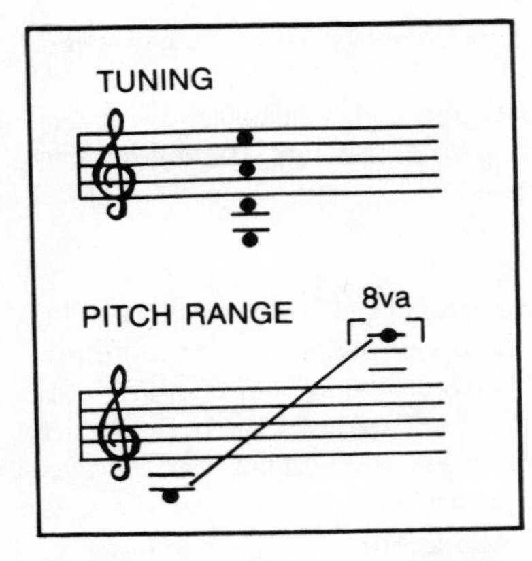

Fig. 1-2. Tuning and pitch range of the violin.

of waves per second, the higher the frequency, that is, the higher the pitch. *Treble* tones have a much higher frequency that *bass* tones.

The fundamental or basic tone is sometimes called the *first harmonic*. However, it is the harmonics or overtones of the fundamental that supply the identifying characteristic of a sound and which enable us to distinguish between two tones having the same fundamental frequency but played on different musical instruments.

We don't always refer to the pitch of a sound, but may group the pitch into general classifications such as *bass, midrange* and *treble*, an arrangement quite commonly used in connection with high-fidelity systems.

The sounds produced by the tympani are high pitched; those made by the longer pipes of the organ are low pitched. Instead of referring to an entire range, such as bass, we can specify a single pitch. A above *middle* C on the piano has a frequency of 440 Hertz—that is, 440 complete cycles or waves per second (Fig. 1-3).

The full range of any musical instrument, such as the piano, is from the pitch or frequency of the lowest tone it can produce to that of the highest. The range of human hearing, our ability to hear from the lowest pitch to

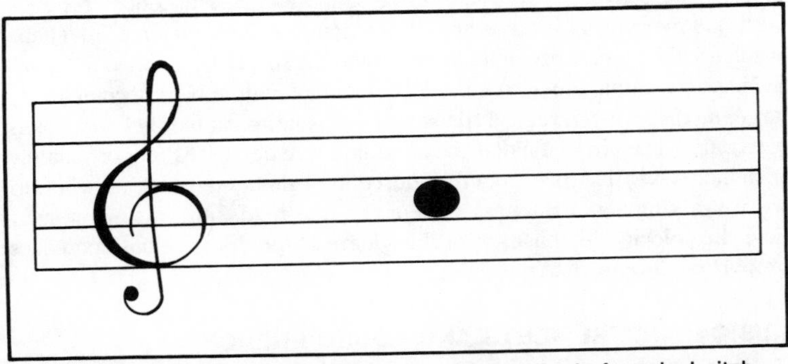

Fig. 1-3. A above middle C is considered the standard of musical pitch.

the highest, encompasses all musical instruments plus the human voice. This doesn't mean sounds do not exist outside our hearing range. They do, but their pitch may be so low or so high that we cannot hear them. And not all human beings have the same hearing range. Human hearing has an average range of about 10 octaves.

HEARING AND AGE

As we get older all of us lose some amount of hearing ability. The amount of loss depends on sex and sound frequency. For a man there will be an average loss of 7 dB at 500 Hz between the ages of 35 and 65. In the same age bracket, the hearing loss will be 10 dB at 1 kHz. In the higher frequency range, at 4 kHz, the hearing loss will be 37 dB between the ages of 25 and 65. For women the hearing loss will be 10 dB between the ages 25 and 65 for frequencies up to 1 kHz. At 4 kHz in the same age bracket, the hearing loss will be 18 dB. These are average figures only and can vary with individuals. Hearing loss can be accelerated by extensive listening to sound at very high levels.

WHAT HAPPENS TO THE SOUND?

As a sound leaves its source, such as a musical instrument, it spreads out, or *diffuses*. The entire region of sound could be called a *sound field* with the microphone immersed in that field. The space occupied by the microphone compared to the total sound field is small. Further, only a small section of the microphone is receptive to sound or is affected by it.

The sound that leaves a source uses up most of its energy in heating the air through which it moves. The amount of heat is small and the volume over which it is distributed is rather large. You cannot detect any heat difference by inserting your hand between the sound source and the microphone. Sounds having a higher pitch diminish much more quickly over a given distance than do sounds of a lower pitch. The greatest amount of sound energy exists at the source. As the sound moves away from that source its energy diminishes rapidly.

We can conveniently characterize sound as consisting of two fields, with one merging into the other. When sound leaves a source, its shape is spherical and, because of its proximity to the source, it is called a "near" field. As the sound moves outward, the radius of this sphere becomes very large, so that we can regard the sound propagation as parallel and this is designated as a "free" field. The sound pressure decreases proportionately with increasing distance from the source. Its velocity depends on whether we are talking about the near field or the free field. In terms of the near field the velocity decreases with the square of the distance but decreases proportionately in the free field.

MUSICAL INSTRUMENTS AND SOUND ENERGY

There is a common misconception that musical instruments produce sound energy but such is not the case. In the case of stringed instruments, for example, all the instrument can do is to change the mechanical energy supplied by the fingers of the musician into sound energy. Energy can be neither created nor destroyed, and so a musical instrument is simply a *transducer*, a device for converting one form of energy to another.

Tones characterized by having a low pitch, such as bass notes, require a higher mechanical energy input on the part of the musician than treble tones. Because of this, sounds that have a higher pitch diminish more rapidly over a given distance than sounds of a lower pitch. Or, put another way, treble tones have a lower energy content than bass tones.

FREQUENCY

When sound energy is converted to its equivalent in electrical energy, the electrical output is an alternating current or alternating voltage waveform. The wave, known as a *cycle* (Fig. 1-4), consists of a pair of alternations, one of which is regarded as positive, the other as negative. Each alternation is half of a full cycle. The *frequency* is the total number of *completed* waveforms or cycles per second, including both the negative and positive half-cycles. At one time, frequency was measured in cycles per second, or cps. The Hertz, as indicated earlier, is now designated as the cycle per second. Abbreviated as Hz, 30 Hz means 30 complete cycles per second. The letter k is used to indicate a multiplier of 1000, so 1000 Hz can also be written as 1 kHz.

The fundamental ac waveform, called a *sine wave*, is shown in Fig. 1-5. The horizontal line drawn through the waveform separates the upper or *positive* (+) half from the lower or *negative* (−) half. With an increase in frequency, there are more cycles per second. A *pure tone*, a tone having no harmonics or overtones, appears as a sine wave.

Frequency is always related to time. As frequency increases, as shown in Fig. 1-6, there are more complete waves or cycles per second. The time reference isn't always specified. Thus, 30 cycles implies 30 cycles per second. The abbreviation Hz (Hertz) eliminates any time ambiguity for it means "cycles per second". Sixty Hertz is 60 cycles per second, so in using Hz the time element is always included automatically.

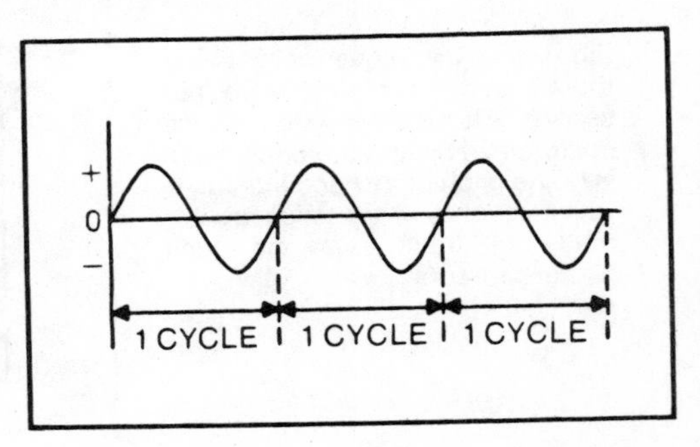

Fig. 1-4. A cycle is measured from the start of one wave to the beginning of the next. Frequency is measured in cycles per second. If three cycles are completed in one second, the frequency is three cycles per second or three Hertz (3 Hz).

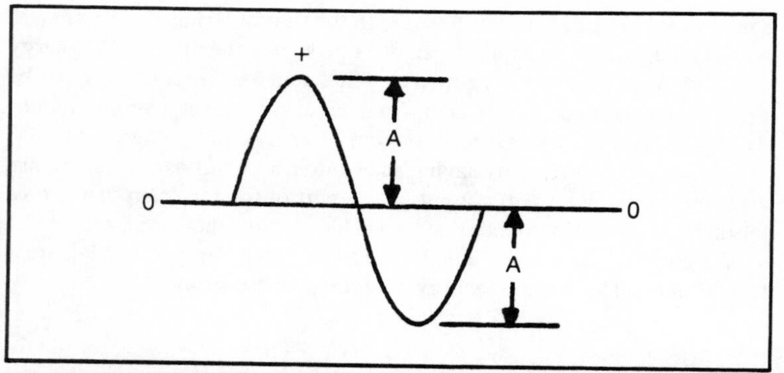

Fig. 1-5. Basic sine wave. The letter A indicates the maximum strength or amplitude of each half cycle.

FREQUENCY AND WAVELENGTH

The frequency of a sound, measured in Hertz, is one way of describing this characteristic. Another is the wavelength which is the distance from one point on a wave to a corresponding point on the following wave. Frequency and wavelength are inversely related. Thus, the higher the frequency—the shorter the wavelength, or, conversely, the lower the frequency—the longer the wavelength. The chart shown in Table 1-2 supplies spot frequencies ranging from 20 Hz to 20 kHz (20,000 Hz) and corresponding wavelengths in feet and in inches.

HARMONICS

Harmonics or *overtones* are multiples of the fundamental frequency (Fig. 1-7). The tone of a musical instrument having a fundamental frequency of 250 Hz could have a second harmonic or overtone at 500 Hz, a third harmonic at 750 Hz, a fourth harmonic at 1 kHz, and so on.

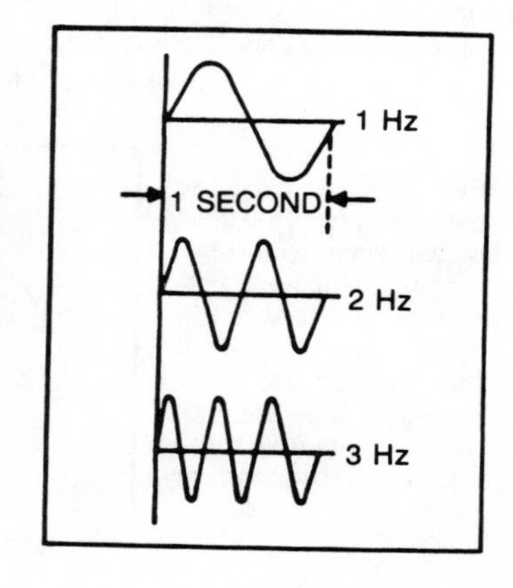

Fig. 1-6. As the frequency increases there are more complete waves per second. The top drawing is 1 Hz, the center is 2 Hz, and the bottom is 3 Hz. The pitch of a sound increases with frequency. Bass tones are low frequency; treble tones are much higher in frequency.

Table 1-2. Wavelengths of Sound.

(1130 ft/sec, in air, at 20 degrees C; 32 degrees F)

Frequency (Hz)	Wavelength (feet)	Frequency (Hz)	Wavelength (feet)	Frequency (Hz)	Wavelength (feet)
20	56.50	140	8.07	380	2.97
25	45.20	150	7.53	400	2.83
30	37.67	160	7.06	420	2.69
35	32.29	170	6.65	440	2.57
40	28.25	180	6.28	460	2.46
45	25.11	190	5.95	480	2.35
50	22.60	200	5.65	500	2.26
55	20.55	210	5.38	525	2.15
60	18.83	220	5.14	550	2.05
65	17.38	230	4.91	575	1.97
70	16.14	240	4.71	600	1.88
75	15.07	250	4.52	650	1.74
80	14.13	260	4.35	700	1.61
85	13.29	270	4.19	750	1.51
90	12.56	280	4.04	800	1.41
95	11.89	290	3.90	850	1.33
100	11.30	300	3.77	900	1.26
110	10.27	320	3.53	950	1.19
120	9.42	340	3.32	975	1.16
130	8.69	360	3.14	990	1.14

Frequency (Hz)	Wavelength (inches)	Frequency (Hz)	Wavelength (inches)	Frequency (Hz)	Wavelength (inches)
1000	13.56	9000	1.51	16000	0.85
2000	6.78	10000	1.36	17000	0.80
3000	4.52	11000	1.23	18000	0.75
4000	3.39	12000	1.13	19000	0.71
5000	2.71	13000	1.04	20000	0.68
6000	2.26	14000	0.97		
7000	1.94	15000	0.90		
8000	1.70				

As we get beyond one of the upper-order harmonics, such as the fifth, the amplitude or strength of such overtones is quite small. The fundamental and its harmonics combine to produce complex waveforms. It is these waveforms that give each tone its particular character. They enable us to distinguish between a tone produced on a piano and a tone of identical frequency made by a guitar or some other instrument.

RANGE OF MUSICAL INSTRUMENTS

The fundamental range of musical instruments, as shown in Fig. 1-8 and Table 1-3 is quite limited. At the low frequency end, there are very few musical instruments that have the capability of producing tones below 50 Hz. The human voice does not go much below 70 Hz, while at the high-frequency end all musical instruments and the voice are below 5 kHz in fundamental frequency.

While it is the fundamental frequency that determines the pitch of a tone, it is the harmonics that add richness and quality. Pure tones (such as the tone produced by a tuning fork)—tones consisting of a fundamental only—make tiresome listening. The number of harmonics produced depends on a variety of factors, whether the instrument is percussive or wind, for example, and also on playing technique. A violinist controls harmonic content by the movement of his fingers. In the case of the flute, playing it softly results in an almost pure tone, as shown in the top drawing of Fig. 1-9. With louder tones, however, there are more harmonics. So our enjoyment of a particular tone depends not only on the number of harmonics produced, but on the variation in the quantity of harmonics.

TIMBRE

That character of a sound which enables us to distinguish between different musical instruments, including the voice, is called *timbre*. Even if two instruments are playing the same tone—that is, each is playing notes having the same frequency and at the same loudness level—the notes have a different sound. Each musical instrument has its own particular pattern of overtones.

Overtones or harmonics (also called *partials* or *partial tones*) are classi-

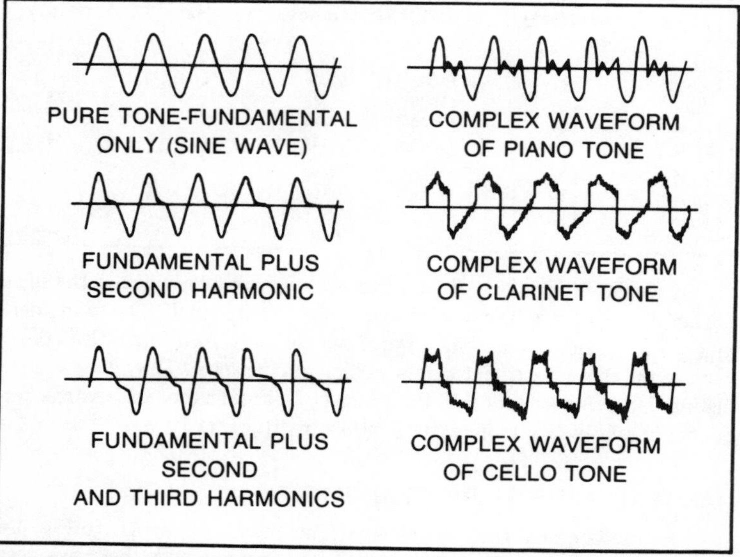

Fig. 1-7. Various types of waveforms. Harmonics are multiples of the fundamental frequency. They may be either odd- or even-numbered harmonics.

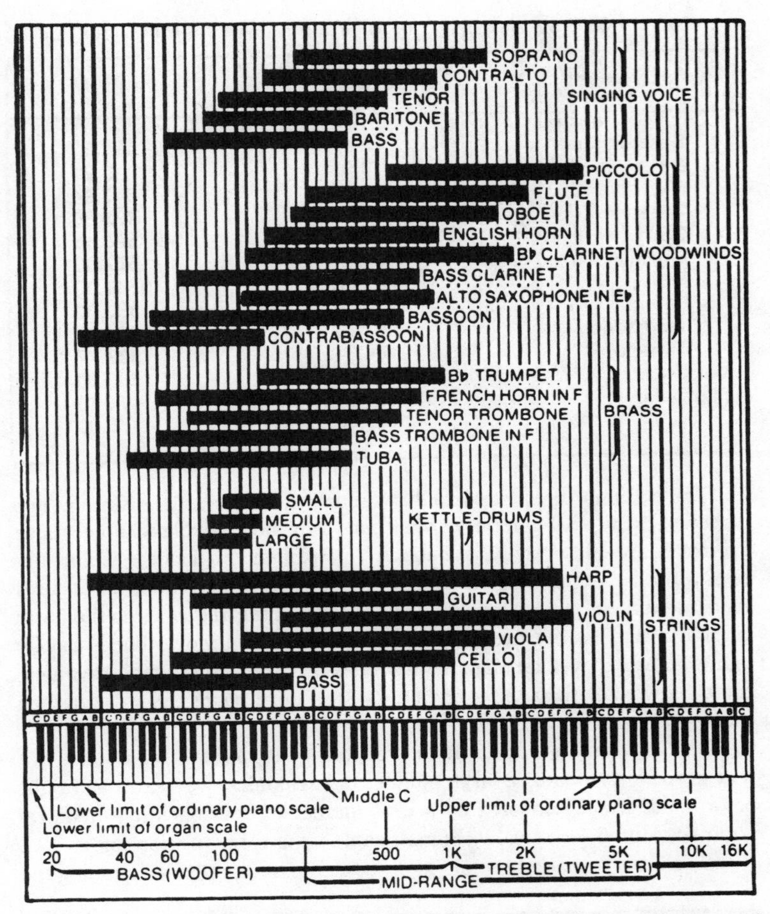

Fig. 1-8. The frequencies of music including the ranges of the fundamental frequencies of instruments and voices. The harmonic frequencies generated by instruments and voices extend off the right side of the chart, though at volume levels far below those of the fundamental frequencies shown. The A above middle C is usually set at the standard tuning pitch of 440 Hz. (Copyright by Ziff-Davis Publishing Company. Reprinted from the April 1980 issue of Stereo Review by permission.)

fied into two groups: *odd* and *even*. An even harmonic is one that is an even-order multiple of the fundamental frequency (f) of the tone. A tone of 440 Hz has its second harmonic at 880 Hz. The next even-order harmonic would be four times the fundamental, or $4 \times 440 = 1760$ Hz. If we call the fundamental $f1$, the second harmonic would be $2 \times f1$ or $2f1$, the fourth harmonic would be $4f1$, and so on. The odd-order harmonics would be $3f1$, $5f1$, and so on. If the fundamental is 400 Hz, the third harmonic, $3f1$, is $400 \times 3 = 1200$ Hz.

Table 1-3. Frequency Ranges of Various Musical Instruments.

Instrument	Low Hz	High Hz
Bass clarinet	82.41	493.88
Bass tube	43.65	349.23
Bass viol	41.20	246.94
Bassoon	61.74	493.88
Cello	130.81	698.46
Clarinet	164.81	1567.00
Flute	261.63	3349.30
French horn	110.00	880.00
Trombone	82.41	493.88
Trumpet	164.81	987.77
Oboe	261.63	1568.00
Violin	130.81	1174.70
Violin	196.00	3136.00

The various tones produced by musical instruments differ in two respects: in the total number of overtones they yield, and in whether those overtones are odd or even, or both. An instrument such as the violin supplies a fundamental plus odd and even overtones. A trumpet produces a fundamental plus odd overtones. The fundamental frequency is the basic pitch. If a note has five overtones, each higher-frequency overtone is usually (but not always) weaker than its predecessor. Thus, the second harmonic could be weaker than the fundamental, the third harmonic weaker than the second, and so on.

The total number of harmonics supplies the *character* of a tone. If a tone is accompanied by a large number of harmonics, we hear it as bright or brilliant; if accompanied by a few number of overtones it sounds restrained, muted, or mellow. Some people even refer to it as dull.

This characteristic of sound gives us a clue for aurally determining the

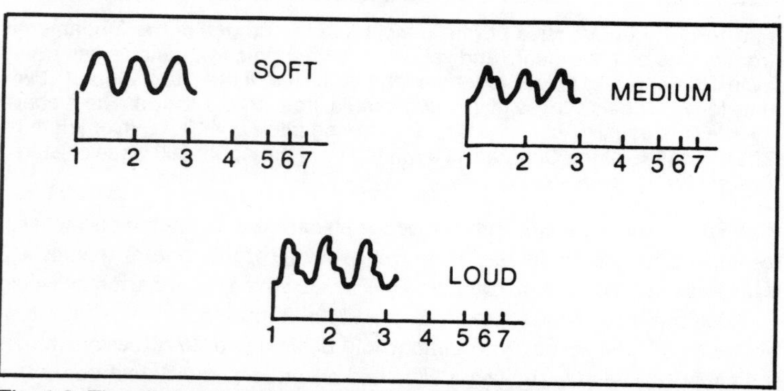

Fig. 1-9. The playing techniques determine the harmonic content of the tone produced by a flute. The louder the tone, the greater the number of harmonics.

functioning of a microphone. If the preamplifier and power amplifier following the microphone have an inadequate frequency response—(good-quality amplifiers go beyond the outermost limits of human hearing)—then our inability to distinguish between wood instruments and string instruments is an indication that the overtones are either not being reproduced or are being distorted.

LOUDNESS

Our ears aren't linear devices. We are most sensitive to tones in the middle frequencies, with decreasing sensitivity to those having relatively lower and higher frequencies.

Loudness and volume are not the same, as evidenced by the fact that a high-fidelity receiver will have both a loudness and a volume control. A volume control is used to adjust the overall sound level over the entire frequency range of the audio spectrum. A volume control is not frequency or tone selective, or at least it shouldn't be. When you advance the volume control in a receiver all tones are increased in level.

FLETCHER-MUNSON CURVES

Fletcher-Munson curves, or equal loudness contours, show the response of the human ear throughout the audio range of 20 Hz to 20 kHz. These curves reveal, as shown in Fig. 1-10, that more audio sound power is required at the low end and the high end of the sound spectrum to obtain sounds of equal loudness. Our ears are more responsive to sound between 3 kHz and 4 kHz than they are at other frequencies.

There are two extremes in this graph. One is identified as curve A and the other as curve B. Curve A represents a very high sound intensity level and is between 110 dB and 130 dB, regions in which sound becomes painful. Curve B is sound that can barely be heard, sound that is at the threshold of audibility. In curve B it takes about 65 dB more audio power at 20 Hz to produce an equivalent hearing sound level compared to 4 kHz. As another example of the tremendous variation in our hearing ability with respect to sound frequency is that the power needed to produce audible sound at about 50 Hz is about a million times greater than that required at 3 kHz.

A more correct name for the loudness control on a hi/fi receiver would be "physiologically correct loudness contour compensation." This control is used to compensate for our hearing—a hearing that is relatively insensitive to bass and treble tones when the overall volume is very soft. The loudness control, or switch, overcomes this hearing characteristic by boosting these extreme sound ranges at low volume settings. The loudness control isn't required when listening at normal or high volume levels.

This situation applies not only to recorded music, but to live music as well. You can easily observe the difference in sound quality of an approaching marching band, depending on how near or far the band may be. At

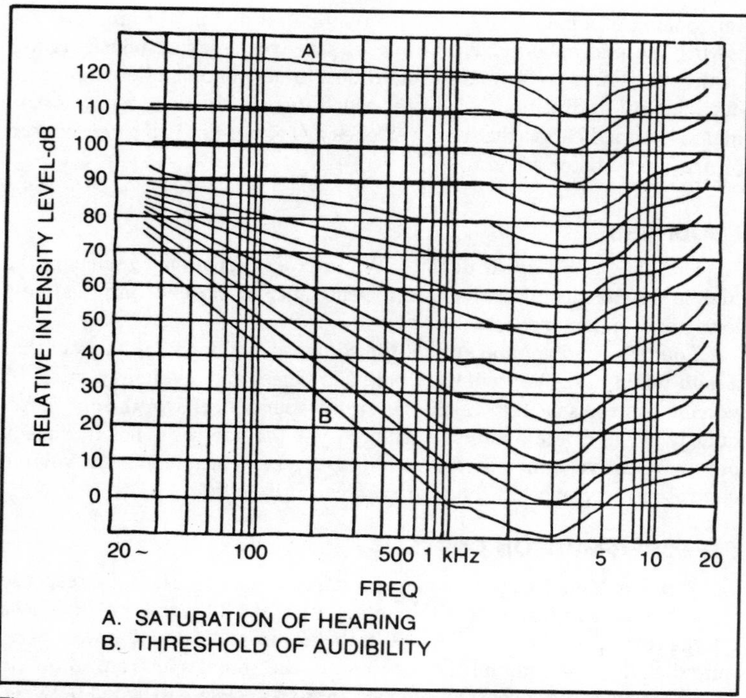

Fig. 1-10. Equivalent loudness or Fletcher-Munson curves. These curves show that more audio power is needed at low and high sound frequencies than those in the midrange to obtain sound of equal loudness.

a distance, the music may sound somewhat flat, but it is much richer sounding when the musicians parade directly past you.

Volume has two important characteristics. The first is its *dynamic range*, extending from the threshold of hearing to the threshold of pain. The other characteristic is the *relationship of sound to time*. The chart in Fig. 1-11 is a comparison of the relative volume levels of ordinary sounds. The threshold of hearing is zero decibels (abbreviated dB) and the threshold of pain is 130 decibels. The decibel is a unit of comparative measurement, explained in more detail in Chapter 7.

Figure 1-12 shows the curves of the thresholds of audibility and feeling. The encircled area is called the *auditory-sensation* area and includes all audible tones of any frequency and intensity. The curves also show that the thresholds for music are more restricted than those for speech.

THE PHON

The *phon* is used for measuring the loudness level of a pure tone. The drawing in Fig. 1-12 is a plot of the frequencies of music and speech in

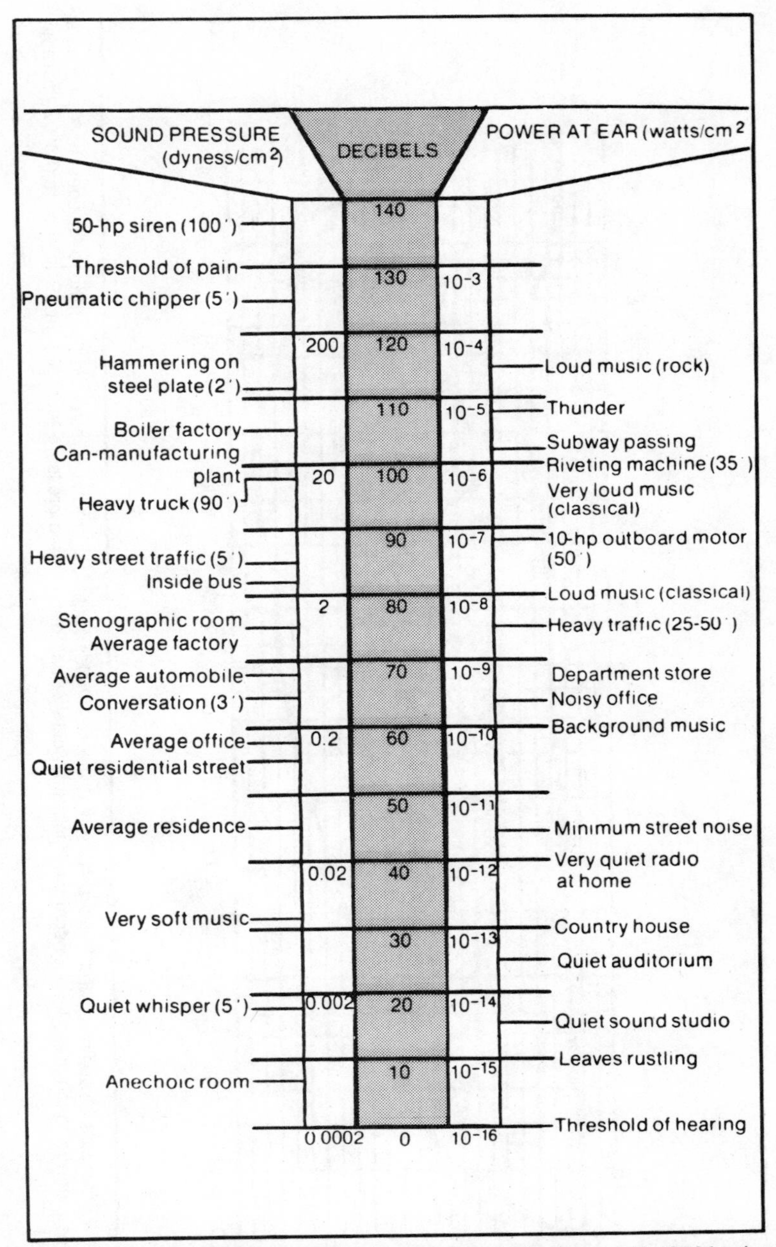

Fig. 1-11. Relative loudness levels of common sounds. (Copyright 1976 by the Ziff-Davis Publishing Company. Reprinted from Jan. 1976 Stereo Review by permission.)

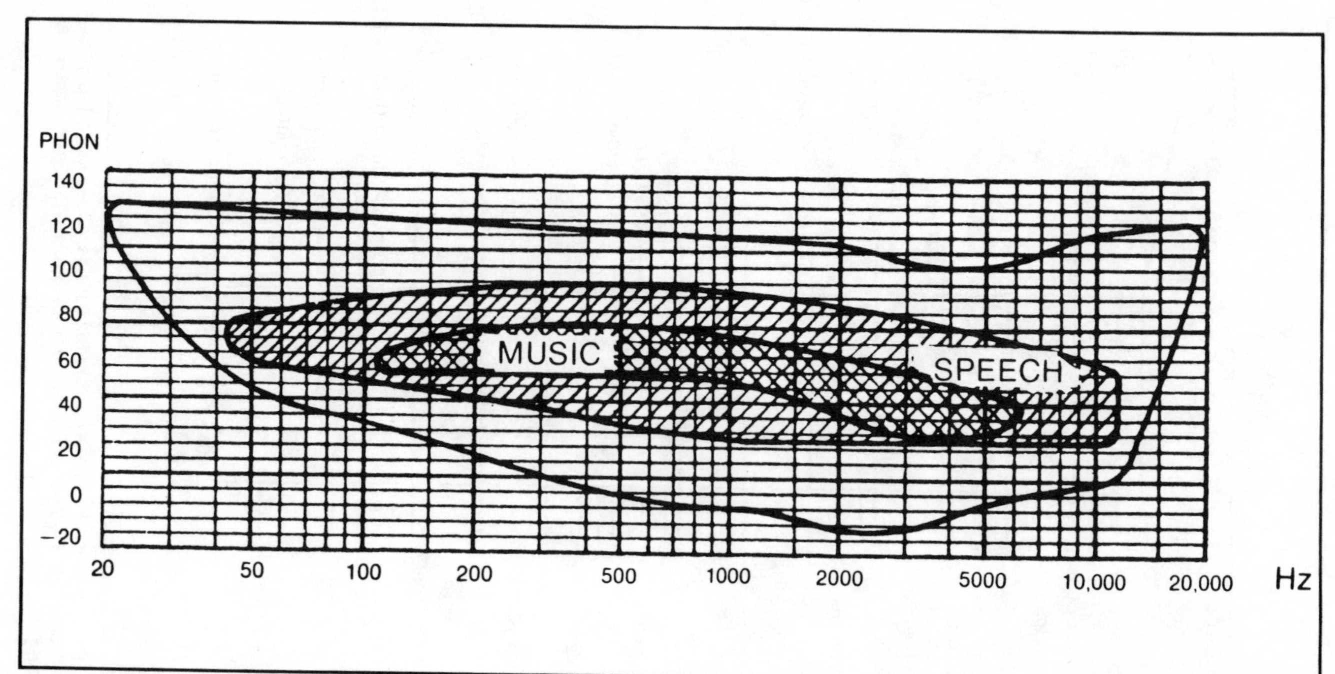

Fig. 1-12. Thresholds of audibility and feeling. Note that the thresholds for music are more restricted than those for speech. A phon is a unit for measuring the loudness level of sounds. It is numerically equal to the sound pressure level SPL in dB relative to 0.0002 microbar of a 1 kHz tone.

phons. We do not hear linearly. If the intensity of a sound is doubled this does not mean that the intensity of sound as we perceive it is doubled.

COMB FILTER

When we hear a sound we perceive its location essentially by sensing the differences in loudness and in arrival time of the sound at each ear. This information is further enhanced by the human brain's ability to sense a "comb filter" effect, a series of notches or dips in the frequency response of the ear itself that affects all sounds heard. The comb filter is so-called because a graph of the response (Fig. 1-13) looks like a comb. The notches are caused by outer-ear resonance and diffraction effects (interference of direct and reflected sound waves) and partially by resonances within the ear canal itself. The comb-filter effect is a natural phenomenon that gives us specific information about sound source localization. We are never aware of the filter per se, but we can certainly hear a difference when it is altered.

THE SONE

The *sone* is a unit of loudness and is used for the measurement of the characteristics of the human ear. One sone is equal to the loudness of a 1 kHz tone 40 dB above the threshold of hearing. Thus, a tone that is twice as loud would be two sones, one that is three times as loud would be three sones, and so on. While the sone is the basic unit, decimal fractions are often used. Thus, a millisone is a thousandth of a sone.

THRESHOLDS OF AUDIBILITY AND FEELING

What you can hear depends on your age, sex, and the physical condition of your ears and brain. While the audio spectrum is assumed to have a range of 20 Hz to 20 kHz, few of us have a hearing capability that goes

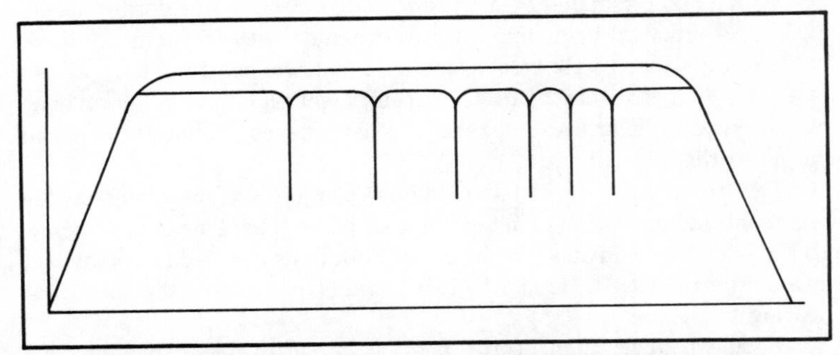

Fig. 1-13. The ideal "flat" frequency response indicated by the straight horizontal line, is not what we actually hear. The anatomy of the ear creates a "comb filter" which produces peaks and dips in the frequency response, as shown in the drawing, which we accept as "normal hearing."

down to 20 Hz, and equally few can really hear as high as 20 kHz. The pipe organ has a possible range below 20 Hz, while the bottom frequency of the contrabassoon is just a bit above 30 Hz; but even so, very little music is written for such low frequencies. Natural sounds include hardly any low frequencies and, as a matter of fact, what you will find among the very low frequencies is just disturbing noise. At the high end, the piccolo and the violin have fundamentals that are lower than 5 kHz, but the harmonics of musical instruments are substantially higher.

USEFUL FREQUENCY RANGE FOR MICROPHONES

Consequently, the useful frequency range for microphones seems to be from about 50 Hz to 15 kHz. You will sometimes see a specification indicating a response of 20 Hz to 20 kHz, but most microphones are within more practical limits.

As you reduce the sound level more and more you will reach a level at which sound perception will stop. This is the threshold at which you will hear no sound, referred to as the *threshold of audibility*. The threshold of audibility depends on frequency; the sound pressure at the threshold of audibility differs substantially.

Going to the other extreme, sound can be made so loud that our perception of it turns into feeling. This upper limit is called the *threshold of feeling*, also known as the *threshold of pain*. Some rock concerts are capable of approaching this level.

THE OCTAVE

A doubling of frequency is called an *octave*. From 30 Hz to 60 Hz could be called an octave, since 30 × 2(or 30 doubled) equals 60. We could regard 60 Hz to 120 Hz as still another octave, and so on. Figure 1-14 shows notes that form a single octave on the piano keyboard.

We do not start with 0 Hz, for this is actually a direct current or voltage such as that supplied by a battery, for example. Considering music, 32 Hz is a practical beginning; but we can start with 16 Hz as a bottom limit. If we select 16 Hz as our starting point, we can then have 10 octaves, up to approximately 16 kHz. Insufficient pickup by a microphone of certain octaves, or overemphasis of other octaves, will alter the sound output of the microphone.

The 10 octaves in Table 1-4 are of particular interest, since they roughly represent the range of human hearing capability. There are sounds below 16 Hz, of course, and above 16 kHz, but most of us cannot hear them and, as a matter of fact, 16 Hz and 16 kHz could be considered the outermost hearing limits.

What we have, then, is the possible range of sound pickup by a microphone—10 octaves. This is a tremendous range. A lens in a camera works with light instead of sound, but both light and sound are forms of energy. Yet the lens can confine itself to just one octave—from 4000 Ang-

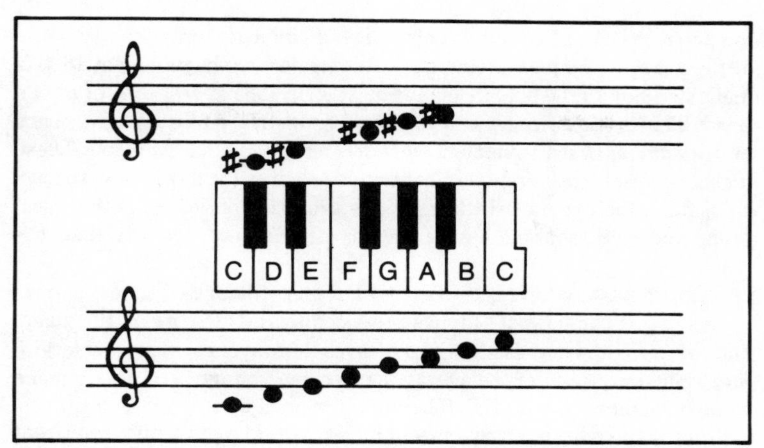

Fig. 1-14. A selected octave on the piano keyboard.

stroms (the Angstrom is a unit of light measurement) to 8000 Angstroms. Motion-picture sound works with a range of about six octaves, and for telephone conversations the range is even narrower, just three to four octaves.

A lens, or course, is just a collector of light, just as the microphone is a collector of sound. But the similarity ends there, for the microphone is a *transducer*, something the lens is not. The microphone must not only collect sound, but transduce or convert it to an equivalent form of electrical energy. A lens receives light, but its output is still light. Although a microphone is sometimes compared to a lens, the microphone has the far more difficult job. For high-fidelity use, a range of 9 to 10 octaves is quite an accomplishment.

CHARACTERISTICS OF THE SOUND SPECTRUM

When all musical instruments are considered the total overall range

Table 1-4. Octaves and Frequency Ranges.

Frequency Range, Hz	Octave
16 to 32	first
32 to 64	second
64 to 128	third
128 to 256	fourth
256 to 512	fifth
512 to 1024	sixth
1024 to 2048	seventh
2048 to 4096	eighth
4096 to 8192	ninth
8192 to 16,384	tenth

is approximately 10 octaves. This range, as indicated in Fig. 1-15 starts at the very low frequency end, the subcontra octave beginning at 16.4Hz and extending to the six-line octave at 16 kHz. The lowest tone of the piano is 27.5Hz but the organ can go lower—to 16.4Hz. At the high frequency end, a number of instruments have a greater range than the piano. These include not only the organ, but instruments such as the xylophone, the piccolo, flute, the oboe, clarinet and violin. In terms of voices, both the soprano and mezzosoprano have a greater high-frequency reach than the piano.

The first two octaves, 16 Hz to 64 Hz, contain the bass tones. Sounds in this region can consist of tones produced by the pipe organ, the piano, and the harp. It is in these first two octaves that we encounter trouble with *hum*. The frequency of the average power-line voltage is 60 Hz (in some foreign localities it is 50 Hz). The hum produced by a defect in a fluorescent fixture can have a frequency of 60 Hz; but if it isn't a pure tone, that is, one without harmonics, then you may be hearing the second harmonic at 120 Hz. Turning off the lights is one solution to this difficulty if you notice hum during playback.

Bass tones supply richness, depth, and power. If a bass tone has a very

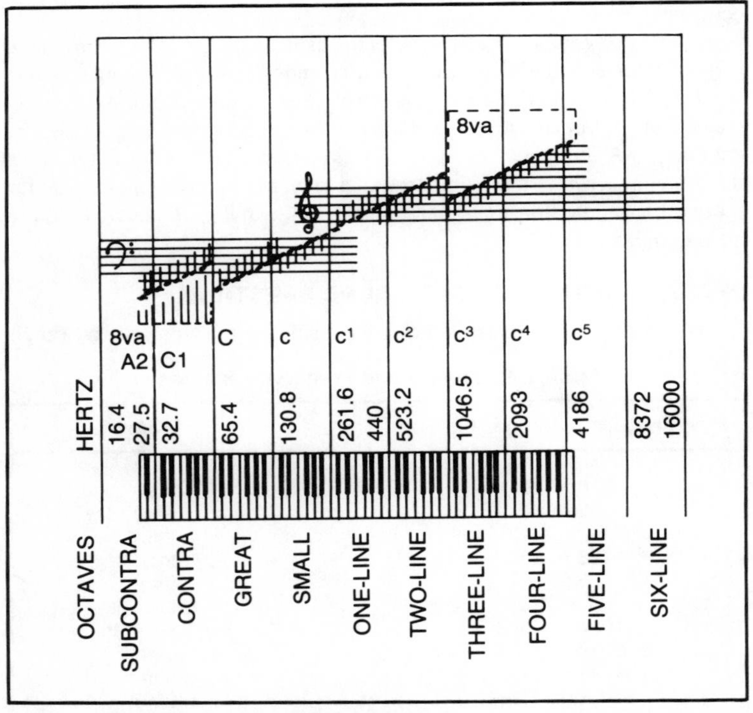

Fig. 1-15. The 10-octave range of music.

low frequency and is strong enough, we may *feel* rather than hear it. Even when bass tones occur only momentarily, they manage to set up a pleasing balance with higher frequency tones. Without them, higher tones, those in the upper midrange and treble range, will seem stronger than they really are.

While noise can occur anywhere in the audio spectrum, certain noises make their home in the bass tones. Street noises, various noises inside the home, and other sounds we normally disregard and aren't conscious of hearing, can show up in the first two octaves.

The third and fourth octaves, from 64 Hz to 256 Hz, contain the frequencies that supply musical tempo. It is in these frequencies that you will hear orchestral rhythm. Drum beats and piano tones appear in these octaves. The fifth, sixth, and seventh octaves are known as the midrange frequencies, or more commonly just as the *midrange*. In this frequency range, 256 Hz to 2048 Hz, is that part of the sound spectrum to which the ear is quite sensitive. *Middle C* on the piano comes near the bottom of the midrange, and the majority of musical instruments have either their fundamental tones or first harmonics here.

The remaining sounds are in the 8th to 10th octaves. Male speech is about 3 kHz to 6 kHz, while female tones are higher by an octave.

CHARACTERISTICS OF SOUND

A sound does not reach its maximum level instantaneously. It takes some time for it to do so, however short that time may be. And after reaching its maximum level, neither does it decrease immediately to zero. The sound may be sustained for a while, if the sound-energy source remains active. Finally, it takes time for the sound to decay, that is, to approach and reach a zero sound level. Figure 1-16 shows this triple characteristic. The *attack* time is the time it takes the sound to reach its peak. *Sustain* is the level maintained by the sound source. (Note there may be a dropoff from the

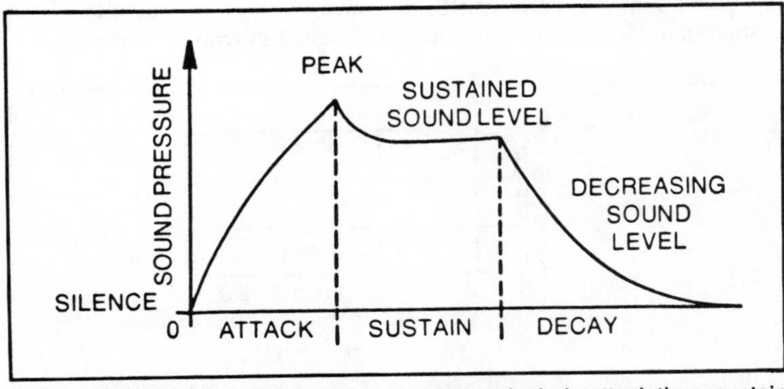

Fig. 1-16. Three characteristics of a sound wave include attack time, sustain time, and decay time.

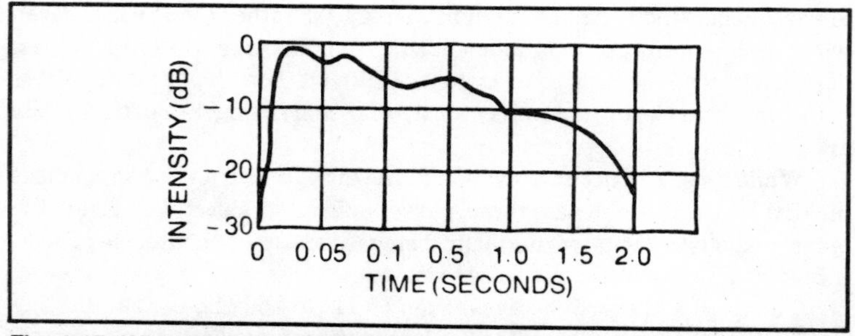

Fig. 1-17. With the sustain pedal depressed, a piano tone can last for a long time. In this example it is almost two seconds.

peak). Finally, there is a *decay* time, during which the sound drops to zero.

Sustain time can be very short or it can be prolonged. With a piano, for example, using the pedal can sustain a tone after the piano key has been struck and released. Figure 1-17 shows quite a prolonged sustain time (in this example, almost two seconds). The human voice can sustain a tone, but the decay time will usually be short. The decay time of a guitar (Fig. 1-18) is rather long. It is these three characteristics—attack, sustain, and decay—that contribute to musical-instrument personality and makes each musical instrument so different.

One of the most beautiful aspects of the human voice is its ability to alter any one of the three characteristics of sound. It can reach maximum level quickly or slowly. It can sustain a tone for a longer or shorter time. And it can make decay time gradual or sharp.

Figure 1-19 contains graphic representations of the ways the human voice can control sound. In the first drawing (A), the attack time is very short. The voice tone reaches its peak rather quickly, with sustain time and decay time merging into each other. In this instance the sound decreases smoothly and gradually.

In the center illustration (B), the voice reached its peak slowly, but sustain and decay are quite rapid. In the final drawing (C), the peak is

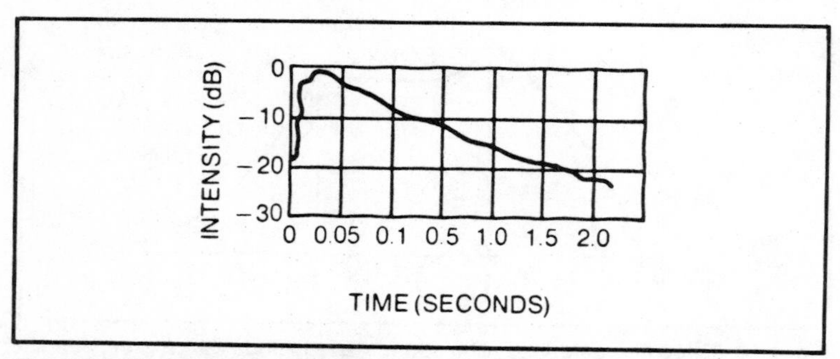

Fig. 1-18. Decay of a guitar tone is rather long.

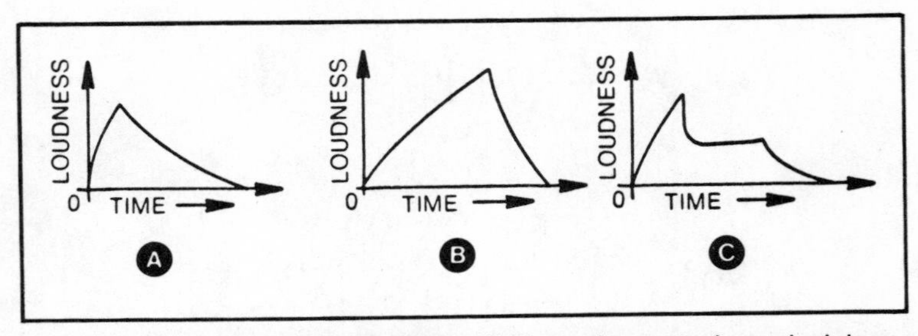

Fig. 1-19. Three variations in the wave of a tone: sharp attack, gradual decay (A); gradual attack, sharp decay (B); sharp attack, sharp partial decay, followed by sustain and then gradual decay (C). Many other variations are possible.

reached quite quickly, followed by a quick decrease. The note is then sustained for a time, and then decays gradually.

Percussive instruments are characterized by very sharp attack times, with the result that the tone reaches its peak extremely quickly. Like the guitar, the piano has a rather long decay time, as shown in Fig. 1-20.

POLARITY

Consider the battery shown in Fig. 1-21A. It has two terminals, one marked plus (+) or positive, the other minus (−) or negative. In describing this battery we can say that terminal B is positive with respect to terminal A, or that terminal A is negative with respect to terminal B.

We can put a variable resistor (Fig. 1-21B) across the terminals of the battery. Since the resistor is now in shunt with the battery, the same polarity considerations will appear across the resistor. If the variable arm of the resistor is at the exact center of the resistor, we will have divided the voltage of the battery into two equal parts. If, for example, the battery is 12 volts, then six volts will appear across one half of the resistor and another six volts across the other half. For this reason a resistor set up in this manner is sometimes called a voltage divider.

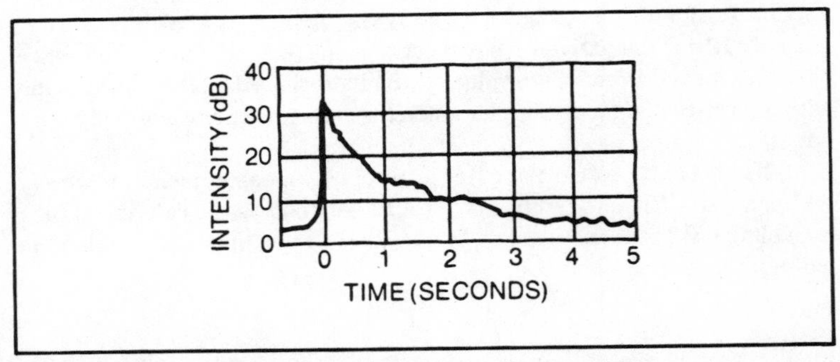

Fig. 1-20. Piano tone is percussive with an extremely short rise time.

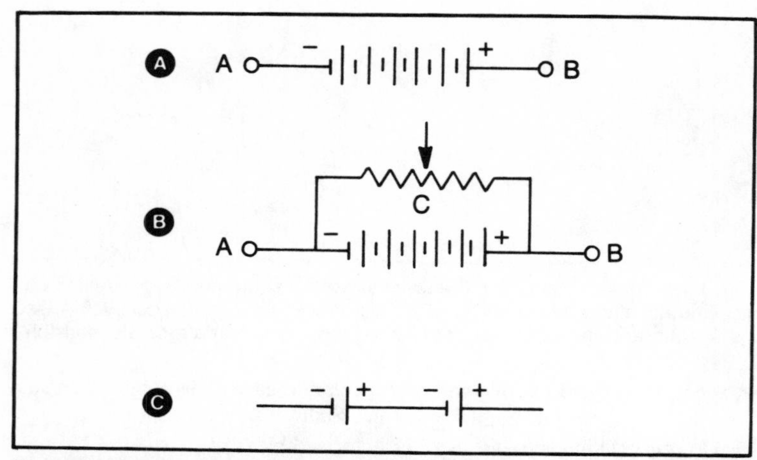

Fig. 1-21. Voltages are always measured across two points. One of the points is used as a reference. In voltage measurements reference points aren't necessarily fixed, but can be moved as required.

The center point of the variable resistor is identified by the letter C. Terminal C is positive with respect to point A, but is negative with respect to point B. The fact that a particular point is negative and positive at the same time is no cause for concern since it is an arrangement that is commonly used. In Fig. 1-21C we have two cells connected in series. At their junction point the plus terminal of the cell at the left is wired directly to the minus terminal of the cell at the right.

While we sometimes refer to voltage *at a point* this is sloppy thinking but is used that way because it is convenient to do so. A voltage is always measured across two points and that is the only way in which such a measurement can be made. One of the points is a reference.

When you make a measurement with a yardstick, the zero point or the left hand edge of that ruler is the reference. If, in Fig 1-21A we say that point B is 12 volts positive, it is 12 volts positive only with respect to point A. If, still using Fig 1-21 we say that point A is 12 volts negative, it is only 12 volts negative with respect to point B. Note that our reference point has moved from one terminal of the battery to the other. Measuring with a yardstick is much simpler since the reference point normally does not shift.

The idea of time does not enter into polarity measurements. If we say that point B is 12 volts positive (with A understood as the reference) then the voltage will remain 12 today, tomorrow, or a year hence, assuming the battery remains in good condition.

PHASE

The subject of phase isn't simply a mental exercise but has practical importance in recording. When the direct sound wave produced by a vocalist, instrumentalist or orchestra strikes a reflecting surface, part of that wave passes through that surface, part of it is absorbed by the surface material, part of it is converted to heat, and the rest of it is reflected as a sound wave. The direct wave and the reflected wave may be wholly or partially in phase, in which case they reinforce each other, or they may be wholly or partially out of phase, in which case the resultant sound may be wholly or partially canceled.

In Fig. 1-22 we have a circle divided into four sections called quadrants. Each quadrant forms an angle marked A,B,C, and D, and each of these angles is 90° (90 degrees). Since there are four such right angles the total enclosed by the circle is 360°. The size of the circle doesn't affect the angular measurement and so every circle is 360°. If we extend radii from the center of the circle we can divide the circumference of that circle into 360 equal segments with each representing 1°.

If we assume that the circle is made of spring steel and we snip the steel at one point, we will then have a straight section or line. The line is now divided into 360° and whether the line is short or long, made from a small circle or a large one, is of no consequence. See Fig. 1-23.

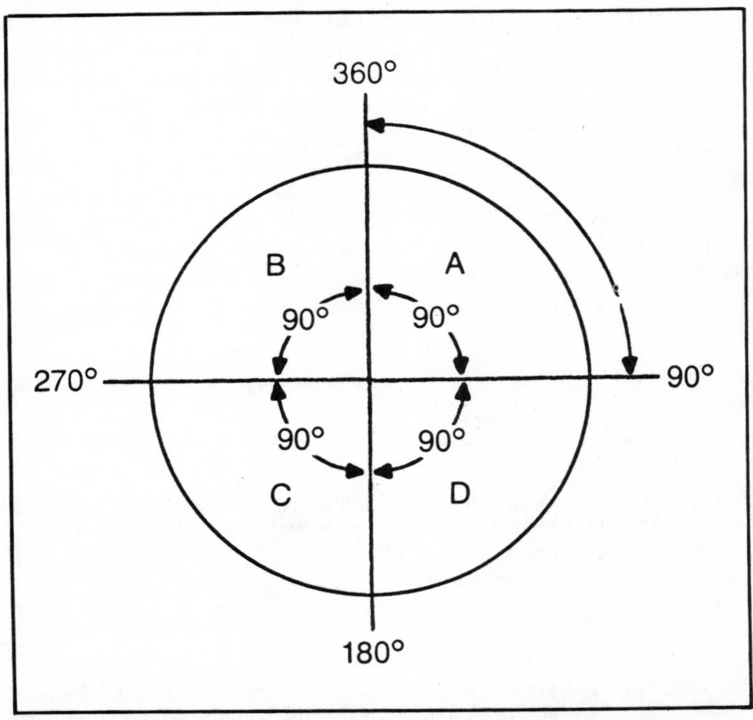

Fig. 1-22. Every circle, regardless of size, can be divided into 360°.

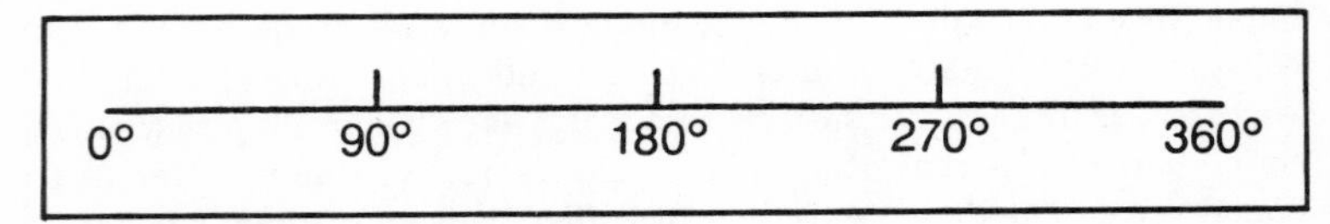

Fig. 1-23. The line is divided into 360°.

In Fig. 1-24 we have two sine waves using the 360° line as a reference. The wave starts at zero degrees, reaches a peak, drops to zero again at 180° and then increases once more to 360°. If the wave continues the same process will be repeated.

It takes time for a wave to reach completion and so the reference line, marked in degrees, is also our time base. The zero line, then, is the reference line for the positive and negative pulses of the ac voltage and is also the time base. See Fig. 1-25.

Phase is a comparison of the time difference or the angular distance of two or more waves. A wave whose frequency is constant is said to be single phase, as shown in A in Fig. 1-26. In drawing B we have two waves displaced by 90°. The two waves are 90° out of phase with each other. Since wave B started later than wave A it lags wave A. We could also say that wave A leads wave B. A similar situation exists in drawing C but here we have three waves that are out of phase with each other by 120°.

A wave can change phase with respect to itself. Figure 1-27A shows a wave having a constant frequency. If the same wave shifts its frequency, as shown in Fig. 1-27B, it can be out of phase with itself.

It is possible for two waves of different amplitude to start and stop at exactly the same time. These waves are then in phases and are additive. It is also possible for two waves to be 180° out of phase. If they both start and stop at the same time, if the waveforms are identical, and if one becomes positive while the other becomes negative, the overall result is wave cancellation and the voltages represented by the two waves equals zero.

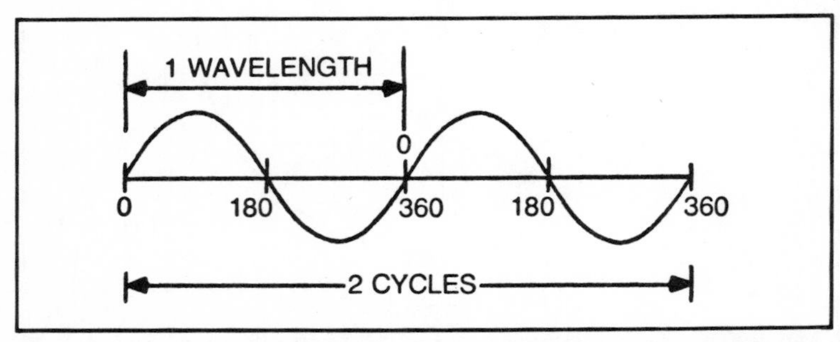

Fig. 1-24. Two cycles of a sine wave superimposed on straight lines, each of which is 360°. The start of each wave is 0°, the center point of the wave is 180°, and the completion point is 360°.

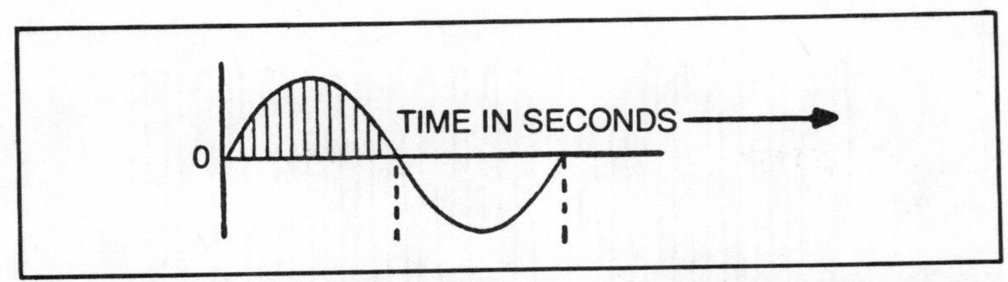

Fig. 1-25. The 0 axis is also the time axis. If it takes 1 second to complete this wave, the halfway point will be 180° or one-half second and the quarter way point will be 90° or one-fourth second. The vertical lines represent instantaneous values, that is, the amount of voltage at that particular time.

Between these two extremes are many possibilities. Waves can be partially in phase or partially out of phase, and so we can have partial voltage additions or cancellations.

ENERGY CONTENT OF MUSICAL TONES

Musical tones not only differ in frequency but in energy content as well. The greatest amount of energy is contained in the bass tones and a small

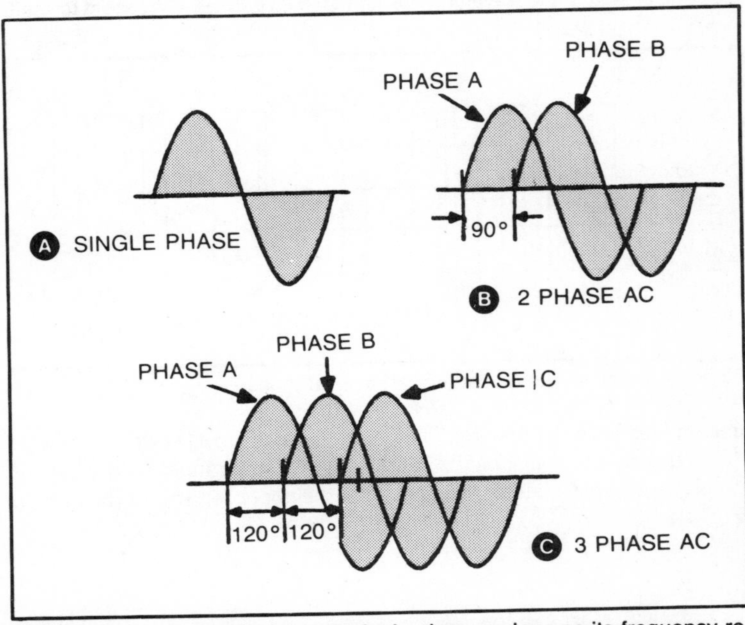

Fig. 1-26. The wave shown in A is single phase as long as its frequency remains constant. The drawing in B shows two waves, 90° out of phase with each other. Wave B lags wave A because its time of starting is later. We could also say that wave A leads wave B. In C there are three waves out of phase by 120°.

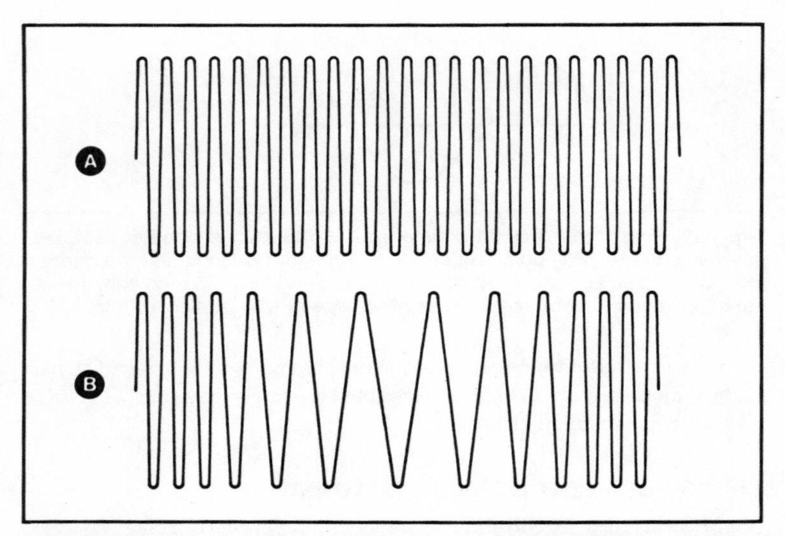

Fig. 1-27. The sine wave shown at the top has a constant frequency. We could also say that its phase remains constant. The lower drawing shows a frequency shift and so part of the waveform has changed its phase with respect to itself.

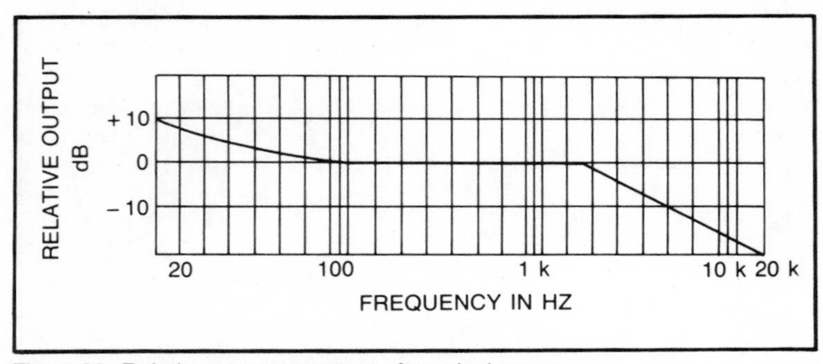

Fig. 1-28. Relative energy content of musical tones.

amount less in the midrange. There is a gradual rolloff starting at about 2500 Hz continuing to 20 kHz. It takes more energy input on the part of the player of a bass fiddle than in playing the upper register of the piccolo (Fig. 1-28).

Chapter 2
Acoustics

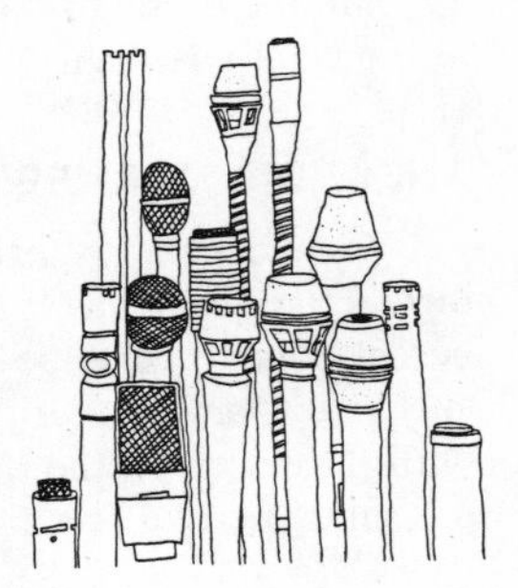

Acoustics is defined by The American Heritage Dictionary of the English Language as the "scientific study of sound, especially of its generation, propagation, perception and interaction with materials" and is further described as "the total effect of sound, especially as produced in an enclosed space." Acoustics, though, is as much an art as it is a science. There are many subject headings for acoustics: psychoacoustics, engineering acoustics, architectural acoustics, electroacoustics, and musical acoustics, but for our purpose acoustics is the effect of space and materials on sound.

DRY SOUND VS WET SOUND

Sound heard directly from a source such as instruments or the voice is known as *dry* or direct sound. Someone sitting in the first row listening to an orchestra outdoors probably hears dry sound only. *Wet* sound is reverberant sound, sound that is reflected from a surface such as a floor, walls and ceiling. Dry sound and wet sound combine to form the composite or total sound heard by an audience.

The sound reaching a microphone is not only that which is directly in front of the microphone. It is a combination of direct sound from a vocalist, or instruments, plus reflected sounds from the same source, entering the microphone after it bounces off the walls, floor, ceiling, chairs, people, and possibly even a lectern. All surfaces of all things reflect sound, comparable to the light reflected by various objects. The amount of sound reflection will vary depending on how smooth (hard surfaces) or how soft (rough, textured, or material covered) the surfaces are. The human skin reflects sound and does so much better than cloth. And strange as it may seem, there will be more sound reflection in an auditorium filled with women wearing miniskirts than in that same auditorium occupied by men and women wearing pants. Microphone pickup also includes direct sound from voices and instruments which are located to the sides of the microphone.

DRY RECORDING

Dry recording is a technique in which the individual performers are recorded separately and need not be in the studio at the same time. As a result, the music of the combined group does not exist until at the time of the final mixdown.

ANECHOIC CHAMBER

We can have two extremes. The first would be an anechoic chamber in which we have direct or dry sound only and in which all possible reverberant sound energy is totally absorbed. In an outdoor situation or an anechoic chamber, the *sound-pressure level* (spl) decreases by one half, a 6 dB drop, every time the distance from the sound source is doubled. If the spl is 90 dB 6 feet from the sound source, at 12 feet it will be 90 dB − 6 dB = 84 dB.

Technically, sound measurements should be made with no individual in the chamber. Anechoic chambers do not resemble real life and are constructed solely for the purpose of making dry sound measurements. We do have a comparable setup in outdoor concerts where there is no backdrop and a minimum of reflecting surfaces. Music in an open air environment is usually less pleasing than the same composition played by the same orchestra in an enclosed environment, making allowances for noise distractions produced by pedestrian or vehicular traffic. The inference is that the human ear finds a correct combination of dry and reverberant sound to be more pleasing. Why this should be so more properly falls in the realm of psychoacoustics.

As a microphone is moved from the anechoic chamber to an open field outdoors, to an outdoor stage, or into a closed auditorium or room environment, it progressively sits in a reverberant envelope of direct and indirect sound. No microphone has a completely dead side. Sound reaching the microphone may be suppressed from the side and rear, but it still generates audible sound, however weak.

ECHO CHAMBER

The other extreme would be the recording or reproduction of sound in an echo chamber, a room completely free of any sound absorbing substances, with walls, floor and ceiling made of some hard materials and their surfaces having maximum sound reflectivity (Fig. 2-1). Anechoic and echo chambers are used for testing microphones and speakers, but in real recording and playback situations it takes a finite amount of time for sound to be absorbed and reflected, with the time dependent on frequency and the kinds of acoustic materials used.

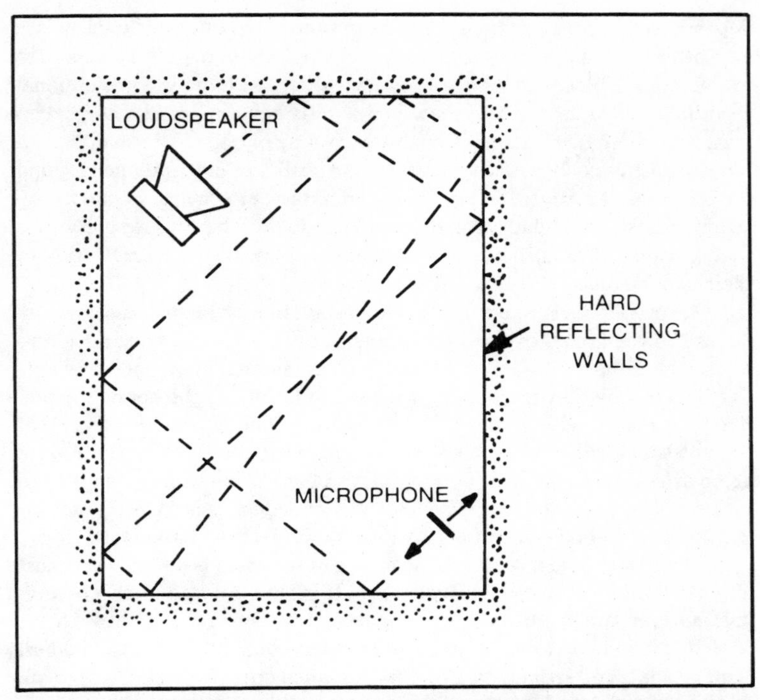

Fig. 2-1. Possible arrangement of speaker and mic in an echo chamber. Very little direct sound from the loudspeaker reaches the mic.

ROOM ACOUSTICS

The acoustics of a room, whether intended for recording or playback, depends on the size and shape of the room, the materials used in it, the furniture, wall coverings and number of people. Some materials, such as cloth and carpeting, are sound absorbers in varying degrees; hard surfaces, such as walls made of plasterboard or wood are highly sound reflective. Usually a room, especially one used in the home for sound reproduction, has a mix of all kinds. Walls, floor and ceiling supply the greatest amount of reflected sound. Clothing worn by people tends to be absorptive, but the total surface area of the clothing of a person is small compared to that of the floor, walls and ceiling. In the studio these can be treated to minimize or prevent interference from outside noise. No two rooms, unless identical in every respect, including occupants and their positioning, will affect recording or playback in the same way.

WHY BOTHER WITH ACOUSTICS?

Since this is a book about microphones, it would seem that a study of acoustics would be an extraneous topic. The answer is that you cannot depend solely on microphones, for no matter how high their quality and no matter how suitably they are placed, the end result may be far less than satisfactory. The acoustics of a room, whether that room is a professional recording studio, or the hi/fi recording and listening room in a home, can mean the difference between a quality recording or one of low quality. And room acoustics do more than just affect the quality of the reproduced sound for acoustics can affect the playing mood of the performers. Vocalists, instrumentalists, or orchestral groups will not do their best if they are aware that, because of acoustics, the end result will not be a true reflection of their capabilities.

The proper study of acoustics is important in both the professional studio and in the home. In the studio, musicians expect, and have a right to expect, that the acoustic treatment of the studio will more accurately reflect the sounds they produce. If a recording is done in the home, inattention to acoustic treatment is a "double curse." The reason for this is that the acoustics will affect the sound as it is being recorded and will once again affect the sound as it is being played back if the same room is used for recording and playback. Thus, a study of acoustics is as important for the amateur in-home recordist, as it is for professional studios.

Whether we like it or not the acoustics of the room is possibly the most important member of the musical group. It is always present for a recording, is never late or absent for any reason. It cannot be ignored with the hope that it will go away, for that it will not and cannot do. For most homes, acoustics is an afterthought. For a professional studio, concert hall or auditorium, acoustic treatment is as important as any other architect's blueprint.

Poor acoustics in a studio will not be cured completely, regardless of the microphone used, unless the studio is acoustically treated. However, the selection of the right microphone can, with proper application, improve a poor situation. If the studio suffers from noise transmission through the walls, or if the room is acoustically "live," the resulting noise or bright, "pingy" sound reflected from wall to wall will be picked up by the microphone.

While a well-designed cardioid microphone, described in detail later, will tend to reject much of this sound, the problem should be resolved by acoustic deadening of the room. However, working relatively "tight" with a good cardioid microphone will allow voice levels to override ambient noise for a better signal-to-noise ratio.

ANNOUNCER'S BOOTHS

Because announcer's booths often have very little sound absorption quality, sound, particularly in the voice spectrum, tends to resonate annoyingly. Not only are booths poor for voice, but mechanical noises from air conditioning and other equipment are reflected and "magnified" by flat walls and windows, creating resonance related problems.

REVERBERATION

It is possible to hear sound after the original sound source has stopped functioning. The sound energy supplied by a source does not go out of existence just because that source has been turned off. The sound energy of the source radiates into space and, if there are no reflecting surfaces, will continue on outward until its energy is completely dissipated. That sound energy, however, isn't "lost," for we can neither create energy nor destroy it. The sound energy continues until all of it has been converted into heat energy.

If the sound energy from the source strikes an object, such as a wall, some part of that energy will be reflected. If our ears are in the path of the direct sound we will hear that direct or "dry" sound. And if our ears are in the path of the reflected sound we will hear it also. The reflected sound or reverberant sound can reinforce the original dry sound, or may be out of phase with it, causing complete or partial cancellations. The reason your voice may sound so rich in the shower is due to reverberant sound from the tile walls that surround you.

ADVANTAGES AND DISADVANTAGES OF REVERBERATION

There are advantages and disadvantages to reverberant sound. It can make the original dry sound appear to be richer. Music with a certain amount of reverberation seems to be more satisfying. Our musical taste calls for a certain amount of reverberation. If there isn't enough of it, music appears to be "muddy," that is, ill defined. If there is too much reverberation, the speech of someone whether using or not using a microphone will be partially or completely unintelligible. So, because of these effects on the sound, it is important for us to be able to control reverberant sound, to increase or decrease it if and when necessary.

As the working distance between the sound source and the mic is increased there is a greater pickup of reverberant sound, adding more ambience to a recording. Consequently, the sound becomes less dry and has added warmth, but this technique has its limitations. A mic that is too far from the sound source will reproduce ambient sound whose level could be too high compared to the original sound, possibly overwhelming it. The recorded sound will now be overwhelmed with echoes and background noise. This effect will be particularly noticeable if playback takes place in the same room used for recording.

Unwanted sound becomes more recognizable during playback. Consequently, a first basic rule in recording is to make sure that most of the sound reaching the mic is direct sound.

ECHOES

It takes a certain amount of time for dry sound to reach our ears. That dry sound, reflected from some surface also takes time to reach our ears in the form of a reverberant sound. Since the reverberant sound wave takes a longer time to reach us there is always a small time displacement between dry and reverberant sound. If this out of phase condition is small, and it usually is, our ears cannot distinguish the time separation, and so we hear the dry and reverberant sound as a composite.

However, if the room in which we are listening is large enough, the reverberant sound will take so long to reach our ears that we will be conscious of that sound as though it was supplied by a separate sound source. It will appear to be weaker than the original and we will recognize it as an echo. The time displacement of the echo depends on the size of the room and our position with respect to the reverberant sound. In some places, such as large railroad stations, the effect of the reverberant sound is to make the dry sound wholly or partially unintelligible.

If the room producing the echoes has hard surfaces, ceiling, floor and walls, it is possible for the echoes to supply multiple reverberations. In that case we will hear not one, but a series of echoes that will continue, until the sound energy is dissipated. A series of sound reverberations of this kind are termed flutter echoes.

An echo is a member of the family of reverberant sounds and so is a sound which has been reflected, but its strength and time of arrival at the ears is such that it is recognizable as a repetition of the original sound. Ordinarily, reverberant sound forms an indistinguishable composite with direct sound. The delay time of a sound for the formation of an echo is about 50 milliseconds or more. We can also get multiple echoes when the echo is repeatedly reflected from other surfaces.

DIFFUSION

Not only the material of which a room is made but its shape as well, can affect room acoustics. A sound field is said to be diffuse if the intensity of the sound is about the same in all parts of the room. A stage having a hard, highly reflective concave back wall will tend to beam the sound and so will a ceiling having a dome shape. If a room has areas where the sound is very strong or weak, then the sound diffusion characteristic of that room is not as good as it should be.

Good diffusion is helpful either in the home room used for recording and playback or in a recording studio. It makes positioning of microphones and performers less critical and makes it easier to get good recordings.

SOUND REFLECTION AND OBJECT SIZE

Whether an object will reflect sound also depends on the size of the object compared to the frequency of the sound striking it. The frequency of a sound is inversely proportional to its wavelength and so high pitched sounds have a short wavelength; low pitched sounds have a long wavelength.

As indicated in Fig. 2-2, if a low-frequency wave strikes an object whose dimensions are smaller than the wavelength of the sound, the sound will

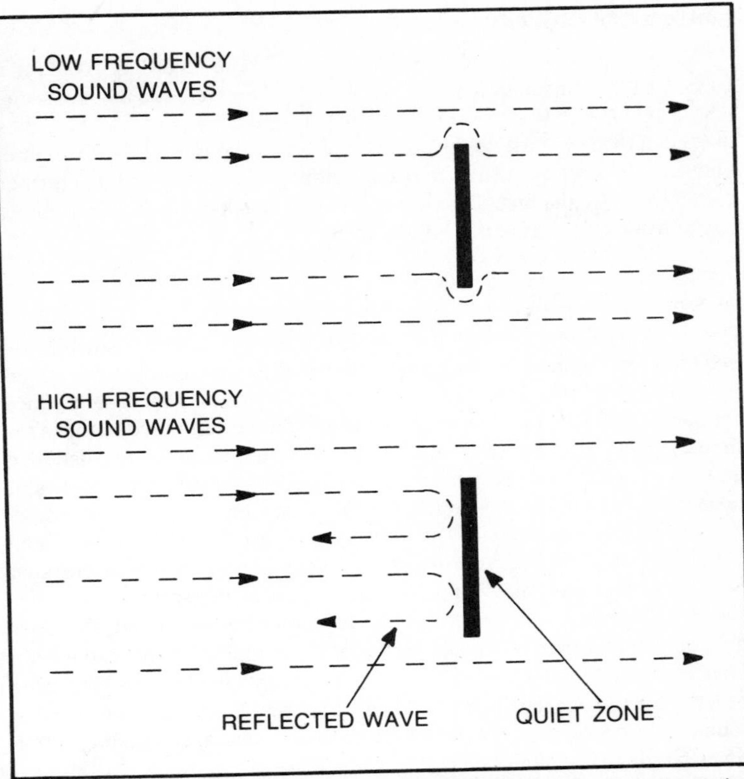

Fig. 2-2. One of the factors in the amount of sound reflection depends on the size of the reflecting object compared to the wavelength of a particular sound.

travel around the object. However, if a high-frequency sound (one having a short wavelength) strikes an object whose size is greater than its wavelength, the sound will be reflected. Sound behind the object will be heard faintly, if at all. Most of the sound will be reflected. Whether the sound will be returned to the source depends on the angular relationship of the object to the source.

The wavelengths of sound in air range from about 1 inch for treble tones to about 50 feet for bass notes. Reflective surfaces in a recording studio or in the home have a large variety of sizes, and in many instances may not be flat. A concave surface, for example, will tend to focus sound waves, but here also whether sound waves will be reflected depends on the relationship of sound frequency to object size. In the case of a concave surface, the diameter of the reflector must be greater than the wavelength of the incident sound. In the case of a convex surface, possibly used for diffusing reflected sound, such elements function when the surface dimensions are greater than the sound wavelengths.

SOUND COLORATION

A professional recording studio, or an in-home recording room, or a concert hall, or an auditorium all have acoustic characteristics that are somewhere between that of the anechoic and the echo chambers. The acoustics of each of these will affect the quality of the sound picked up by the microphones. Thus, we get sound coloration right at the outset and it is further complicated by the fact that we also get sound coloration in the listening room in which the sound is reproduced.

SOUND DEFINITION

When we listen to sound it is the direct, not the reflected sound, that indicates the location of the sound. The location of sound is determined primarily by the sound that reaches the ear first, that is, direct sound. Once we have determined its location, the sound source can move but it takes time for us to become aware of the new location. If, as we are listening and have determined just where the sound is, the sound source moves, it takes time for us to become aware of the new location, an acoustic phenomenon known as *precedence effect* or *Haas effect*. Precedence effect, for example, takes place when a vocalist or instrumentalist performs at one spot on a stage and then moves to some other part of the stage.

While our ears can determine the location of direct sound, this does not apply to reverberant sound. Such sound is supplied by areas which are huge compared to that covered by the original dry sound source. For some psychoacoustic reason, this combination of direct, location identifiable sound, and reverberant, non-location identifiable sound, is pleasing, within reverberation time limits.

However, when music is recorded, both direct and reverberant sounds are recorded at the same time. In the playback process what we hear is a combination of direct and reverberant sound coming from a pair of speakers usually located in front of the listener. This is an unnatural situation and does not represent the sound as it is produced in the studio or a concert hall. A better and more realistic method would be to extract the ambience from the recording and feed it to a separate pair of rear positioned speakers.

REVERBERATION TIME

When sound strikes a surface, such as that of a wall, some of the sound is reflected, some is converted to heat energy, and some of the sound passes through. A hard smooth surface will supply optimum sound reflectivity; a material that is acoustically transparent, such as the grille of a speaker, will permit passage of most of the sound. Aside from its frame, an open window is acoustically transparent.

Some of the sound is reflected and absorbed by the person hearing it but the area of the human body is small compared to the combined surfaces of walls, floor, and ceiling and it is from these that we get the greatest amount of reflected sound. Further, the clothing worn by people tends to be more absorbent than reflective. The continuous reflection of sound in an enclosed space is called reverberation or reverberant sound and the time it takes for this sound to decrease by 60 dB, one millionth of the original sound source intensity (after the input signal has been cut off) is known as reverberation time. Reverberation time is sometimes written as T_{60}. The number 60 refers to 60 dB.

Reverberation time is a function of room volume and sound absorption. It is directly proportional to the volume of an enclosed space, such as a recording studio or an in-home listening room and inversely proportional to the total amount of sound absorption in that enclosure. Reverberation time is not the same for all frequencies. The formula for calculating reverberation time, developed by professor Wallace C. Sabine in 1895 indicates average reverberation time and does not take frequency into consideration.

$$T_{60} = \frac{0.05\ V}{S_a}$$

In this formula, T_{60} is the reverberation time in seconds; V is the volume of the room in cubic feet and S_a is the total equivalent of sound absorption in sabins per square foot of surface material.

When sound moves through air some of its energy is dissipated in the form of heat which moves the air molecules. The total sound energy leaving a sound source, such as an instrumentalist, vocalist, or orchestra, never reaches the reflecting surfaces. Consequently, the reflected sound is not as strong as the direct sound, assuming there is no resonant condition that could conceivably strengthen the sound. In movement through the air, high frequencies are energy losers more than lower frequencies. Little sound energy is lost in the form of heat by air at frequencies below 1500 Hz.

The basic Sabine formula can be rewritten to take into account the conversion of sound energy to heat energy in its passage through the air. Using the metric system, the formula can be written as:

$$T_{60} = 0.16V/(A + S_a)$$

In this formula, the reverberation time, T_{60}, is still in seconds. V is the volume of the room in cubic meters, S_a is the total amount of sound absorption in square meter sabins. A is the absorption coefficient of air. The absorption of sound by air, the loss of sound energy in the air in the room, is significant only for concert halls or auditoriums. For the usual recording studio or in-home setup, A can be ignored. The formula then reduces to:

$$T_{60} = 0.16V/S_a$$

and, except for the coefficient, (0.16) is the same formula previously supplied for calculations in the English system of measurement. In both formulas we are concerned only with the volume of the enclosed space and the amount of sound absorption. Note that neither formula takes the velocity of sound into consideration.

TOTAL SOUND ABSORPTION

To calculate the value of sound absorption, it is necessary to multiply the area of each room surface by its coefficient of absorption. The total value is the sum of all these products, plus any absorption in the room due to chairs, rugs, drapes or curtains and people. The acoustics of an in-home listening room, hall, auditorium or recording studio is subjective. We generally tend to associate long reverberation times with large enclosures; cathedrals, churches, and auditoriums. Long reverb time conveys an impression of spaciousness. Conversely, you would expect chamber music,

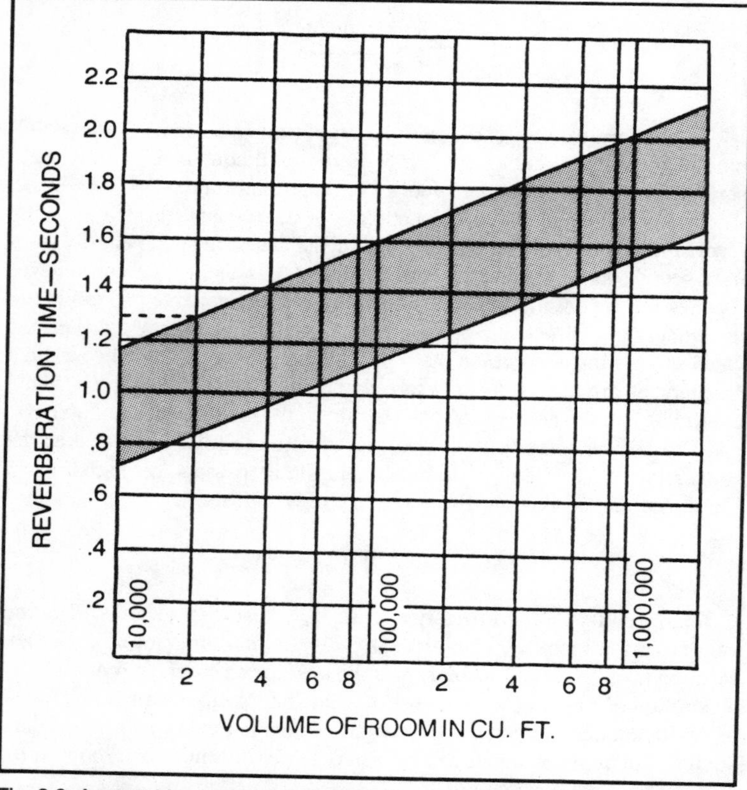

Fig. 2-3. Acceptable reverb times for rooms of different volumes. The dark area indicates acceptable reverberation time.

as its name implies, to be played in a small, intimate room. The tinted area in the graph shown in Fig. 2-3 is an indication of acceptable reverb times. Acceptable reverberation time isn't a single fixed number, but as the dark area of the graph in Fig. 2-3 indicates, has upper and lower limits. For example, to find the reverberation time of a room having a volume of 20,000 cubic feet, locate the number 2 at the bottom of the graph. This digit 2 represents 20,000 cubic feet. Move up along the line marked 2 until you reach the bottom edge of the shadowed section. This is on a horizontal line marked .8. The minimum reverberation time, then, is 0.8 second. Move up along the vertical line represented by 2 until you reach the top of the shadowed area. If you will look to the left you will see that you are on a line represented by 1.3. The reverberation time of this room should be between 0.8 and 1.3 seconds.

SOUND ABSORPTION COEFFICIENTS

The upper and lower limits of sound absorption coefficients are 0 and 1. The number 0, or 0% means that the material does not absorb sound at all. This means that the material either allows all sound to pass through, or absorbs all of it. A large open window that permits all sound to pass through without reflecting any of it can be said to have a sound absorption coefficient of 0. An ideal echo chamber would have all its surfaces made of materials having a 0% sound absorption coefficient or as close to that ideal as possible.

At the other extreme, a sound absorption coefficient of 1 or 100% indicates a material that absorbs all sound directed at it and reflects no sound at all. Such a material would be the ideal substance for an anechoic chamber. In practice most substances are somewhere between 0% and 100%.

SOUND ABSORPTION COEFFICIENT VS FREQUENCY

The sound absorption coefficient of a material isn't a single fixed number but varies with frequency. For some materials the sound absorption coefficient remains fairly fixed. For others the coefficient increases with an increase in frequency, while it may decrease for different materials. Table 2-1 shows how the sound absorption coefficient can vary with frequency. However this is for a selected material and does not apply to any other.

Table 2-2 may be helpful in making calculations of the reverberation in rooms. The coefficients given are for a frequency of 512 Hz. A more useful table is Table 2-3. Here we have commonly used materials but the sound absorption coefficients are supplied in terms of their behavior on an octave basis. The lowest frequency given is 125 Hz, the next octave is 250 Hz, then 500 Hz and so on. Note that a glass window has a decreasing coefficient with an increase in frequency while lightweight drapes have an opposite effect. Another point of interest in Table 2-3 is that a sound absorption coefficient may increase and then decrease with an increase in frequency. Heavy drapes for example, have an increasing sound absorp-

Table 2-1. Variation of Sound Absorption Coefficient with Frequency for 3/16" Plywood.

Frequency (Hz)	Sound Absorption Coefficient
125	0.35
250	0.25
500	0.20
1,000	0.15
2,000	0.05
4,000	0.05

tion coefficient until 1 kHz and then the coefficient gradually decreases.

SOUND ABSORPTION COEFFICIENTS OF BUILDING MATERIALS

The absorption coefficients of various building materials are often supplied as an overall figure without taking frequency into consideration, such as that shown earlier in Table 2-2. According to this table an upholstered seat could have a sound absorption coefficient of 0.05. This indicates that

Table 2-2. Sound Absorption Coefficients of Various Building Materials.

Material	Sound Absorption Coefficient
Linoleum on Concrete Floor	0.03 to 0.08
Upholstered Seats	0.05
Ventilating Grilles	0.15 to 0.50
Painted Brick Wall	0.017
Unpainted Brick Wall	0.03
Unlined Carpet	0.20
Felt Lined Carpet	0.37
Light Fabric, 10 oz. per square yard	0.11
Medium Fabric, 14 oz. per square yard	0.13
Heavy Fabric, 18 oz. per square yard	0.50
Concrete or Terrazo Floor	0.015
Wood Floor	0.03
Linoleum on Concrete	0.03 to 0.08
Glass	0.027
Marble	0.01
Ventilating Grilles	0.15 to 0.50
Plaster on Brick	0.025
Plaster on Lath	0.03
Rough Finish Plaster	0.06
Wood Paneling	0.06

Table 2-3. Sound Absorption Coefficients of Commonly Used Materials at Various Frequencies.

Material	Frequency in Hz					
	125	250	500	1000	2000	4000
Glass window	0.35	0.25	0.18	0.12	0.07	0.04
Lightweight drapes	0.03	0.04	0.11	0.17	0.24	0.35
Heavy drapes	0.14	0.35	0.55	0.72	0.70	0.65
Wood floor	0.15	0.11	0.10	0.07	0.06	0.07
Carpet	0.02	0.06	0.14	0.37	0.60	0.65
(on concrete)						
Brick, glazed	0.03	0.03	0.03	0.04	0.05	0.07
Carpet, heavy, on 40-ounce hairfelt or foam rubber	0.08	0.24	0.57	0.69	0.71	0.73
Concrete block, coarse	0.36	0.44	0.31	0.29	0.39	0.85
Concrete block, painted	0.10	0.05	0.06	0.07	0.09	0.08
Fabric, medium velour, 14 ounces per square yard, draped to half of its flat, un-draped area	0.07	0.31	0.49	0.75	0.70	0.60
Floors						
concrete or terrazzo	0.01	0.01	0.015	0.02	0.02	0.02
linoleum, asphalt, rubber, or cork tile on concrete	0.02	0.03	0.03	0.03	0.03	0.02
wood	.15	.11	.10	.07	.06	.07
Glass						
large panes of heavy plate glass	0.18	0.06	0.04	0.03	0.02	0.02
Gypsum board, 1/2 inch, nailed to 2 inch × 4 inch lumber, centers 16 inches apart	0.29	0.10	0.05	0.04	0.07	0.09
Marble or glazed tile	0.01	0.01	0.01	0.01	0.02	0.02
Plastr, gypsum or lime, rough finish or lath	0.02	0.03	0.04	0.05	0.04	0.03
Plywood paneling, 3/8 inch thick	0.28	0.22	0.17	0.09	0.10	0.11
Water surface, as in a swimming pool	0.008	0.008	0.013	0.015	0.02	0.025

the individual seat will absorb 0.05 or 5% of the sound and will reflect 95%. Note how widely different these materials behave. A heavy fabric has a coefficient of 0.50 or 50%, reflecting only half the incident sound. Marble has a coefficient of 0.01 or 1% reflecting 99% of the sound.

To calculate the total equivalent sound absorption of materials in sabins per square foot, multiply the sound absorption coefficient by the area. Thus, if a cement floor with an absorption coefficient of 0.015 measures 40′ × 30′, the total area in square feet is 40 × 30 = 1200 square feet. 1200 × 0.015 = 18. This floor, then, has a total absorption of 18 sabins.

Table 2-4. Calculation of the Sound Absorption of a Hall.

Material	Dimensions (feet)	Area (sq. feet)	Absorption Coefficient	Equivalent Absorption (sabins)
Floor, cement	56 × 112	6,272	0.015	94
Walls, wood panel	8 × 336	2,688	0.06	161
Plaster	20 × 336	6,720	0.025	168
Ceiling, plaster	56 × 112	6,272	0.03	188
Curtain, velour	39 × 20	780	0.5	390
Total absorbing power, bare room				1,001
Plus 800 unupholstered seats at 0.25 sabin				200
Total absorbing power, no one present				1,201

CALCULATION OF REVERBERATION TIME

To calculate the reverberation time of a room it is first necessary to determine the equivalent absorption in sabins. As an example, assume we have a hall measuring 112′ × 56′ × 28′. The volume of the room is approximately 175,000 cubic feet. Measurements of the floor, walls and ceiling are supplied in Table 2-4. Each area in square feet is multiplied by its absorption coefficient. We are assuming the room contains 800 seats which are not upholstered. Table 2-4 shows the measurements of this hall and its total absorbing power in sabins with no one present. The Table shows that this is 1,201 sabins.

To calculate the reverberation time, we refer to the formula supplied earlier:

$$T_{60} = \frac{0.05\ V}{S_a}$$

V is 175,000 and S_a is 1201. Substituting these numbers in the formula we get:

$$\begin{aligned} T_{60} &= (0.05)\ (175{,}000)/1201 \\ &= 8750/1201 \\ &= 7.29 \text{ seconds} \end{aligned}$$

The absorption, in sabins, per person, depends on the individual and on the kind of seat they are using, but it ranges from about 3.0 to 4.3. If we assume an absorption of 4.05 sabins, we can compute the total absorption in sabins and the reverberation time in seconds. Table 2-5 shows the reverberation time for an audience of up to 800 persons. Note that the sound absorption increases and the reverberation time decreases. However, dou-

Table 2-5. Effect of an Audience on Reverberation Time.

Audience (number present)	Absorption (sabins)	Reverberation Time (seconds)
0	1,201	7.3
200	2,011	4.3
400	2,821	3.1
600	3,631	2.4
800	4,441	2.0

bling the size of the audience does not mean doubling the absorption power, nor is the reverberation time cut in half.

If you will now go back to Fig. 2-1 you will see that even with the room filled to capacity the reverberation time is too high. As a result the audience will have some difficulty in understanding a speaker. A remedy in this case would be to increase the absorption by applying absorbent acoustic materials to the walls and ceiling to bring this hall down to an acceptable reverberation time. For this size hall the reverberation time should be in the range of 1.2 to 1.7. A good figure would be somewhere between these two limits, possibly 1.5.

Our formula for reverberation time, supplied earlier, is:

$$T_{60} = \frac{0.05\ V}{S_a}$$

The volume (V) is 175,000 cubic feet and the desired reverberation time (T_{60}) is 1.5 seconds. S_a is the number of sabins required to achieve this result. If we substitute these numbers in the formula we will have:

$$1.5 = \frac{(0.05)\ (175{,}000)}{S_a}$$

We can transpose S_a and 1.5 to get:

$$S_a = \frac{(0.05)\ (175{,}000)}{1.5}$$

= 5,833 sabins. We can round this number off to 5,830.

This amount of sound absorption is for an average audience of 400. Naturally, it is impossible to determine in advance just how large the audience will be, but 400 can be regarded as a practical number. Referring to the value for an audience of 400 (Table 2-5) the necessary increase in absorption is:

$$5{,}830 - 2{,}821 = 3{,}009 \text{ sabins (rounded off to 3,010)}$$

The area of treatment, in square feet, can be found by dividing the necessary added absorption (in sabins) by the coefficient of the material used.

That is:

$$\frac{\text{necessary absorption (sabins)}}{\text{absorption coefficient}} = \text{area of treatment}$$

We can arbitrarily select a number of different materials having a fairly high absorption coefficient: Thus:

$$3010/0.40 = 7{,}525$$
$$3010/0.60 = 5{,}017$$
$$3010/0.80 = 3{,}762$$

The areas required to give the added absorption of 3,010 sabins using coefficients 0.40; 0.60 and 0.80 are:

Coefficient	Area Required (square feet)
0.40	7,525
0.60	5,017
0.80	3,762

The location of the absorbing material is governed somewhat by the size and shape of the room, but as a rule the ceiling and the rear wall are used. Side walls are used when the ceiling is low.

STUDIO REVERBERATION TIME

As another example, consider the calculation of the total sound absorbing characteristics of a studio measuring 40′ × 30′ × 10′. The total volume is the product of these numbers and is 12,000 cubic feet. The total absorption is shown in Table 2-6. This is the total absorption value of the studio, with no one present. You can allow 4.05 sabins (see Table 2-7) for each person who will be in the studio. With ten persons present S_a will be 1074.5 plus 40.5 or 1115 sabins.

The reverberation time of this studio will be:

$$T_{60} = \frac{0.05\ V}{S_a} = \frac{0.05 \times 12{,}000}{1115} = \frac{600}{1115} = 0.54\ \text{second}$$

The chart (Fig. 2-1) shows that for a room of this volume the reverberation time is not acceptable. The wall to wall carpeting is the main factor

Table 2-6. Sound Absorption in a Studio.

Material	Dimensions (feet)	Area (sq. feet)	Absorption Coeff.	Sa
Cement Floor	40 × 30	1200	0.015	18
Carpeting	40 × 30	1200	0.57	684
Wood Panel Walls	40 × 30 × 4	4800	0.05	240
Plaster (ceiling)	40 × 30	1200	0.025	30
Velour curtain	20 × 10	200	0.5	100
Seats(10)			0.25	2.5
				1074.5

contributing to this result. The large velour curtain is another cause of low reverb time.

In this case, using a carpet measuring 30′ × 20′ instead of the original size would reduce the sound absorption of the carpet to:

$$30 \times 20 \times 0.57 = 342 \text{ sabins}$$

The total absorption would drop to 732 sabins which we can round off to 730. Using the formula for calculating reverberation time would increase the reverberation time to 0.82 second, a more acceptable figure.

Neither formula for calculating reverberation time (the one using the metric system and the other the English system) takes frequency into consideration. To get a more accurate representation of reverberation time, the formula should be worked for a series of frequencies starting at about 125 Hz and continuing through 4,000 Hz. A curve can then be plotted and drawn to show the effect of frequency on reverberation time. To be able to do this we would need to know the sound absorption coefficients of all the materials in the studio for each spot frequency from 125 Hz to 4,000 Hz. In a room that is properly treated acoustically, all sonic frequencies will have the same reverberation time, that is, the decay time of the reverberant sound will be independent of frequency. The room can then be con-

Table 2-7. Sound Absorption of Seats and Audience at 512 Hz.

	Equivalent Absorption (in sabins)
Audience, seated, units per person, depending on character of seats, etc.	3.0-4.3
Chairs, metal or wood	0.17
Pew cushions	1.45-1.90
Theater and auditorium chairs	
Wood veneer seat and back	0.25
Upholstered in leatherette	1.6
Heavily upholstered in plush or mohair	2.6-3.0
Wood pews	0.4

sidered to have a "flat" response.

In a room having hard, smooth, rectangular walls, it is possible for sound reflections to travel back and forth, but with the absorption of sound always at some specific groups of frequencies. This selective absorption results in the presence of some sounds after the rest have been completely absorbed, giving the room a distinctive sound coloration. The effect is most prevalent in small rooms. The goal should be to have the reverberation time the same for each octave of sound.

A wood paneled room is more sound energy absorbent for bass tones; a carpet for treble. When you have a situation using pure sinusoids, a room that is both carpeted and paneled will not distort the sound. Except for very quiet notes from the flute, music consists of a fundamental plus a number of harmonics. A carpeted room will not only absorb treble tones, but the harmonics of lower frequency notes. This results in an unclear bass since it is the harmonic content of a tone that gives it a distinctive character. Such bass tones sound "unclear" or "muddy."

A common misconception is that if a room is carpeted, for example, the solution is to increase bass output to compensate for the higher rate of absorption of bass tones. Equalization can be used in this example to make up for bass attenuation in the recording or playback room due to acoustics, but it does not affect decay time. Sound pressure level (spl) in the bass region has been altered by the equalization process, but reverberation time still remains a function of frequency.

Reverberant sound isn't "single source" sound, but is a mixture of all reflected sounds, supplied by areas which are huge in comparison with dry sound sources. The ear cannot identify the location of such sound and so, for reproduced reverberant sound to approximate the original, it too must be "location unidentifiable." This highlights one of the problems in sound reproduction which, in the home, is typically given by two speakers. Both direct and reverberant sounds are supplied by these speakers, generally positioned in front, and since such sound is position identifiable, even if only approximately, there is a lack of realism in the reproduction. A better arrangement would be to have the front speakers supply direct sound, with pairs of speakers at the sides and the rear furnishing reverberant sound. The character of this sound should then supply the necessary ambience, that is, the sum total of reverberant sound, supplying a sonic illusion as to the kind of hall or auditorium in which the music was performed.

In a concert hall we are aware of the source of direct sound, for not only is the dry sound stronger than the reverberant, but we can see the musicians. However, we are unaware of the exact location of reverberant sound. Not only are the reflecting surfaces large, but they can vary in their reflective abilities. We also hear multiple reflections randomly as the sound is bounced from one surface to the next.

NONLINEARITY

Any acoustic environment is nonlinear. Unlike an amplifier capable of supplying a flat frequency response, not only over the audio range but well below and above it, reverb sound suffers treble loss and so this sort of sound has a much narrower band than dry sound. Further, the outer ear, the pinna, is shaped so as to favor direct sound, hindering the reception of sound from the rear. Reverberation also has a considerable effect on the intelligibility of speech. If the reverb time is too long, someone close to the speaker may understand the articulation but there will be places in the listening room where the speech will sound garbled. The same effect takes place in music with instruments seeming to lack clear definition.

The shape of the reflecting surface has an effect on the distribution of reverberant sound. A surface that is concave focuses the sound and so this kind of shape is unsuitable for recording situations. A shape that is convex distributes the sound over a large area, much larger than that of a comparably sized flat surface. The amount of sound absorption not only depends on the type of material, but on its surface area. The larger the area, the greater the absorption.

USING A ROOM FOR SPEECH

The criteria of a room used for speech are different from those required for a room used for musical sound reinforcement. With speech, the objectives are clarity and distance. This means that listeners throughout the room must hear every word of a speaker clearly and distinctly. Furthermore, the sound must reach the final row of seats with sufficient level.

With speech only, reverberation can be a nuisance since sound cancellation can lead to dead or weak areas, or where sound reinforcement can possibly produce garbled speech.

The problem, though, may start right at the mic. Most speakers haven't the faintest idea of how to use a mic, whether hand held or stand mounted. Some prior practice and some instructions by the recordist are often desirable. The recordist may also want to experiment with the setting of the gain control of the sound amplification system and make allowances for the sound absorption of an audience.

In many instances, though, a room or a hall or an auditorium will be used for both speech and music, as in a church, or even in a concert hall when announcements are to be made. In a church, the use of an organ and choral music means a relatively long reverb time is needed. This is a sharp contrast to speech that requires a short reverberation time. What is required, then, is a combination of compromise and adjustment. In the case of a church, a sound reflector positioned directly behind the pulpit will be a help in reinforcing the direct sound. The use of a cardioid is indicated as a means of cutting down on audience noise and side and rear generated noise. It may be necessary to use two loudspeakers positioned somewhat differently than those for music.

USING A ROOM FOR MUSIC

While it is possible to calculate the reverberation time of a hall and while this is a step in the right direction, the fact remains that determining the optimum acoustics for an auditorium or a hall is as much of an art as it is a science. The Metropolitan Opera House in Lincoln Center in New York City has a reverberation time of 1.6 seconds measured at a frequency of 500 Hz.

One of the problems in music hall acoustics is that reverberation time is not the same for all types of music. There is also a further subdivision for orchestral music, for example, since the reverberation time demands will vary depending on the composer. It is not the same for Mozart as it is for Brahms.

ARTIFICIAL REVERB

Artificial reverb can be produced by a system of vibrating springs, through the use of a reverberant chamber, by an endless magnetic tape loop and by analog or digital electronic methods. The purpose of artificial reverb in recording is to create an ambience that is missing, to create the illusion of a concert hall, auditorium, cathedral or any other space associated with particular kinds of music. However, some modification of the ratio of dry to reverberant sound (although not to such a great extent) can be made without resorting to artificial reverb. Whether a recording will sound dry or reverberant depends not only on the size and shape of the room and its total absorption properties, but on other factors such as the working distance of microphones, amount of sound reinforcement, settings of tone or equalizer controls and level controls. Thus, reducing the working distance of the mics will increase the ratio of dry to reverberant sound.

Artificially induced reverberation, whether barely discernible or cavernously reverberant, has become indispensable to recording studios, broadcast stations, theatres or individual performers. Its many uses range from mix auditioning to final mastering, to enhance or "sweeten" individual instruments, vocalists or an entire group, or even in live performances. It can create variations in depth and fullness of sound throughout a spectrum of controlled spaciousness.

ROOM RESONANCES

Any enclosed space that has reflective surfaces is capable of supporting not only one, but a number of resonances. Resonances at the fundamental frequency can cause sound pressure levels to increase by about 10 to 20 dB. Resonances are due to reverberant sound having an in-phase condition with the direct sound. We can also have dips where they are out of phase. Resonances can occur at even order and odd order harmonics of the fundamental frequencies.

Sound cancellation, or *nodes*, and sound reinforcement, *antinodes,* become less significant as values of reverberation time become lower and do not appear at all in anechoic chambers. At nodal points in the room we have relative freedom from the vibratory motion of air molecules.

Room resonances, also called *standing waves* or *eigentones*, can result in hum pickup in the studio if a microphone is positioned at an antinode. What is regarded as 60 Hz hum is generally more likely the second harmonic at 120 Hz.

Room resonances can occur when walls are parallel, with the amplitude of standing waves at a maximum if the walls are spaced some multiple of a half-wavelength apart (Fig. 2-4).

Standing waves can be modified or eliminated by room furnishings: chairs, couches, tables, lamps, and even by people in the room. No object in the room should be permitted to vibrate and while furnishings are helpful in breaking up resonance conditions, a room having a non-uniform shape is better: a bay window, an alcove, walls not parallel to each other. To eliminate flutter effect (a repeated echo caused by strong transients) try using decorative materials having a high absorption coefficient; scatter rugs or wall coverings. Nodes and antinodes can be close to each other in frequency. Room resonance varies directly as the velocity of sound in air and inversely as the various dimensions of the room multiplied by two.

VELOCITY OF SOUND

The average velocity of sound in air can be considered as 1117 feet/second or 766 mph. Sound velocity can be calculated from:

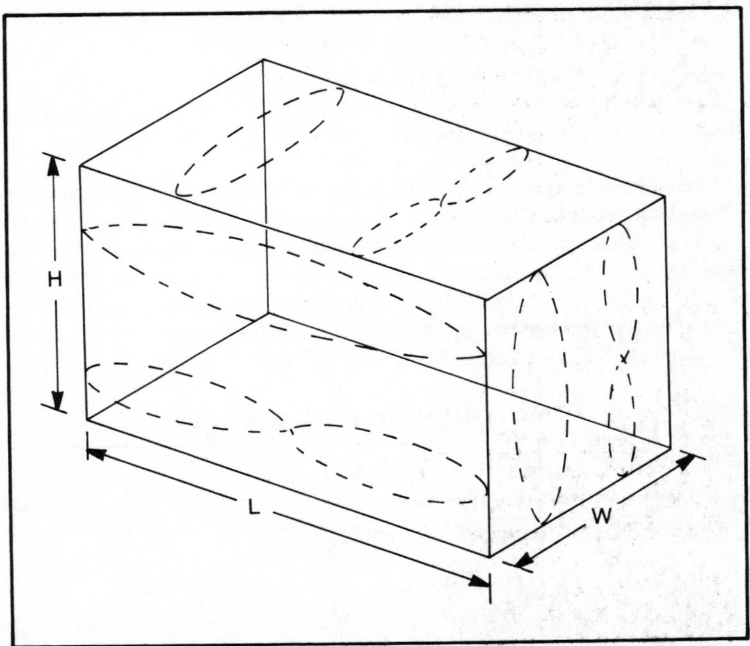

Fig. 2-4. Parallel walls a half wavelength apart and the resulting eigentones.

$$V = 49 \sqrt{459.4 + {}^\circ F} \text{ feet/second}$$

$$V = 20.06 \sqrt{273 + {}^\circ C} \text{ meters/second}$$

V is the velocity in feet/second or meters/second; °F is the temperature in degrees Fahrenheit; °C the temperature in degrees Celsius.

The velocity of sound isn't affected by frequency, somewhat slightly by humidity, much more so by temperature, and varies considerably depending on the material through which it moves. In solid substances such as brick or steel the velocity of sound is far greater than in air. At the freezing point, 0 °C or 32 °F, the velocity of sound in air is 1087 feet per second, increasing to 1127 feet per second at 20°C or 68°F. See Table 2-8.

The average velocity of sound in air, 1117 feet/second or 340.5 meters/second can be regarded as the time lag of dry sound. Someone positioned 11.17 feet from the original sound source would hear the sound 0.01 second later than someone positioned right at the source (Table 2-9). For larger distances the time lag of dry sound is proportionately greater.

Room resonances can be calculated by dividing the average speed of sound in air by the length, width and height of room surfaces. These dimensions are multiplied by 2 since the resonances occur at one-half wavelength. Further, there are also resonances at the harmonics of the calculated resonant frequencies.

In a room measuring 40 feet long by 30 feet wide by 10 feet high it would be possible to have these resonance conditions:

$$\frac{1117}{2 \times 40} = 14 \text{ Hz} \qquad \frac{1117}{2 \times 30} = 19 \text{ Hz} \qquad \frac{1117}{2 \times 10} = 56 \text{ Hz}$$

Since resonances also occur at the harmonics of these frequencies, we would have for the first example: 14 Hz, 28 Hz, 42 Hz; for the second we would have 19 Hz, 38 Hz, 57 Hz etc. and for the last we would have 56 Hz, 112 Hz, 168 Hz, and so on. We then have a series of resonances: 14 Hz, 19 Hz, 28 Hz, 38 Hz, 42 Hz, 56 Hz, 57 Hz, 112 Hz and 168 Hz. Also note that the frequencies of resonance are rather close to each other. Resonances were calculated only through the third harmonic. When room

Table 2-8. Velocity of Sound in Air.

Deg. F	Speed (ft.sec)	Deg. C	Speed (meters/sec)
32	1087	0	331.32
50	1107	10	337.42
59	1117	15	340.47
68	1127	20	343.51
89	1147	30	349.61

Table 2-9. Time Required for Sound to Reach a Reflecting Surface and then the Ears of a Listener. This is Applicable Only to Single, not Multiple, Reflections.

Total Distance Traveled (feet)	Distance from Ear to Barrier (feet)	Approximate Time Required (seconds)
11.2	5.6	0.01
22.4	11.2	0.02
33.6	16.8	0.03
44.8	22.4	0.04
56.0	28.0	0.05
63.2	31.6	0.06
78.4	39.2	0.07
89.6	44.8	0.08
100.8	50.4	0.09
112.0	56.0	0.1
224.0	112.0	0.2
336.0	168.0	0.3
448.0	224.0	0.4
560.0	280.0	0.5
632.0	316.0	0.6
784.0	392.0	0.7
896.0	448.0	0.8
1008.0	504.0	0.9
1120.0	560.0	1.0

dimensions are supplied in meters, some resonances can be calculated from:

$$f = \frac{17025}{L}$$

This simplified formula is derived from:

$$f_{abc} = 1/2\ V \sqrt{\left(\frac{n_a}{L_a}\right)^2 + \left(\frac{n_b}{L_b}\right)^2 + \left(\frac{n_c}{L_c}\right)^2}$$

The numbers for n are the modes of sound vibration and can be any digit such as 0, 1, 2 . . . etc. V is the speed of sound and is taken as 340.5 meters/second. The letters L refer to room dimensions, with the subscripts, a, b, and c specifying a particular dimension, such as length, width and height in a room, in meters.

For calculating the lowest resonance of any single surface, we can substitute 0 for n_a for n_b, and 1 for n_c. To find the lowest resonance of some other surface we can represent n_a, n_b, and n_c as 0 1 0 or as 1 0 0. For other resonant modes we can have combinations such as 0 1 1, 0 0 2, 0 2 0, 2 0 0, 0 2 2, 2 0 0, etc.

If we are interested in determining the lowest mode of vibration for any single dimension only, we can assign values of zero for the other measurements. The equation then becomes:

$$f_a = 1/2\ V \sqrt{\left(\frac{1}{L_a}\right)^2 + \left(\frac{0}{L_b}\right)^2 + \left(\frac{0}{L_c}\right)^2}$$

Those terms with the number 0 in the numerator drop out and the equation becomes:

$$f_a = 1/2\ V \sqrt{\left(\frac{1}{\Lambda_\alpha}\right)^2}$$

and this, of course, is:

$$f_a = 1/2\ V \frac{1}{L_a} = 1/2\ (340.5) \frac{1}{L_a} = 170.25 \frac{1}{L_a} = \frac{170.25}{L_a}$$

corresponding to the equation previously supplied.

The resonant frequencies, sometimes referred to as natural or *eigen* frequencies, are for a rectangular room having smooth, hard walls. The numerator, 170.25, is one half the velocity of sound in meters or 340.5/2 while the denominator is the length, width, or height of a reflecting surface. A room 40′ × 30′ × 10′ would be 12.195 m × 9.146 m × 3.04 m.

$$f = \frac{170.25}{12.195} = 14\ \text{Hz}$$

$$f = \frac{170.25}{9.146} = 18.61$$

$$f = \frac{170.25}{3.04} = 56\ \text{Hz}$$

You can expect slight differences in results between the English and metric systems, but these aren't significant. Since standing waves are the result of sound reflections, they become weaker as sound absorption is increased.

Because recording studios often must make use of available commercial space not particularly acoustically desirable, it is frequently not easy to make acoustic modifications which can produce substantial changes in reverberation time. Even when a hall, theatre, or auditorium is constructed with an effort at obtaining the best acoustic results, the finished product can sometimes be a sonic disaster. And, unlike a bottle of wine, the acoustics of any given enclosure do not improve with age.

REVERB COMPONENTS

The advantage of reverb units is that they can change what is essentially a fixed quantity, the reverb time of a room, to a variable. Further, with a reverb unit, time delay can be made to accommodate particular recordings. For non-professional use, in-home reverb devices offered in the past did not become popular because of the distortion they supplied. That is changing and with improvements in reverb technology, we now have semi-professional and professional recordists making more use of time delay devices.

One of these, for semi-pro use, is Sansui's Model AX-7. It has four input channels with built in reverb circuits. The amount of reverberation added can range from 0 at minimum to as much as 3.2 seconds at maximum. Reverb can be added in toto or to individual instruments. As an example, in the case of an instrumental trio plus vocalist, reverb can be added to the singer's voice but not to the instruments simply by connecting the singer's microphone to the channel 1 input of the unit and setting the reverb control selector to this channel. The reverb control is then advanced until the desired amount of presence or depth is obtained. With this method a singer's voice can be made to sound richer and deeper. This is unusual since it is an effect that cannot be achieved by relying solely on the natural reverb qualities of a recording room.

AKG's BX-5 portable stereo reverberation unit is designed for use in small sound studios and broadcasting stations. It works on the principle of a torsional transmission line which (TTL) uses a series of springs whose transmission properties are controlled. The unit does not contain any dry input signal at its output. Balanced reverb is obtained at any decay-time setting. Decay time is approximately 1-1/2, 2-1/2 or 3-1/2 seconds. The frequency range of the reverb signal is 50 Hz to 8 kHz.

In recording studios, the unit can be used as a primary reverb source. It can be used with "live" voice to increase average modulation level, station "loudness" and to increase coverage area. It can also provide enhancement and special effects in the production of commercials.

AKG's Model BX25E/BX25ED, AKG's reverberation unit, employs the same patented TTL principal. To increase channel separation, separate spring boxes are used for each channel. An optional digital delay gives the user additional features including pre-delay and two individual echoes per channel. This allows for more accurate creations of acoustical environments.

The spring reverb is a three-part electromechanical device consisting of a transducer which converts the sound input into mechanical vibrations, one or more stretched coil springs, and an output transducer for converting these vibrations into an electrical signal (see Fig. 2-5).

The purpose of the spring is to simulate the behavior of sound in an enclosed space in which the sound not only reverberates through reflections from the walls, ceiling and floor, but also from objects such as chairs,

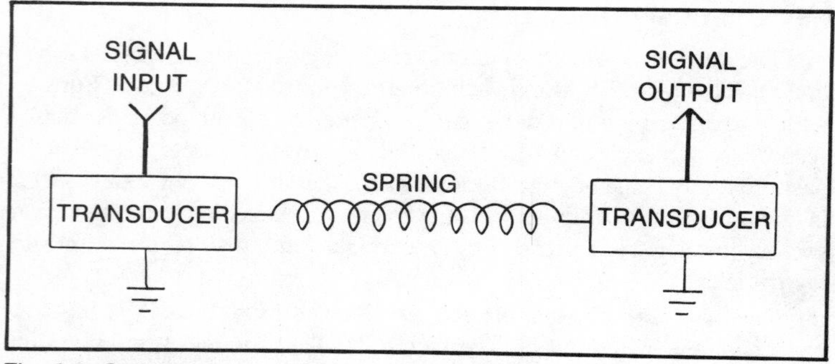

Fig. 2-5. Spring reverb.

pictures, statues, railings, people. The original dry sound can also produce one or more resonances. The sound, in passing through the springs, is time delayed with respect to the original sound. To prevent spurious responses, the springs must be well enclosed and shielded from external vibrations.

As in the case of microphones, spring reverbs range from minimum quality to professional. In some of the lower quality units, the reverb action favors a limited sound bandwidth while a professional component may contain a parametric equalizer. The input signal transducer may be preceded by a limiter circuit to prevent excessive sound signal input. Single spring reverbs are wholly inadequate for professional use. High grade reverbs have multiple springs and separate transducers for each.

Various techniques are used for the enhancement of the output reverb signal. The reverb signal may be fed back, in phase, from the output of the signal output transducer to its input. This not only emphasizes the reverb but increases the amount of time delay. Another technique is to mix the dry and wet signals.

The spring is not simply a spring mounted between two fixed points. It may be supported at various points along its length or damped at different points to limit spring action. The spring coils aren't uniformly spaced, and may be compressed or stretched along various parts of its length.

A study of acoustics is based on the assumption that the recording and/or transmission of sound will take place in a studio. But when live outdoor broadcasts are made, the sound engineer is left without the facilities of the control room to which he is accustomed. Contrary to the normal practice of controlling recordings through speakers, the sound engineer must then do level control work with headphones.

Chapter 3

Noise

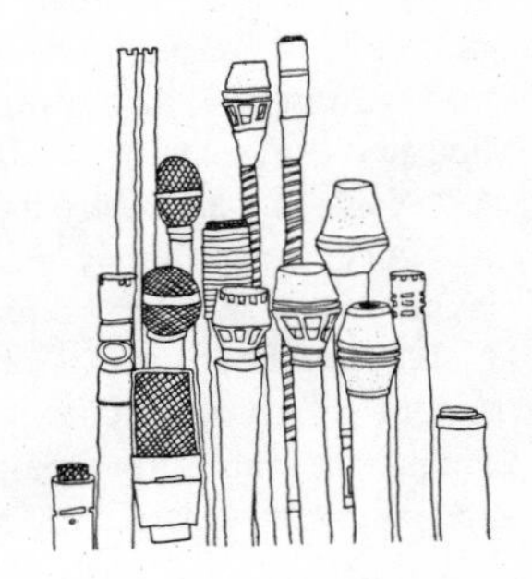

Everyone knows what noise is but none one can define it to everyone's satisfaction. One possible definition is that noise is random sound containing frequencies which are not harmonically related. Another possible definition of noise is to call it unwanted sound. If there is noise during a recording session you can be sure that some of it will find its way into the recording.

But there is no definition of noise that cannot have holes punched in it by someone determined to do so. The problem is that whether a sound is noise or not is subjective. In some instances noise is musical. Recordings have been made and sold of noise. In the 1950s Gilbert Briggs, then head of Wharfedale Wireless Works in England, demonstrated some noise recordings in New York's Carnegie Hall.

Some noise is soporific; some will set your teeth on edge. Noise is usually given a bad character reference since it is so undesirable in many applications. Yet we are so accustomed to it that a completely silent world is unthinkable.

MASKING EFFECT

Oddly, we may be conscious of noise only after it is removed. We may not be aware of the noise when it is present, but whether we are aware of it or not, listening to noise produces fatigue.

When we listen to sound with noise included we hear the sounds in their relative levels with the loudest dominant and the weakest subordinated and not heard consciously. Known as masking effect, it means that low level noise or hiss will not be heard in the presence of loud program sound. But if the program level drops to zero or close to it the sound of the noise will become noticeable. If we assume a constant level of hiss as our example of noise, it is our perception of it that varies. Playback at high levels does not solve the problem of hiss; it merely obscures it.

APPARENT NOISE LEVEL

Because the human ear is more responsive to midrange sounds, noise in that part of the spectrum can have a higher apparent loudness than a comparable level of noise at or near the extremes of the audio frequency range. Noise impairs speech intelligibility and interferes with our enjoyment of music. The average home has a noise level of about 40 dB, an office may have about 55 dB while a factory may run to 80 dB or more.

Noise is completely natural for we live in a sea of noise. It surrounds us every day of our lives on a 24-hour basis. Noise is so pervasive that it is only when we try to eliminate it that we first begin to realize how completely persistent it is. Getting rid of noise completely is an impossibility except on a heavenly body that has no atmosphere. There is no noise on the moon, and, in the complete absence of an atmosphere, there never is any sound, even when the moon is struck by an object such as a meteorite.

WANTED VS UNWANTED NOISE

Offhand, it might seem as though all noise is unwanted and that all noise interferes with the recording process. However, under certain sound reinforcement conditions, certain kinds of noise are desirable since it adds to realism. Thus, background audience noise might be wanted prior to a symphonic recording. The strong silence that follows the cessation of audience noise can be used to dramatize the opening movements of a concert.

Unwanted noise, however, can interfere with a recording. The problem is that such noise becomes more conspicuous following the recording than before it. It is possible to become so habituated to the presence of noise that a recordist may be completely unaware of it. To become conscious of a noise, it may become essential to listen for it specifically with the sole intent of trying to identify it. Unwanted noise may include air conditioning noise; office machinery noise; nearby conversations; traffic noise; airplane noise; acoustic feedback; indoor plumbing noise; recording and film equipment noise; wind noise; sounds of nature; and people noise. People noise can include not only conversation, but the noise made by people moving, shuffling, or clothing noise.

SOURCES OF NOISE

Noise has two basic characteristics giving us some clues to identifying and eliminating it. One of these factors is that there are many more sources of noise than music or speech. The other is that noise has the ability to hide in and among music waveforms.

There are various ways of producing noise, and one very common method is impact noise, a noise produced by mechanical means. The striking of one object against another, the slamming of a door or footsteps are all examples of impact noise. Impact noise doesn't stay away from high fidelity recordings either. The path followed by a stylus in the grooves of a phono record is an impact noise producer.

Electrical noise voltages can be produced by the flow of an electric current through a wire, or a coil, or a resistor, or a transistor. Microphones, preamps, power amps, and speakers are all electrical noise makers. And so are tuners and receivers.

Thermal noise can be produced or augmented by the presence of, or an increase in, heat. The thermal agitation of molecules produces noise. As an object is heated, the accelerated motion of its molecules increases the noise level.

Microphone noise is a function of temperature and is present in all types of mics. In the dynamic mic, it is the result of the thermal agitation of electrons in the microphone's moving coil.

The amount of noise output can be determined with a peak reading voltmeter and a weighting network (an equalizer arrangement). The weighting network simulates the response of the human ear allowing the voltmeter to supply an indication of the subjective effect of the noise.

A microphone is an energy converter, an action which involves the generation of noise. To keep noise from being perceived, its level must be kept low enough so it is inaudible or else it should be masked by having an adequately high level of wanted sound. At a given acoustic sound pressure it is desirable for the microphone to produce as high an electric output level as possible, that is, to have high sensitivity.

SENSITIVITY OR RESPONSE COEFFICIENT

Sensitivity is the output level in millivolts (mV) with the mic positioned in a sound wave whose pressure amounts to 1 μbar. In German standard definitions, mic sensitivity is termed *response coefficient* to indicate that mic sensitivity has been measured in a free sound field with the mic open circuited, that is, not terminated with a loading resistor. Response coefficient is ordinarily measured at 1 kHz.

The voltage output of a mic should be a direct function of the sound pressure, regardless of the frequency of that sound. This is a design problem for we are concerned here with the ratio of two variables, pressure and frequency. This ratio, though, should be a constant. It is this ratio that is referred to as the response coefficient.

NOISE AND RECORDING

In recording or in sound playback, noise is any sound not produced by the signal. Much has been made of the random nature of noise, that it is percussive, but noise can be sinusoidal as well. During a recording session our objective is to reduce the ambient noise level, and the noise produced by our recording instruments, including microphones, as much as possible. We need not worry that a noise-free recording will sound unnatural, for there is more than enough ambient noise present during playback to supply the "natural" noise to which we are accustomed.

THERMAL NOISE

All audio components, and that includes microphones, no matter how well made or efficiently designed, produce noise because of heat. And heat means any temperature above absolute zero – 459.72 °F or – 273.13 °C. As temperature rises, random molecular movement increases and with it the movement of electrons, whether free or atomically bound. It is this random electron movement that produces electrical noise signals that are also random. We hear this noise as hiss. There is no way in which thermal noise can be eliminated at its source.

IMPEDANCE VS NOISE

The random movement of electrons is current flow and this current (I) moves through the impedance (Z) of the conducting substance. This produces a noise voltage, $E_{noise} = I \times Z$. We can reduce the current to zero by dropping the temperature to absolute zero (obviously impractical). But we can reduce Z and if we can make Z equal zero, the noise voltage will drop to zero. As Z increases, so does the internally generated hiss voltage. A low impedance mic produces less hiss than one that has a higher impedance. But here, as in many other types of audio components, there is a tradeoff. As a condition of our using a higher impedance mic we also get a higher signal level. But the signal-to-noise ratio of both types of microphones (low impedance and high impedance) will remain the same.

NOISE AND HEALTH

Noise can injure the delicate structure of the inner ear, a structure which enables us to detect different sound frequencies and then send them along to the auditory center of the brain. Permanent hearing loss can be caused by loud noise and the louder the sound the less exposure time required for such loss. Hearing damage starts at about 75 dB, but exposure to a higher level, such as 90 dB, can result in hearing loss in just a few years. Ordinary conversation takes place at about 60 dB.

It's a good idea to keep sound levels low when wearing headphones since the sound source is so close. Since sound pressure varies inversely as the square of the distance, consider that distance as a protective cushion. Remember, though, it's the sound pressure level at your ears that does the damage.

WEIGHTED NOISE

Weighting curves for noise are needed because human hearing is nonlinear and at low volume levels has a low and high end rolloff. Noise voltages can be sent through a weighting filter (Fig. 3-1) and then measured. This makes noise comparisons more meaningful, not only for a more correct comparison of noise levels, but the noise filter output corresponds more nearly to what the listener hears.

NOISE AND DYNAMIC RANGE

Dynamic range is defined as the sonic distance between the softest and loudest sounds, in dB. If in a noisy environment the noise floor is 40 dB and the top of the range of wanted sound is 60 dB, then the useful dynamic range is the difference of 20 dB. 60 dB doesn't sound like much when some live sounds can conceivably hit 120 dB, but 60 dB is about tops for the average phono record or tape for in-home use. Dynamic range in a live situation is superior before any of the recording electronics is put to use.

HUM

Noise is so often aperiodic that a cyclical type, such as hum, is often not regarded as a noise problem. But noise it is and it can be as exasperating and fatiguing as any other type. Hum can be prevalent in a high fidelity system for any number of reasons. It can be injected into the system by way of a ground loop by not using correct component grounding techniques. Power supplies can be hum prone when filter capacitors age or develop leakage. An improperly shielded phono cartridge can be a notorious hum producer.

Hum is generally thought to be 60 Hz since this is the fundamental frequency of the power line throughout much of the U.S. Generally the

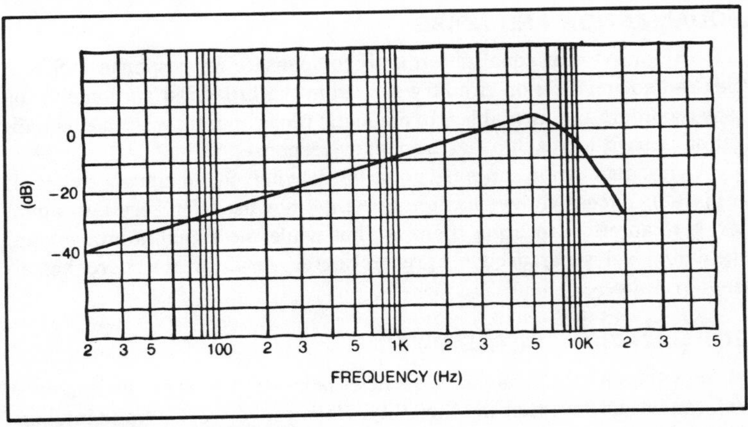

Fig. 3-1. Weighting filter curve.

hum nuisance is quite possibly the second harmonic or 120 Hz. Fortunately, the cure in many cases is easy. A better grounding connection, a tighter cable, better shielding, a new location for components are all just a few of the commonly recognized and successful hum elimination techniques.

HUM SENSITIVITY

External hum fields are produced by various sources, including power cables, light installations, electric motors, tools, etc. Microphone sensitivity to such hum fields is expressed as output voltage per induced field strength.

Hum, either 60 Hz or its harmonics, can be picked up by a microphone voice coil, transformer, or by an ungrounded or unshielded case. A hum-bucking coil greatly reduces pickup of magnetic hum and careful attention to grounding and shielding in the design of the mic reduces hum pickup through the case.

Magnetic hum pickup is specified as sound pressure equivalent, expressed in dB SPL, from a 1-millioersted (mOe) hum field. For instance, a hum pickup of 17 dB equivalent SPL means that the microphone's hum output will be the same as from an acoustic source of 17 dB SPL: comparable to a soft whisper about 10 feet away. A 1-millioersted field roughly corresponds to the hum field found in a typical studio environment.

THE NOISE BAND

Noise may not only cover a wide passband, but may consist of a series of discontinuities over a given frequency spectrum. These may have a number of high amplitude spikes and can produce either a narrow band or a spot frequency punching effect, producing dropouts over the musical range.

EQUALIZATION AND NOISE

The purpose of equalization is to compensate for response deficiencies and so equalization circuitry is used in the broadcast station and for tape recorders and turntables. In magnetic recording, for example, equalization is used to reach a flat frequency response.

Over-equalization can raise noise levels when sound signals are weak (Fig. 3-2). Excessive equalization is sometimes used for very low signal levels in an effort to bring them up, but while the signal is brought up, the noise level is increased to a greater degree, resulting in a poorer signal-to-noise ratio.

EQUALIZATION ON THE MIXER

For speech amplification, frequencies below 200 Hz are of little interest and can be cut for better intelligibility. To improve presence and intelligibility of a singer's voice, boost the upper midrange in the 4 kHz to 7 kHz range.

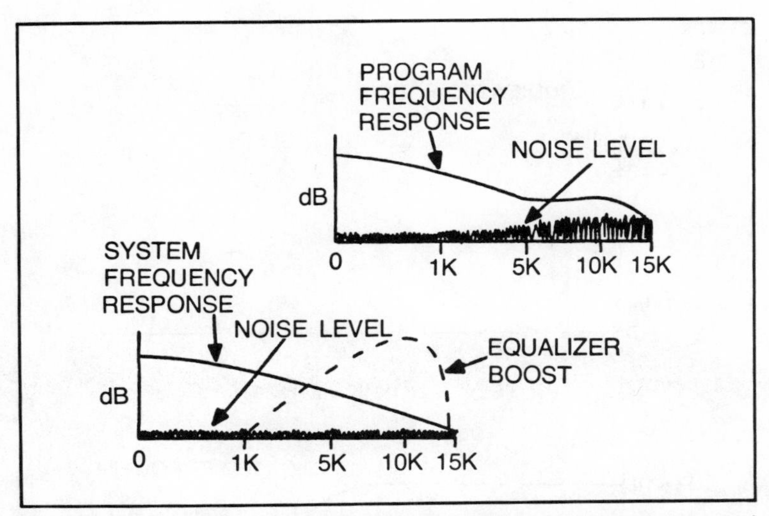

Fig. 3-2. Excessive equalization can raise the noise level. The noise is boosted in the frequency range covered by the equalizer.

SHELVING

All equalizers use various arrangements for boosting signals to a desired level. Two of these are known as a low-frequency shelf and a high-frequency shelf. Figure 3-3B illustrates a low-frequency shelf. Note that the amplitude remains constant from 0 Hz (actually dc) to 160 Hz, but at the indicated frequency there is a rolloff to zero with the zero point reached at 500 Hz. The significance of this graph is that all frequencies below 160 Hz are boosted to the amount of the indicated frequency.

Figure 3-3A shows a high-frequency shelf. This drawing illustrates that all signals above 5 kHz are boosted to the same extent. The letter "Q" is used to indicate the rate of increase or decrease of signal level in terms of decibels per octave. The slope or Q of the low and high-frequency shelves of Fig. 3-3 is 6 dB/octave.

While a boost may sometimes be required over a band of frequencies it may sometimes be wanted over a rather limited range. In that case a peaking type equalizer can be used. Figure 3-4 shows the increase in level in dB. In this case the selected frequency is 1 kHz. Curve C shows the optimum boost is at the indicated frequency with a slope of 18 dB/octave. Curve B has a Q of 9 dB/octave while curve A is 6 dB/octave. The steeper the slope, of course, the narrower the frequency range of the boost.

CALMING NOISE

The challenge of noise is not so much that we want to eliminate it completely but rather that we would like to be able to control it. An environment that is too quiet can be disturbing. Some noise sounds, such as the

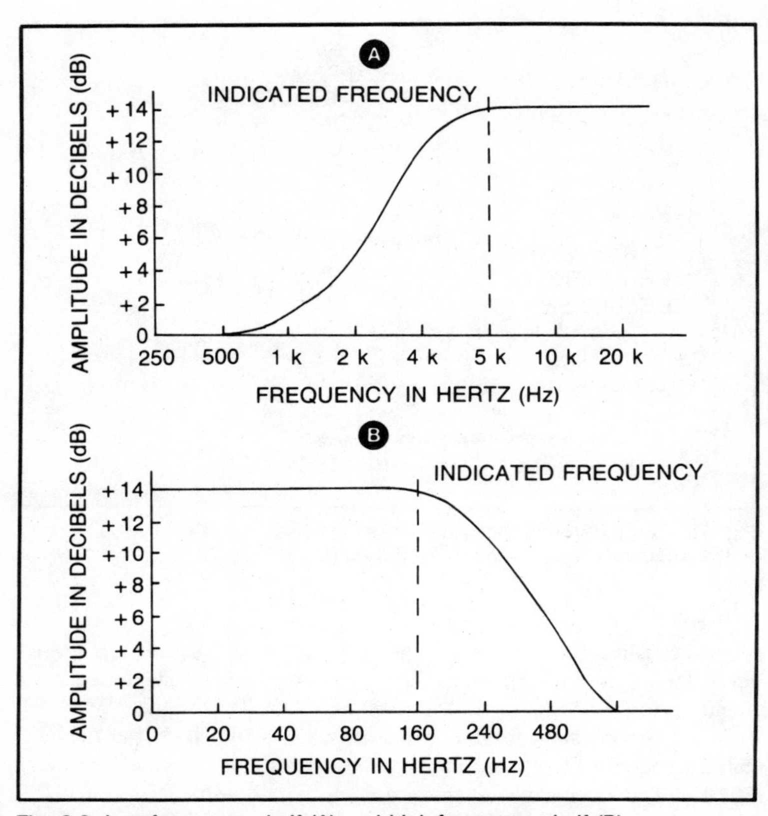

Fig. 3-3. Low-frequency shelf (A) and high-frequency shelf (B).

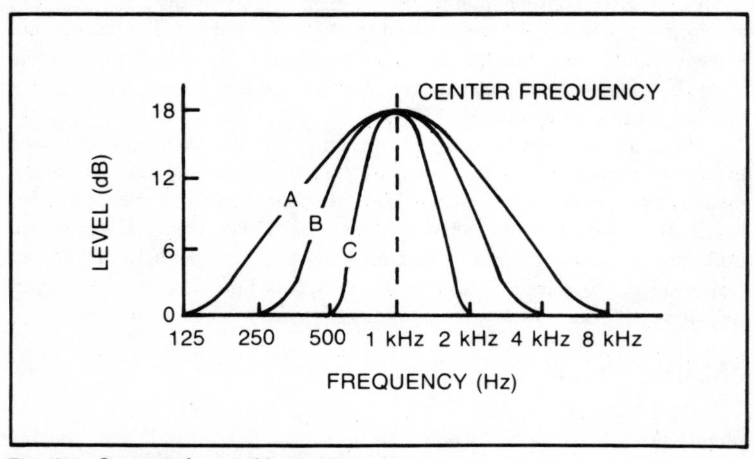

Fig. 3-4. Curves of a peaking equalizer.

patter of rain, or wind sounds, are rhythmically pleasing, functioning as a mask over irritating sounds, making meditation, sleeping, mental concentration, relaxing, or listening to music, much more pleasurable. Phono records are available which supply calming noise for environments which do not have it. Unfortunately, the word "noise" carries an unhappy connotation, implying that all noise is undesirable.

TAPE NOISE

There are three major contributors to tape noise—software, hardware and the user. If a tape isn't properly made, if its particles have an incorrect length to width ratio, if the particles are uneven or aren't polished, if the tape particle density is poor or uneven, noise will be present. The amount of noise produced by the tape is dependent on the chemical composition of the particles, on the amount of bias current, and on the bias frequency.

Tape noise includes modulation noise and clicking caused by improper tape editing. All tapes produce audible hiss, although low noise tapes have less. Wow and flutter are sometimes regarded as noise, but technically they are not. Wow and flutter take place when tape speed varies during playback to produce a wavering in the music or sound pitch. Wow is a slow rate of speed variation; flutter is a rapid rate.

MODULATION NOISE

Modulation noise, also called asperity or behind-the-signal noise, can be caused by non-homogeneous magnetic particles and by vibrations across the tape as it slides past the head gap. A non-homogeneous coating means that there is clumping or clustering of particles, as well as low particle density areas. Modulation noise also depends on the bias setting.

One technique for reducing tape noise is to run tapes at lower speeds. The lower speed reduces treble response, taking some noise along with it. This is a tradeoff of frequency response for lower noise, somewhat equivalent to throwing the baby out with the bath water. Definitely not recommended unless there is no alternative.

Noise has several opportunities for getting in its licks during the sound recording and playback process. Noise is registered during the recording of a master tape, again during the production of a disc master and then again during disc playback.

HISS

The background noise in the production of a master tape is hiss. Hiss is due to the random arrangement of the magnetic particles on the tape, and so, in this sense, every blank tape is noise prerecorded. The act of recording music on tape does not eliminate the noise and the transfer from a master tape to a master disc means a transfer of noise along with the music. The amount of noise level restricts the dynamic range of both records and tapes.

Hiss generally appears somewhere around the high end of the audio spectrum and unlike hum is often a background accompaniment to music. Unfortunately, hiss can be generated in many ways: it can exist as surface noise on phono records, it can be produced by biasing used in tape recording as well as by the noise inherent in all magnetic tapes. It is a byproduct of the multiplex network (MPX) in FM receivers. Unlike hum, which exists as spot frequencies at 60 Hz and 120 Hz, hiss consists of random sound distributed through the mid and high-frequency audio spectrum.

In home hi/fi systems, operators of the system sometimes use the treble tone control to reduce hiss. But this also reduces the high-frequency content of the program material, tending to make it lose clarity and presence, giving it a sort of muddy reproduction effect, most notably with instruments that have a substantial high-frequency content—cymbals, flute, tympani, and female vocalists.

DOLBY AND dBx

One possible way of minimizing the effects of hiss is to use an encoding and decoding technology such as that featured by Dolby or dBx. These function by raising signal level during soft passages in the recording process and by reducing signals to their original level during playback. This process improves the recording signal-to-noise ratio but does not eliminate noise inherent in the original source. The disadvantage is that these systems can be used only when the program source is pre-encoded. While some commercially produced tapes have been treated with such techniques, many phono records, FM broadcasts and tapes are not encoded. A tuner, receiver, or tape deck equipped with Dolby cannot compensate for non-encoded material, although adjustment of the treble tone control is of some help.

PINK NOISE AND WHITE NOISE

The spectra for pink noise and white noise are the same, extending from 20 Hz to 20 kHz. White noise has a constant amplitude across this frequency range while pink noise has a 3 dB/octave slope (Fig. 3-5). The colors pink and white were deliberately selected as an analogous comparison to light. White light consists of all colors in the light spectrum. In that spectrum red is down at the low frequency end. A mixture of red and white produces pink. White noise has a uniform distribution of power with frequency.

NOISE AND SPL

No matter how noise is produced, ultimately it registers on our consciousness (via our ears as sound) as a change in the *sound pressure level* (spl). We measure sound pressure level by:

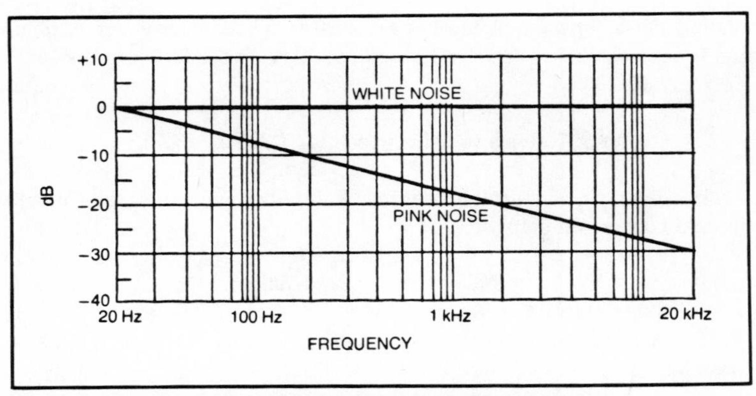

Fig. 3-5. Frequency characteristics of white and pink noise.

$$\text{spl} = 20 \log_{10} \left(\frac{P_2}{P_1} \right)$$

In this formula, P_1, the denominator in the fraction P_2/P_1 is the reference level and is unity in terms of relative sound pressure or 0 dB. This is considered as total silence. Total silence is a mathematical concept only, for even in an anechoic chamber sound pressure level is in the order of 10 dB. A whisper at a distance of four feet is 20 dB, noise in the average home is some 45 dB and average street noise is 70 dB.

The reference level used in the formula is 2×10^{-4} microbar with a microbar corresponding to the pressure of one millionth of an atmosphere. The formula is the same as that used to calculate voltage or current ratios in dB. In your living room the noise level is about 40 dB contrasted with the 20 dB spl of a recording studio. This means that it is ten times the relative sound pressure.

CALCULATING MICROPHONE NOISE VOLTAGES

The determination of circuit noise can be approximated by using a modification of Nyquist's formula:

$$N(f) = 4kTR$$

In this formula N(f) is the spectral density of the noise voltage, k is constant and is equal to 1.38×10^{-23} joules per Kelvin, K, R is the resistance in ohms of the resistor whose noise voltage is being measured and T is the temperature of that resistor in Kelvins.

The Kelvin temperature scale is sometimes called the absolute scale. Zero on the Kelvin scale is equal to −273° on the Celsius (formerly Centigrade) scale and is the temperature at which all molecular motion stops. It is the point we call absolute zero. At that point there is no longer any thermal noise. The actual value of absolute zero in Kelvins is −273.13° C and is equivalent to −459.72° Fahrenheit. To convert degrees Celsius to Kelvins add 273.18.

$$\text{Thus: Kelvins (Celsius absolute)} = {}^{\circ}\text{C} + 273.18$$

However, if you start with degrees Fahrenheit, you must take an additional conversion step.

$${}^{\circ}\text{C} = ({}^{\circ}\text{F} - 32) \times 5/9$$

To calculate noise voltage we can use a modified form of Nyquist's formula:

$$N_v = (4KTRF)^{1/2}$$

A new factor has been added and that is the letter F for the bandwidth in Hertz.

As an example, assume that the room temperature is 70° Fahrenheit and that the equipment is a microphone with an impedance of 600 ohms. The microphone has been in the room long enough to reach room temperature. The formula does call for resistance, but the output impedance of the most widely used microphones for high fidelity sound, and that includes dynamic, condenser and ribbon microphones, is almost purely resistive. This means that the output impedance of these mics remains fairly constant over the entire audio frequency range. As a first step, convert temperature in Fahrenheit to Celsius.

$$\begin{aligned} {}^{\circ}\text{C} &= ({}^{\circ}\text{F} - 32) \times 5/9 \\ {}^{\circ}\text{C} &= (70 - 32) \times 5/9 \\ &= 38 \times 5/9 = 21.11\,{}^{\circ}\text{C} \end{aligned}$$

To convert this to Kelvins:

$$\text{Kelvins} = 273.18 + 21.11 = 294.29 \text{ Kelvins}$$

The spec sheet of the mic will give us the frequency response and in this example has a range of 50 Hz to 18 kHz. To get the bandwidth we subtract the low end of the response from the high:

$$18{,}000 \text{ Hz} - 50 \text{ Hz} = 17{,}950 \text{ Hz or } 17.95 \text{ kHz}$$

We can now solve for the amount of noise voltage produced by this mic by substituting the numbers into the Nyquist formula:

$$N_v = (4 \times 1.38 \times 10^{-23} \times 294.29 \times 17950 \times 600)^{1/2} \text{ volts}$$

This isn't a difficult bit of arithmetic, but it has a high nuisance value. Since the answer will be in volts, we can make an immediate conversion by multiplying by 1,000,000 or 10^6 and get the result directly in microvolts.

$$N_{\mu V} = (4 \times 1.38 \times 10^{-23} \times 294.29 \times 17950 \times 600 \times 10^6)^{1/2} \text{ microvolts}$$

We can simplify by using exponents

$$N_{\mu V} = (4 \times 1.38 \times 10^{-23} \times 294.29 \times 1795 \times 10^1 \times 6 \times 10^2 \times 10^6)^{1/2}$$

And by combining terms that have exponents we get:

$$N_{\mu V} = (4 \times 1.38 \times 10^{-23} \times 294.29 \times 1795 \times 6 \times 10^9)^{1/2}$$

10^9 was obtained by combining 10^1, $\times$ 10^2 $\times$ 10^6. By multiplying 10^9 and 10^{-23} we get 10^{-14}. And so we now have:

$$N_{\mu V} = (4 \times 1.38 \times 10^{-14} \times 294.29 \times 1795 \times 6)^{1/2}$$

Multiplication now gives us:

$$N_{\mu V} = (17{,}495{,}658 \times 10^{-14})^{1/2}$$

The 1/2 outside the parenthesis means we are to take the square root of the numbers inside the parenthesis. The square root of 10^{-14} is 10^{-7} or 0.0000001. The square root of 17,495,658 is 4183.

We now have 4183 × 0.0000001 or 0.0004183 microvolt. This is the minimum noise voltage and is that produced by the microphone only. It is not a constant, not an absolute amount and will probably be higher than this result indicates. But it is a fairly good approximation.

Assuming a mic that is completely passive, we can regard the number obtained for the noise voltage as a reasonably correct one. It is also essential to remember that the noise figure is that developed by the microphone and does not take any noise pickup by that microphone into consideration.

ADDITIVE NOISE

As more tracks are laid down on tape the noise level increases by 3 dB each time as indicated in Fig. 3-6.

EQUIVALENT NOISE LEVEL

The electrical self-noise of a microphone is compared to the theoretical output voltage with a sound pressure level at the threshold of hearing (2×10^{-4} μb or 20 μPa). The measuring unit (dB spl) is related to an output voltage at 20 μPa sound pressure level.

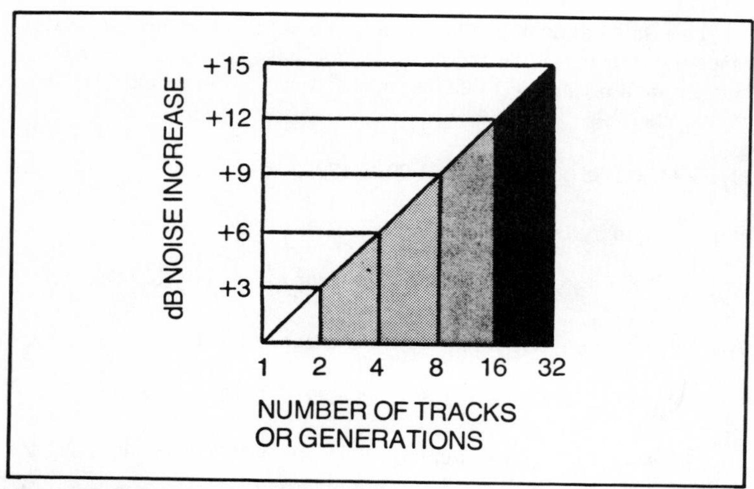

Fig. 3-6. Additive noise chart (courtesy, dbx).

Even with the complete absence of an input signal, as in an anechoic chamber, a microphone will still produce some output due to residual noise internally generated. In the case of dynamic mics (voice coil mics), this noise is caused by the random movement of electrons in the voice coil due to the presence of ambient temperature.

Residual noise can be measured using a test instrument capable of supplying peak voltage values. With the help of a weighting filter (DIN 45 405), certain components of the noise are emphasized. The equivalent noise is then calculated taking into consideration the sensitivity of the mic and comparing it to the threshold of hearing generally accepted as 2 × 0.0001 microbar. Equivalent noise is always included in the microphone spec known as the signal-to-noise ratio (s/n).

Chapter 4
Meet the Mic

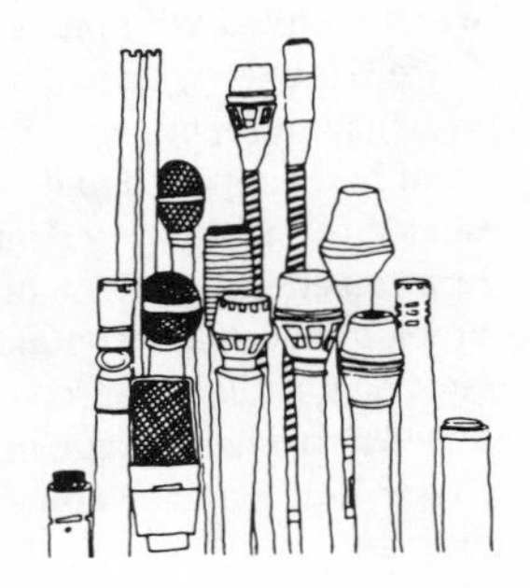

The idea behind the microphone is simple, for all a microphone, any microphone, is required to do is change one form of energy to another. The *Encyclopedia Britannica* defines the microphone as a "device for converting acoustic power into electric power which has essentially similar wave characteristics."

A telephone can meet this requirement, and one of the basic parts of a telephone is a microphone. All of us have some familiarity with the microphone for the one in the telephone is the most widely used. But what we take for granted today was quite a problem more than a hundred years ago. At the Centennial Exposition (Philadelphia, 1876), Alexander Graham Bell demonstrated a liquid microphone. He put a metal wire, about a fraction of an inch thick, into water mixed with acid. At the other end of the wire he attached a diaphragm that could be vibrated by voice waves. The resultant jiggling of the wire in the water varied an electric current that was made to flow through the diaphragm, the wire, and the acid/water mixture.

The patent for the Bell telephone is sometimes referred to as the most valuable ever issued in the U.S. In 1877 Bell offered to sell his patent, No. 174,645 to Western Union for $100,000, but this offer was turned down, quite possibly since serious attempts were being made to get around the patent. The operating heart of the telephone, of course, was a microphone.

Bell was the inventor of the first telephone that could carry the human voice for some distance so it was still intelligible to the listener at the other end. Despite Hollywood, though, he wasn't the inventor of the telephone and didn't even give the component its name. Philip Reis, of Friedrichsdorf, Germany, not only constructed the first telephone (in 1861), but named it as well. And Elisha Gray of Chicago was able to build a short-distance telephone in 1873. Alexander Graham Bell didn't come along with his telephone until 1876. So much for laurels, glory, and who deserves the credit—they all do.

The basic problem in the development of the microphone was that there was no concept that a single vibrating diaphragm could respond to the entire gamut of voice frequencies. In Bell's early experiments, he used, instead of a diaphragm, a series of metallic strips of different lengths. These were mounted with one end fixed but with the rest of the strip free to vibrate. However, during one experiment a single reed became stuck, and so began to function as a diaphragm. Bell was astute enough to recognize what had happened and consequently he moved from the reed concept to the idea of a vibrating diaphragm. Actually, the diaphragm simplified matters considerably. Each of the vibrating reeds had been used to make and break an electrical contact. The diaphragm was not only simpler but was electrical rather than completely electromechanical. As it vibrated it induced a varying voltage in an electromagnet placed behind it. Figure 4-1A shows Bell's first telephone made in 1876 while Fig. 4-1B is a schematic diagram.

Microphones are associated with telephones because you must have a microphone before you have a telephone. The early microphones used by Bell, Reis, and Gray were attic-inventor contraptions. They worked, but you wouldn't want to use any of them to call your home. Credit for the invention of the microphone goes to Prof. D. E. Hughes, an Englishman, who finally came through with one that was eminently practical, and

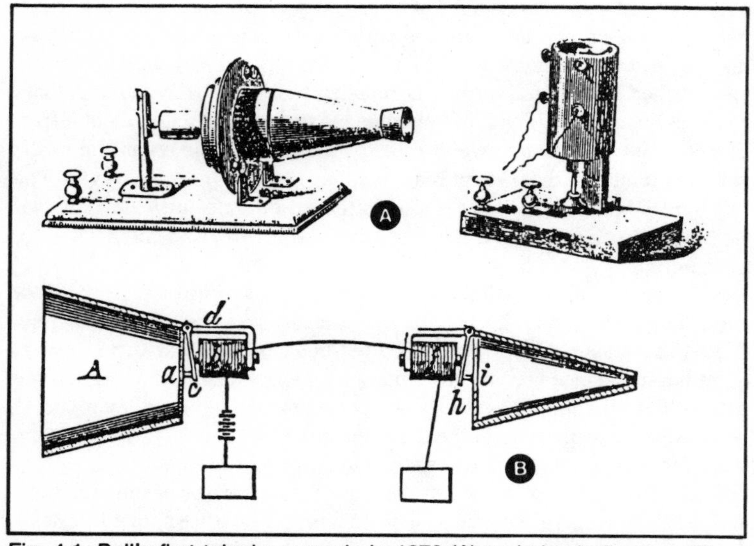

Fig. 4-1. Bell's first telephone made in 1876 (A) and circuit diagram (B).

he is regarded as its inventor. This took place in 1878 and, oddly, microphones had already been in use for almost 20 years.

ELECTROACOUSTIC TRANSDUCERS

Since a mic converts acoustical energy to its electrical equivalent, it is sometimes referred to as an electroacoustical transducer. There are a number of ways in which this can be done: resistance control (carbon mic); piezoelectric (ceramic and crystal mics); electromagnetic (magnetic mic); electrodynamic (dynamic mic); and electrostatic (condenser and electret mics). Of all these, the carbon mic, the mic having the most restricted frequency range, is widely used as the transducer of choice for telephones. For quality sound reinforcement and for recording the electrodynamic and the electrostatic are preferred.

While the microphone is ordinarily thought of in connection with sound recording or broadcasting, its largest use is in telephones, hearing aids, and dictating machines.

THE MICROPHONE AND ITS RELATIVES

The microphone is a strange and wonderful instrument, even more so when you consider its family tree. The microphone is directly related to speakers, for in some cases (as in intercoms), the microphone also works as a speaker. Since the microphone is a *transducer*—a device that changes one form of energy to another—it is first cousin to storage batteries, motors, generators, piezoelectric crystals, and the electric light.

This disparate group is related in the sense that they are all energy converters. A battery changes chemical energy to electrical energy, the motor converts electrical energy to mechanical energy, and so on. What we require of the microphone is quite simple—to change sound energy to its electrical equivalent. Electrical, *not* electronic. We ask it to perform like a battery or a generator, that is, to produce an electric current.

WHAT IS A MICROPHONE?

A microphone is basically a collector of sound, taking acoustical energy input and converting it to electrical energy. The problem is that the acoustical energy contained in our voices or in musical instruments is full of sudden starts and stops. The piano, a percussive instrument, is easily capable of going from a condition of *no sound* to *peak sound* in almost no time at all. And the sound level of some instruments, as well as the human voice, can decrease extremely rapidly. So one of the required characteristics of a microphone is that it must be responsive to rapid changes in acoustical energy. This is somewhat like asking an automobile to move from *rest* to 70 mph, down to 30, up to 45, and down to 10, practically instantaneously. One of the problems the auto has in doing this is a law of Nature which insists that a body at rest prefers remaining at rest and a body in motion prefers staying in motion. More elegantly known as a *law of inertia*, it is as applicable to microphones as it is to cars. The comparison isn't quite fair, however, for the mass of a car is tremendous in comparison with the moving element of the microphone.

As collectors of sound, mics must often meet a number of requirements:

1. We can consider the mic as supplying two levels of sound in the form of electrical voltages. One of these voltages corresponds directly to the self-noise level of the mic. The other voltage is that produced by the conversion of energy supplied by the sound source(s). The self-noise voltage must be small in comparison to the sound signal voltage.
2. The output signal voltage should be undistorted over the frequency range of the sound source. The mic should not be frequency selective, unless the mic is so made for some specific purpose. Further, this operating characteristic must exist over a wide dynamic range, that is, from the softest to the loudest sounds.
3. The polar characteristics of a mic (described in a later chapter) should be the same for all operating frequencies.
4. The microphone should be as unobtrusive as possible so as not to distract attention from the performer(s).
5. The microphone must be able to tolerate repeated connections and disconnections, as well as physical abuse.
6. Particularly in television studios, the finish of the mic should have a very low value of light reflectivity.
7. The output of the mic should be large enough to be able to drive a following preamplifier.

MAKING A MICROPHONE

The simplest microphone you can imagine consists of a pair of tin cans (Fig. 4-2), each having one cover removed. Any pair salvaged from the kitchen will do. Punch a hole in the center of each remaining can cover and insert a length of string. Just tie a knot on the inside of the can so the string cannot come loose. With this arrangement and with the string kept taut, you can transmit your voice over a distance of about 20 feet.

In this setup each tin can works both as a microphone and a speaker. The acoustical energy going into one of the cans causes the bottom cover to vibrate. This vibration is transferred to the string and travels to the other can. The bottom of the second can will be caused to vibrate by the energy it receives from the string. But in vibrating, the can cover will move air back and forth, in step with the sound going into the first can. This air movement, upon reaching the ear, is converted into sound.

What we have here is a transducer, changing sound energy to mechanical energy at one end, then converting that mechanical energy back to sound energy at the other.

Fig. 4-2. The simplest microphone is a tin can with one of its covers removed. Two such cans connected with a length of dry string kept taut can transmit sounds up to about 20 feet.

Granted that this is a very crude system, that the string must be kept dry, that it must be kept tight, that the communications distance is very small, and that the quality of the sound reproduction is terrible. All this is beside the point right now. What *is* important is to realize that this device tells us that acoustical energy can be changed into some other form and that the acoustical energy can be made to work as a control. In this example, the energy produced by the human voice caused the cover of a tin can to vibrate. Not just to vibrate, but to vibrate in step with the variations and changes in the human voice.

There is one other problem presented by the tin-can microphone. It seduces us into taking a completely wrong path toward the development of a microphone, for while it converts sound energy into some other form, there is no way in which we can modify,improve, or amplify that other form. We *can* change or amplify an electrical form, however, so our search for a practical microphone must take a path in which sound energy is converted to electrical energy.

The string telephone is quite possibly the oldest telephone in existence for it was invented in 1667. It predated the tin can which did not appear on the scene until some centuries later and so thin metal plates were used instead. Because of the fact that it could only operate simplex (only one person could talk at a time) and because of its short operating distance, it remained only as an interesting toy.

THE VARIABLE RESISTOR MICROPHONE

Theoretically, at least, it isn't at all difficult to design a microphone which will change acoustic energy to electrical energy. As a first approach, consider Fig. 4-3. Here we have a variable resistor connected to a battery. Whenever an electrical current flows through a resistor, a voltage appears across it. The variable resistor R1 is equipped with a knob, so we can change

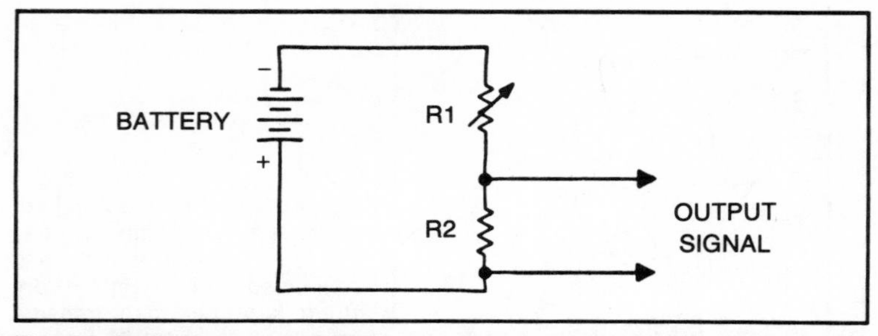

Fig. 4-3. Preliminary steps in making a microphone. R1 and R2 are resistors. The arrow through R1 indicates that it is variable; it has a knob to change its resistance. In turn, this varies the amount of current flow through the circuit. The changing current through R2 produces a changing voltage across it. This becomes the output signal voltage.

the amount of current at will, just as rapidly as we can twirl the knob of R1. But as we do so, the current flowing through the next resistor, R2, will also vary. This produces a changing voltage across resistor R2.

But now we have another requirement. We must be able to turn the knob on R1 exactly in step with the variations in sound. A large current would correspond to a loud sound; a weak current would be equivalent to a weak sound. We could then take the varying voltage produced across resistor R2 and connect it to an audio amplifier as indicated in Fig. 4-4, thereby strengthening the voltage produced by the "microphone," ultimately leading the voltage from the amplifier into a speaker. At the speaker, the reverse process would take place. Our electrical energy would be converted into acoustic energy and we would hear our original sound, considerably amplified.

Theoretically, this microphone would work. From a practical viewpoint it is impossible, however. There is no way in which we could manually rotate the knob on variable resistor R1 so as to follow the extremely rapid speed of changes in sound. However, we could do so if we could manage to make the acoustic energy act as the control element. Instead of turning

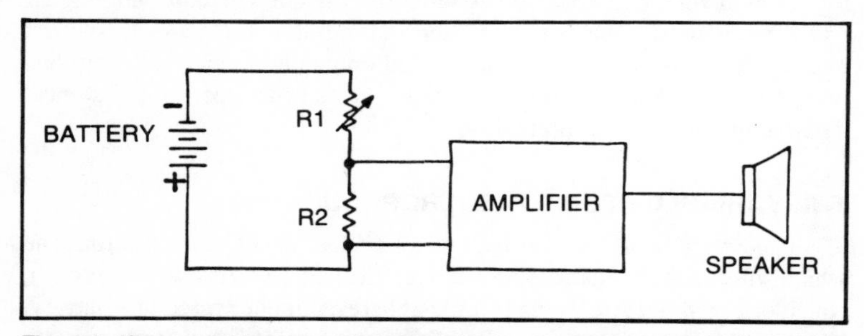

Fig. 4-4. If you could manage to rotate the knob on resistor R1 in step with the sound you would hear an amplified version of the sound from the speaker.

the knob manually, we should let the acoustic energy do the turning. So now that we know the problem, all we need to do is solve it.

THE CARBON MICROPHONE

The two microphones just described—the tin-can version and the knob-controlled type—aren't as ludicrous as they might appear. Based on the earlier definition of microphones, that is just what they are, for they are transducers, changing one form of energy to another. They both have serious faults, however, which preclude giving them further consideration. The arrangement using the variable resistor fails because the current flowing through it must be controlled manually, rather than by acoustic energy. The tin-can microphone fails because it is physical movement only and does not use an electric current. We have no way of using electricity to strengthen the sound signal.

The *carbon* microphone overcomes both of these objections, and while its design and concept are as outlandish as the tin-can microphone and the variable-resistor microphone, it nevertheless does work and has practical applications.

Figure 4-5 shows two views of the general arrangement of this microphone. The unit consists of a small cylinder, known as a *button*, packed with tiny granules of carbon. Pressing against the button containing the carbon granules is a *diaphragm*, a metallic disc supported only around its circumference.

Carbon is a conductor of electrical currents, so we now have a complete current path, as shown by the arrows in the lower drawing. You can trace this current path by starting at the minus terminal of the battery, marked with a minus (−) sign, moving up through the diaphragm, then through the carbon granules, through the resistor marked R, and finally back to the battery. This is a simple, straightforward circuit and the current that flows in it, supplied by the battery, is direct current (dc).

The diaphragm, or microphone element, just a thin, very flexible circular metallic plate, is held tightly in place by a rather thick support. While the plate cannot move at its periphery because it is fastened, the center of the plate can move. It will do so when you bring it close to your mouth and talk into it. The energy of your voice, impinging on the diaphragm, will cause it to vibrate in step with your voice. When it does so, it will move toward and away from the button containing the carbon granules, compressing them when the diaphragm moves inward, giving the granules a chance to separate when the diaphragm moves outward. The action here is exactly the same as that of the variable resistor shown earlier in Fig. 4-3, with one difference, however, and quite a significant difference at that. In Fig. 4-5 the resistance is controlled by the energy in the sound reaching the diaphragm. There is no need for manual control, for now the voice or musical-instrument sounds determine the resistance of the granules in the button. When the granules are compacted, the resistance is lowered; when

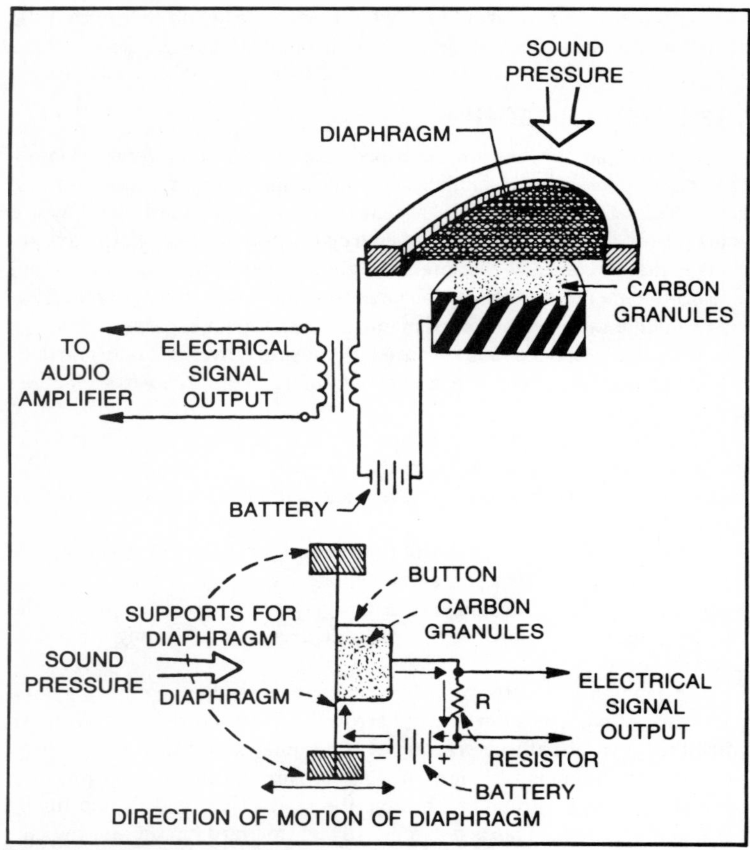

Fig. 4-5. Basic structure of the carbon microphone.

the granules are not pressed together, the resistance increases. And, of course, there are all sorts of variations between these two extremes. But as the resistance varies, so does the flow of current. The higher the resistance, the smaller the current, and vice versa. But this current, in flowing through resistor R, produces a varying voltage across R, a voltage that fluctuates in step with the sounds reaching the diaphragm. Because it follows the variations in sound presented to the microphone diaphragm, we refer to it as an audio voltage. It increases and decreases when the sound pressure increases and decreases. When the sound has abrupt starts and stops, so does the audio voltage.

The audio voltage output of the carbon microphone is high compared to modern dynamic microphones and so this microphone is useful in inexpensive intercommunications systems. However, the pickup range of the carbon microphone is limited and its quality is poor.

The carbon microphone, one of the earliest microphone types, and still widely used today, has one tremendous advantage. It proves that a microphone can be a practical reality. It demonstrates that a microphone can be acoustically controlled. However, it is noisy and will not respond to other than a limited range of sound frequencies. Also, the diaphragm must be tightly stretched. This limits its possible movement, thereby restricting the electrical output level. Still another problem of the carbon microphone is the size of its diaphragm. The diaphragm should be small compared to the wavelengths of sound that reach it. And this is precisely why a loudspeaker makes such a poor microphone. The function of a speaker is opposite that of a microphone, for it must push air instead of being pushed by it. For a speaker, large surface area is important.

IMPEDANCE

The word *impedance* is one you will often hear used in connection with microphones. All it means is total opposition to the flow of current. High impedance means large opposition; low impedance means smaller opposition. Impedance is no indication of quality; rather, it is just a fact of electrical behavior. As a general rule, low impedance means large current flow; high impedance smaller current flow. The basic unit of impedance is the *ohm* (abbreviated Ω), just as the inch is a unit of measurement for a ruler. Actual values of impedance for microphones aren't always specified exactly. Rather, general terms, such as *low impedance* or *high impedance*, are used. Impedance is not a critical value and designations such as low impedance or high impedance usually suffice. (Impedance is covered in greater detail in Chapter 7).

In general, any low-impedance microphone is one which is rated as having an impedance between 50 to 500 ohms. A high-impedance mic is one having an impedance between 10,000 and 40,000 ohms. However, the impedance values given here can be higher or lower than indicated.

The actual impedance of a microphone cannot be measured with a test instrument such as an ohmmeter, the ohmmeter section of a multimeter, or a DUM (digital voltmeter). While resistance and impedance are both specified in ohms, impedance also includes reactance. Reactance, whether inductive or capacitive, varies with frequency and so, unlike resistance, is not a fixed quantity. In the case of microphones, the reactance portion of impedance is substantially larger than resistance.

THE MICROPHONE TRANSFORMER

Still considering Fig. 4-5, we could take the sound signal, now an audio voltage, from across fixed-resistor R, similar to the manner shown earlier in Fig. 4-4. We could feed this voltage into an amplifier and use it to drive a speaker. This supplies us with a problem, however. For maximum transfer of signal the impedance of the microphone should be relatively similar to the input impedance of the amplifier. The carbon microphone, though, is a current-operated device. A rather large current will flow through resistor R in Fig. 4-5. But a large current is synonymous with low impedance, so we look on the carbon microphone as a low impedance device. (Note that we do not supply you with the exact amount of impedance in ohms.)

To be able to get the most signal transferred from the carbon microphone to the input of a following amplifier, we need some sort of interfacing device. This device can be a microphone transformer, as shown in Fig. 4-6. The transformer can be built right into the microphone case or it can be a separate item.

Like other electrical components, the transformer also has impedance, with the impedance depending, to a considerable extent, on the number of turns of wire. The transformer consists of two coils of wire, each would around an iron core, but with each coil physically independent. One coil, the primary, contains relatively few turns, so its impedance, its opposition to the flow of current, is comparatively small. The other coil, the secondary winding, has many more turns. It has a much higher impedance than the primary winding. This makes the microphone transformer an impedance transformation device. The voltage across the primary winding is small, the current comparatively large. Conversely, the voltage across the secondary winding is larger, the current is smaller. Thus, the transformer in this case is not only an impedance changing device. It also supplies a voltage step-up. The higher impedance of the secondary winding matches the input impedance of the amplifier to which it is connected, not precisely down to the last ohm, but at least both impedances, that of the secondary winding and that of the amplifier input, are somewhat close.

The transformer supplies still one more advantage. While we need the current from the battery to actuate the carbon microphone, we do not want

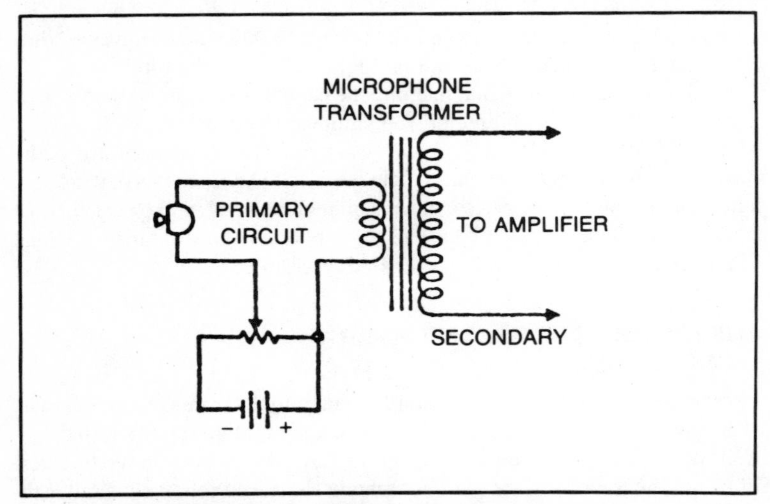

Fig. 4-6. Carbon microphone and microphone transformer.

this battery interfering with or supplying any voltage to the amplifier. Because the transformer primary and secondary windings are distinct and separate, there is no way for the voltage of the microphone battery to get over to the secondary winding and, from there, to the amplifier. Thus, the transformer also supplies a needed isolation for the microphone battery.

ON LINE

The phrase "on line" or "line matching transformer" has nothing to do with the ac power line. In audio technology, on line refers to any component that is part of the recording process. A mic is an on-line component, so is a control console, or a speaker, a tape recorder, a mixer, a reverb unit, or an attenuator pad.

THE CRYSTAL MICROPHONE

Today we have all sorts of microphones. There are some which make all voices, including those of women, sound as though they consisted chiefly of bass tones. However, you shouldn't downgrade a microphone without first considering its function. A crystal microphone is adequate if you are looking for sound output without any consideration for sound quality. The crystal microphone has a high output. This means you can connect such a microphone to an amplifier without worrying whether or not the output voltage of the microphone will be enough to drive that amplifier satisfactorily. That's why such microphones are popular with manufacturers who are willing to supply you with a mic, amplifier, and speaker at what may seem to be bargain prices. If you want a microphone for fun and games or for partying, and you aren't concerned with sound quality, there's nothing wrong with using a crystal microphone.

The heart of the microphone is a crystal of *Rochelle salt*. A diaphragm is held in position around its circumference by a metal ring (Fig. 4-7). While it is fixed along its edge, the diaphragm is flexible enough so the area near its center can move back and forth. A drive pin is attached to the center of the diaphragm with the other end of the pin fastened to a metal plate.

Placed near the diaphragm is a metal sandwich consisting of a pair of metal plates, with a filling consisting of the crystalline substance. The bottom plate is fixed in position, but the upper plate, the plate attached to the drive pin, is free to move. With this arrangement, sound waves from your voice or from musical instruments push against the diaphragm, which in turn, pushes the drive pin, which, in turn, pushes against the top metal plate. The diaphragm is made of spring metal, so the movement of the diaphragm is somewhat in step with the varying pressure produced by the sound.

The crystal material, then, is subjected to a series of varying pressures, depending on whether the diaphragm is moving forward or back, which, of course, depends on what the voice or music pressure is doing at the moment.

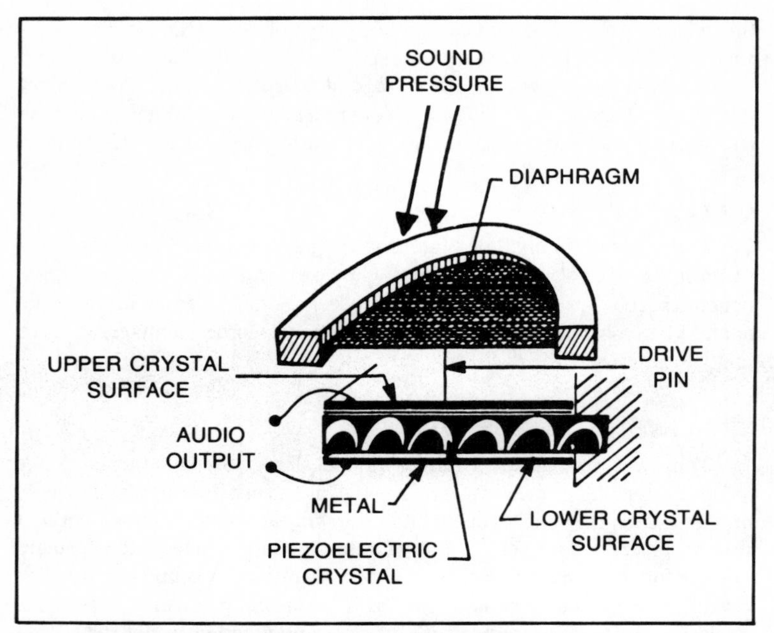

Fig. 4-7. Basic structure of the crystal microphone.

PIEZOELECTRIC EFFECT

The crystal material used in the crystal microphone, has an unusual electrical property known as *piezoelectric effect*. The varying pressure of the metal plate transmitted to the surface of the crystal tends to deform the crystal slightly, permitting the crystal to resume its normal shape when the pressure is removed (Fig. 4-8). This deformation generates an alternating voltage at the rate of sound-pressure change. This microphone, like the carbon type, is basically an alternating voltage generator. Unlike the alternating voltage produced by your local power company, usually a fixed frequency of 60 Hz, the output of the microphone is an entire range of frequencies covering the sound spectrum.

One of the advantages of the crystal microphone is that it supplies a moderately high output-signal voltage for a given sound input. However, the crystal can easily be damaged by high temperatures and high humidity levels. Its frequency response is too poor to make it suitable for recording work, so it is used mostly in voice communications where the objective is intelligibility of sound.

The output impedance of the crystal microphone is high, so it can be connected to the high input impedance of an amplifier without resorting to a microphone transformer. Still another advantage is that crystal microphones can be made quite small, hence are suitable for applications such as hearing aids.

Not all crystal microphones use a diaphragm for exerting sound pressure on the crystal element. In some, known as *sound cells*, the air pressure produced by a voice or instruments is applied directly to the face of the crystal. In the sound cell, a pair of crystal elements are mounted in a housing and are separated by an air space. Both crystals are subjected to sound pressure and both produce a voltage corresponding to the variations in sound. The voltages are *additive*, so this type of design supplies a better frequency response than the diaphragm-driven type. One of the other advantages of the sound cell is that a number of crystals can be used, connected in parallel. This has the effect of lowering the impedance. In areas of high heat and humidity, however, the crystal microphone, no matter how constructed, may declare a holiday and lower its output voltage.

THE CERAMIC MICROPHONE

One way of identifying microphones is through the name of the material used as the transducing element. In the carbon microphone it is carbon, in the crystal microphone it is a substance that has piezoelectric properties, and in the ceramic microphone it is barium titanate. The ceramic mic is a bit better than its crystal counterpart in the heat and humidity departments and it also has a rather high signal output, but it tends to have a frequency response that resembles a roller coaster. What goes in may sound sweet; what comes out may be pickled in brine. It is possible to manufacture ceramic microphones having a smooth, wide frequency response, a response that is fairly uniform. There is a tradeoff, however, and the payment that is exacted is a smaller output signal voltage.

The voltage-generating element in the ceramic microphone has piezoelectric properties; that is, the substance will develop a voltage across its faces when subjected to a pressure and then released (Fig. 4-8). In this respect, then, the ceramic mic is similar to the crystal type. For a given frequency response its output is comparable to that of the crystal microphone.

DYNAMIC MICROPHONES

While both ribbon and moving-coil mics are dynamic, the word *dynamic* has been applied to the moving-coil mic so often that *dynamic* and *moving coil* have become synonymous. Thus, if you order a dynamic mic, it is extremely likely that you will be supplied with a moving-coil type, not a ribbon. Moving-coil mics are sturdy, aren't bothered by humidity and temperature changes encountered in the home or studio, and are sensitive.

The Ribbon Microphone

The ribbon microphone, also known as a *velocity* microphone, consists of a thin, stretched duralumin ribbon approximately 1/4 inch wide and 2 to 4 inches long, suspended between the poles of a permanent magnet. The

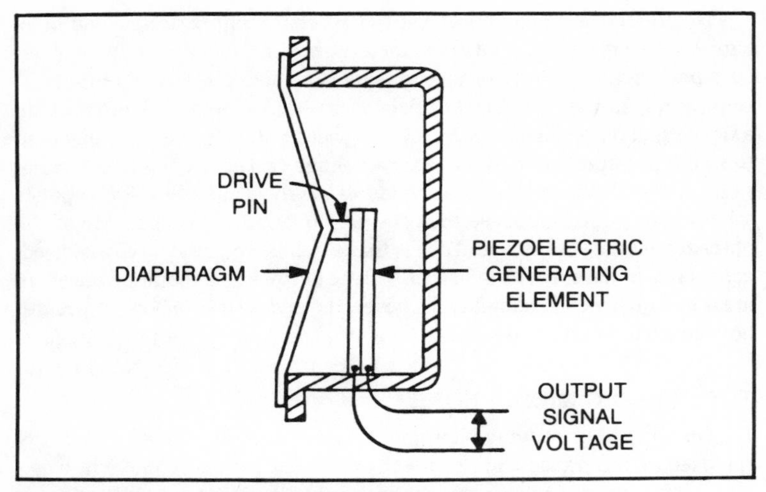

Fig. 4-8. Basic structure of a piezoelectric microphone.

ribbon is clamped at both ends but, except for these ends, it is free to move back and forth (Fig. 4-9).

Any metallic conductor moving in a magnetic field will have a voltage induced across it. This is the basic principle on which your local power company's utility generators work. So the ribbon microphone is first cousin to an electric generator with the magnetic field supplied by a magnet shaped somewhat in U form. Attached to the magnet are a pair of pole pieces that saturate easily with the magnetic lines of force supplied by the magnet. These magnetic lines of force extend from one pole piece to the other, and the ribbon is positioned directly in the magnetic field. When the ribbon is made to move it will produce a voltage. This voltage is ac, not dc, and has a frequency corresponding to the frequency of the impinging sound waves.

Fig. 4-9. Basic structure (highly simplified) of the ribbon microphone. The ribbon is secured at both ends, but is still free to move.

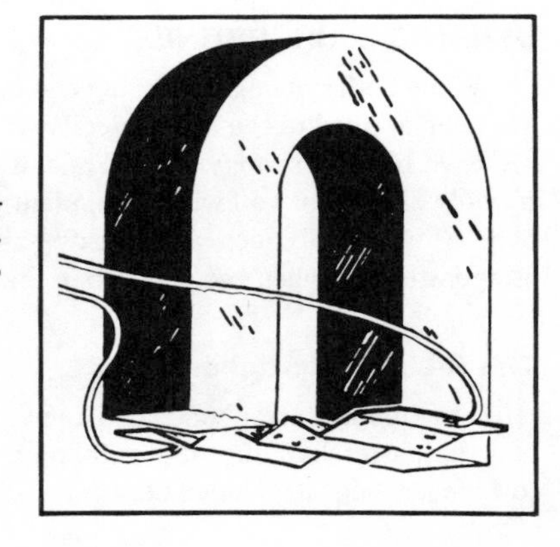

The ribbon inside the microphone, like all metallic conductors, has a certain amount of opposition to the flow of current. The duralumin ribbon inside the velocity mic is no exception. When measured with an instrument such as an ohmmeter, the dc resistance is about 1 ohm, or possibly less. You can consider this equivalent to the impedance of the microphone and it is extremely low. For this reason, such microphones contain a step-up transformer mounted in the mic case. The step-up transformer actually performs a double function. As in the case of the transformer for the carbon microphone, it is an impedance transforming device, enabling the microphone to be connected to the much higher input impedance of a preamplifier. At the same time, the step-up transformer raises the output voltage. With the transformer, the ribbon mic has a typical impedance of about 200 ohms. While this is a single-value impedance, some ribbon mics supply a choice of impedances, such as 30, 150, and 250 ohms. In a typical ribbon mic the frequency response extends from 35 Hz to 18 kHz, plus or minus 4 dB. For other ribbon mics the response may be wider. The ribbon microphone is sensitive only to sounds coming at it from the front or back, and *not* from the sides.

Ribbon mics typically provide a bidirectional or figure-8 response pattern picking up sound from the front and the rear of the mic, while not responding to sounds 90 degrees off-axis. Ribbon mics are very susceptible to proximity effect, popping and other breath noises. (These operating characteristics are described later.)

One of the very severe problems of older ribbon microphones was the extreme fragility of the ribbon. The ribbon could be damaged by blowing across it with the mouth in close proximity to the microphone head. Coughing into such a mic was an invitation to disaster. Used outdoors, a strong wind could literally rip the ribbon apart. Newer ribbon mics use a shorter ribbon element and aren't as fragile as their predecessors. They respond easily to the low sound-pressure level of treble tones. Consequently, this mic has an excellent high-frequency response. Further, the response over the entire audio range tends to be flat. The output voltage, though, is very low. Ribbon mics were widely used in the early days of radio broadcasting and studio recording. They were a wonderful alternative to the carbon mic. But the ribbons that were used then were soft and very light. To be able to produce an adequate amount of output signal voltage, either the ribbon had to be large, or the permanent magnet surrounding it had to be very strong (and this meant very large), or both. But to get good frequency and transient response there was no choice but to make the ribbon small. So the surrounding magnet was made correspondingly larger to supply an even stronger magnetic field. The result was that such microphones were heavy and bulky. In recent years magnet design has enabled manufacturers to reduce magnet size without decreasing magnetic field strength, so you will see some ribbon mics that have a more practical size.

Few manufacturers are now making ribbon mics. Those that are available are more rugged than they used to be. In some you can even replace the ribbon. They generally have a smoother response than the much more popular moving-coil mics. Their treble rolloff is gentle and they do not have much of a popping problem.

The Model M 500 N(C) (Fig. 4-10) is a dynamic hypercardioid ribbon mic by Beyer suited for high sound level applications. Its ribbon measures only 0.85 in. in length and weighs only 0.00034 gram. It contains a four-stage blast filter making possible hand held use with lips almost touching the mic without danger of popping, hissing, or breathing sounds. This mic was developed to cover the special needs of pop vocalists and instrumentalists and is capable of withstanding a sound level of more than 130 dB. The M 500 is unaffected by extremes of atmospheric conditions.

The Beyer dynamic M 160 is a hypercardioid ribbon microphone (Fig. 4-11) utilizing a special double ribbon element. The two aluminum ribbons used in the generating element are each less than one twelve-thousandth of an inch in thickness and are much shorter than those used in conventional ribbon microphones. The short length, in conjunction with a special forming process, makes the ribbons highly immune to damage from overload or mechanical shock and produces a most rugged microphone.

Positioned one above the other, there is a separation of only 20 thousandths of an inch. The low inertia of the ribbon transducer results in an excellent transient response. The frequency response of the M 160 extends from 40 to 18,000 Hz and the microphone's hypercardioid characteristic rejects unwanted sound reflections and extraneous noise coming from behind the microphone, with 25 dB attenuation at 120°.

There are two red dots on the ring around the head of the microphone to mark the longitudinal axis of the two ribbons. When this microphone is used on stands the best possible recording results are obtained with this axis in vertical position.

Unlike other cardioid-type microphones, the M 160 needs no venting on its shaft to achieve its hypercardioid characteristics, and is thus ideal for hand held or stand use. The M 160 is designed to withstand the rigors of professional use, and is unaffected by extremes of temperature and humidity. The matte chromium-plated metal body reflects very little light, and the small size of the microphone makes it suitable for use in television and film studio applications. The M 160 N (C) version features a black chromium-plated finish.

The Moving-Coil Microphone

If you know how a loudspeaker works, you already know how a moving-coil microphone functions, for both use the same basic principles. You can consider the moving-coil mic as a reverse form of speaker. In intercom systems, for example, the dynamic speaker is also used as a dynamic microphone.

Like a speaker, a moving-coil microphone (Fig. 4-12), now better known as a dynamic microphone, contains a coil consisting of a few turns of wire.

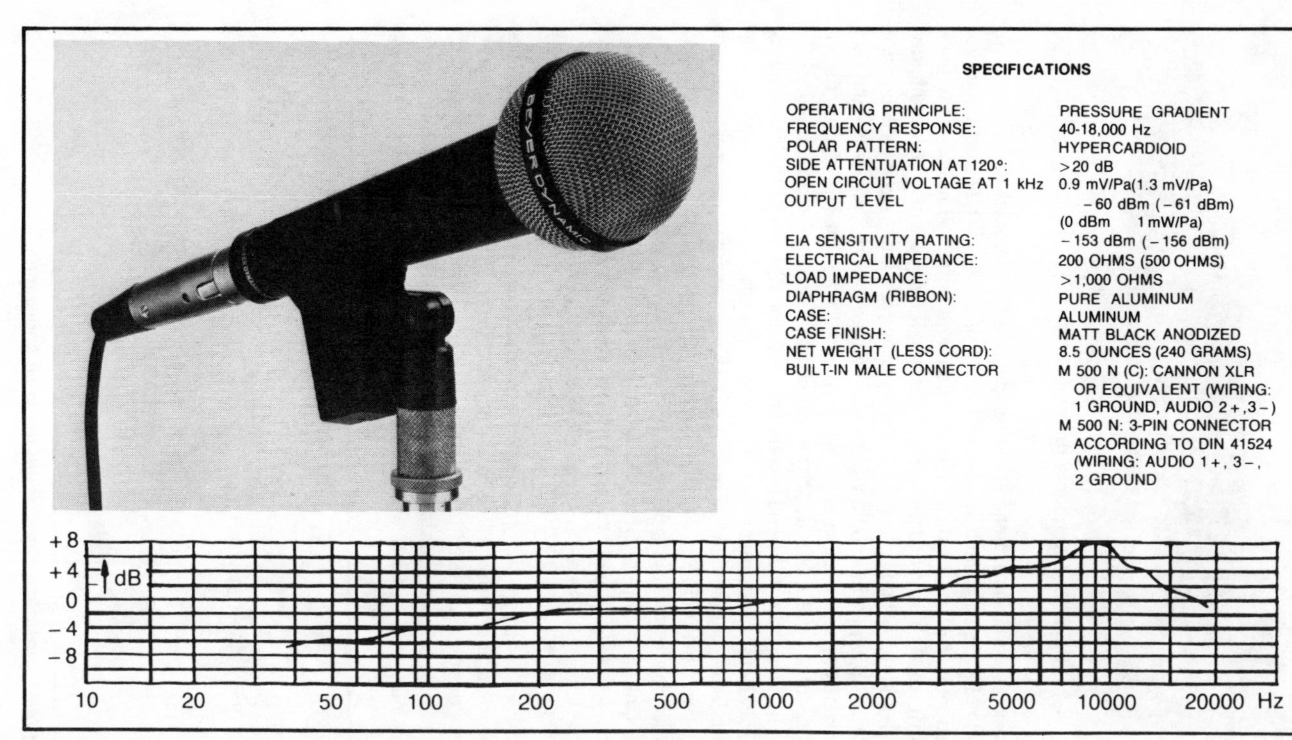

Fig. 4-10. Frequency response curve of a standard Beyer M 500 N(C).

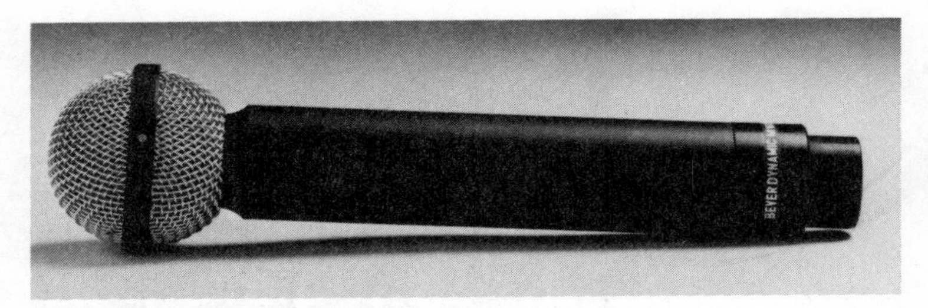

Fig. 4-11. Beyer M 160 double ribbon microphone.

This coil of wire is attached to an suspended by a nonmetallic diaphragm, made of a material such as plastic, Lexan, or Mylar. While the diaphragm is supported along its outer circumference, it is free to move easily back and forth. When it does, it moves the small coil of wire. The coil slides back and forth along one arm of a strong magnet. Once again, a conductor moving in a magnetic field will have a voltage induced across it—and that is exactly what happens now. When a sound wave pushes against the diaphragm, it will move it, at the same time moving the attached coil.

In a way, the voice coil is comparable to the ribbon of the ribbon microphone. Both work similarly, for both consist of conductors moving in a magnetic field. The advantage of the coil, though, is its much greater length compared to the ribbon. Actually, we can consider the coil as a ribbon wound in circular form. This greater length results in a much larger induced voltage and, as a consequence, the moving-coil mic develops a much greater output signal voltage for a given sound pressure input than the ribbon mic.

Bass Reflex Techniques

Bass reflex speaker techniques are sometimes included in dynamic mics to extend and improve low-frequency response. In a bass reflex speaker, ports or vents are used to allow the back pressure created by the moving cone to be utilized, making such speakers more efficient than acoustic suspension types. In other words, in a bass reflex speaker, the air that is nor-

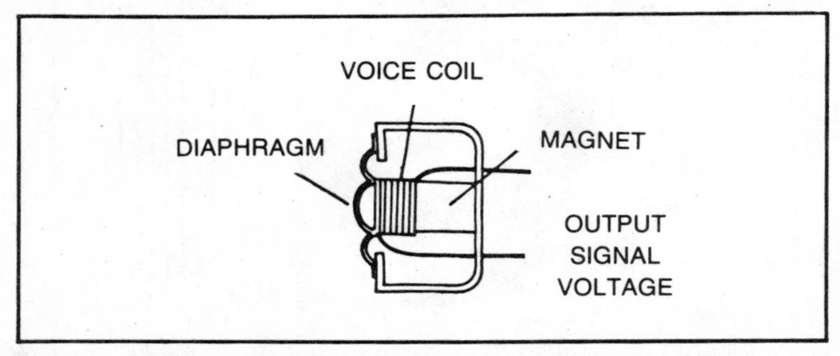

Fig. 4-12. Basic structure of the dynamic (moving-coil) microphone.

mally trapped in the rear of the enclosure is allowed to escape through the front, supplying reinforcement to bass sounds.

In the dynamic mic, the trapped air volume within the microphone body is often used with the mic housing working as a bass reflex enclosure for extending and improving low-frequency response. Unlike bass reflex speakers, however, the dynamic mic can include a bass attenuation control, so the amount of bass response can be adjusted. This is done mechanically by gradually closing the internal reflex port, using a rotating ring below the head, or electronically with a tone control switch.

While microphones and speakers work at opposite ends of sound systems, they are similar in a number of respects. Like speakers, the mass, dimensions, shape, and efficiency of all moving parts of the mic, plus maximum utilization of the magnetic field supplied by the built-in permanent magnet, are all responsible for total response, low distortion, and accurate translation of the sound input into a corresponding signal voltage at the microphone connector pins.

Advantages of the Dynamic Microphone

Of all microphones available the dynamic or moving-coil types (Fig. 4-13) are the most widely used. They are the most dependable, most rugged, and most reliable design for both indoor and outdoor work. These microphones are capable of the smoothest and most extended frequency response when compared to carbon, crystal, and ceramic types. Because they have a happy combination of highly desirable characteristics—good transient response, high reliability, and moderate cost—they are more widely used in both studio and home recording than any other type of microphone. This does not mean that all dynamic mics are automatically good. Whether a dynamic mic does have all these desirable characteristics depends entirely on the manufacturer. A good microphone is like a fine watch. It takes a considerable amount of expertise, know-how and responsibility to make one. The same is true of microphones.

Double-Element Dynamic Microphone

The AKG double-element dynamic microphone (also called a two-way or coaxial mic) uses a separate high-frequency element, a *tweeter*, coaxially mounted over a separate low-frequency element, a *woofer*. The system crosses over acoustically and electrically at 500 Hz, like a two-way loudspeaker system.

CONDENSER MICROPHONES

A basic condenser (Fig. 4-14) consists of a pair of metal plates separated by an insulating material called the *dielectric*. When you connect a *condenser*, more properly called a *capacitor*, to a voltage source such as a battery, the capacitor will take an electric charge. This means there will

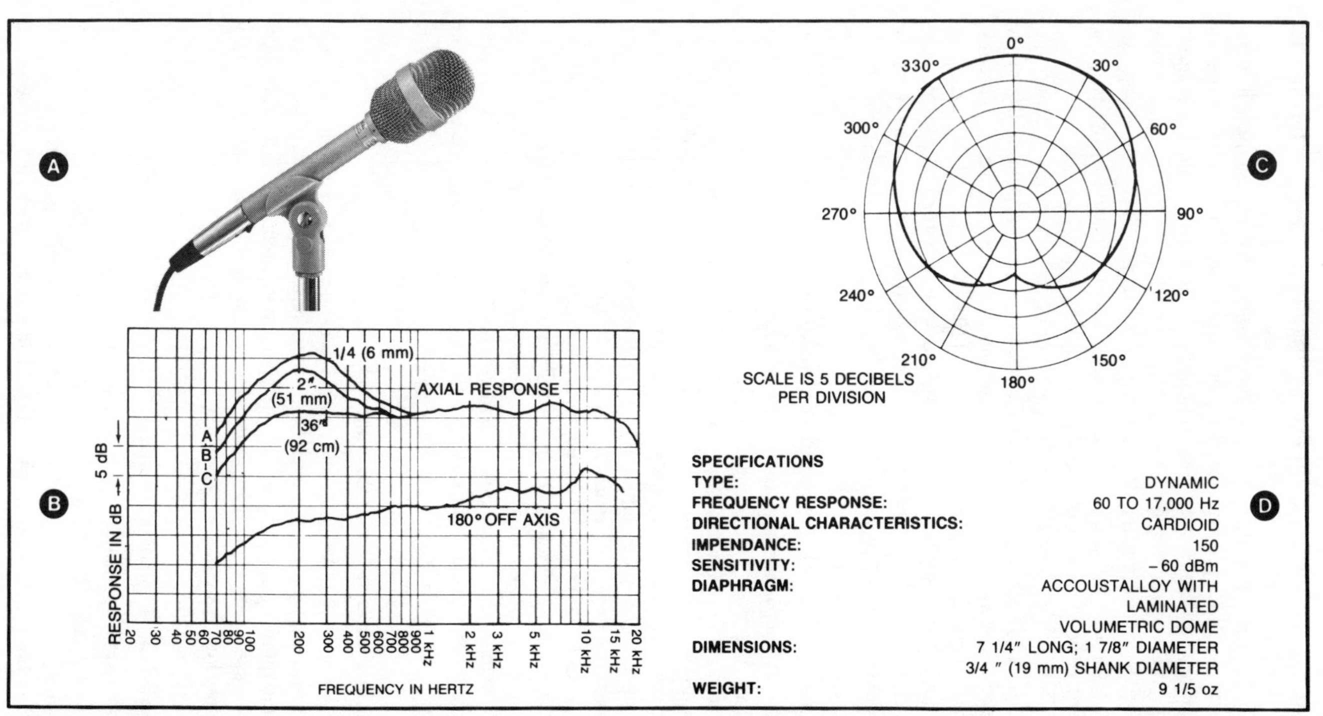

Fig. 4-13. A dynamic microphone. Electro-Voice Model DS35. (A) the microphone; (B) frequency response curves; (C) polar pattern; (D) specifications.

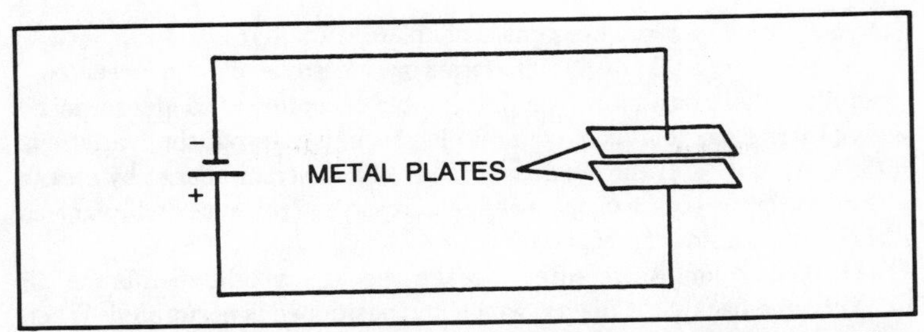

Fig. 4-14. A capacitor (or condenser) consists of two or more metal plates. When connected across a battery, the capacitor will become charged.

be a voltage across the plates of the capacitor. The amount of this voltage will depend on the kind of dielectric used, the area of the plates of the condenser, and how close they are to each other. It is this latter fact that makes the condenser mic possible. If you make one of the condenser plates fixed and the other capable of moving, you can then have a variable voltage device. If you can manage to push one plate closer to the other, the voltage across the condenser will be larger. Conversely, if the moving plate is farther away, the voltage across the condenser becomes smaller (Fig. 4-15).

In the condenser mic we manage to fulfill these conditions. Since one of the plates is able to move, we can use sound pressure to make it do so. In the condenser mic we have a metal-plate diaphragm, tightly stretched but capable of movement. The diaphragm has a very low mass, highly important for the ability to reproduce musical transients. It is often made of some plastic material, such as polyester film, but plastic is a nonconductive material. To make it conductive it is coated with an extremely fine, thin covering of gold. The diaphragm then forms one of the plates of the condenser, and its plastic backing the dielectric. The dielectric faces the

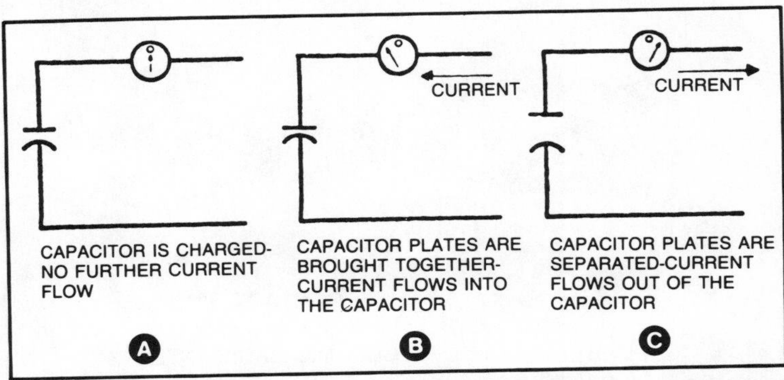

Fig. 4-15. When a capacitor is charged (A), no more current can flow into it. When the plates are brought closer, however, more current can move into the capacitor (B). Conversely, when the plates are separated, current flows out of the capacitor (C).

fixed-position gold-plated ceramic backplate (Fig. 4-16).

Briefly, then, a thin diaphragm forms one electrode of a condenser (now generally called a capacitor with the possible exception of condenser mics) and vibrates close to a fixed counter-electrode, thus producing variations in the capacitance. If the condenser is given a constant charge by means of a very high resistance, the voltage across the condenser will vary as a direct function of the capacitance.

To enable sound pressure to reach the lightweight membrane diaphragm, the backplate against which it is positioned is perforated. When sound pressure reaches it, the ultra thin polymer film moves. Note that this film is not only the dielectric (that is, a nonconductor or a poor electrical conductor) but is also metallically coated. This coating and the backplate form the two plates of a capacitor. What we have here, then, is a variable capacitor, and is a direct relative of variable capacitors used in

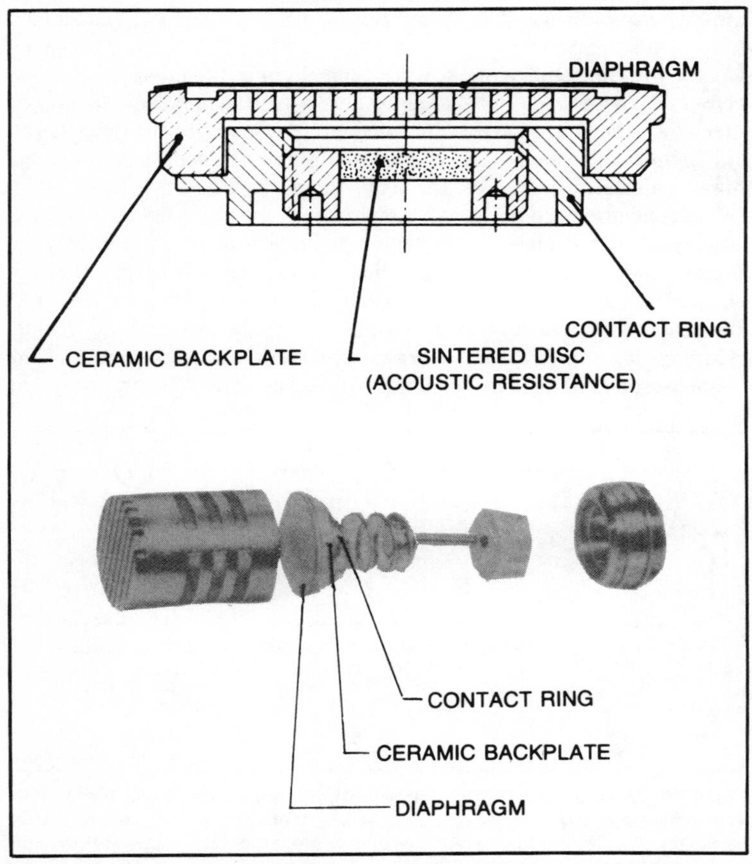

Fig. 4-16. Structure of AKG CK-1 condenser capsule.

electronic equipment. The capacitance is directly proportional to the dielectric constant of the polymer film with the film selected having as high a dielectric constant as possible. The total capacitance is also proportional to the area of the two metal plates and inversely proportional to the distance separating them. However, it is not the total capacitance that is significant as much as the possible amount of variation of that capacitance.

This capacitive element requires a charging voltage, known as a polarizing voltage. The amount of voltage that this capacitor will accept depends directly on its capacitance. Incoming sound varies the separation of the plates (actually only one of the plates moves; the backplate is fixed in position) and this, in turn, varies the electrical charge across the plates. It is this varying charge that is the output signal voltage.

The polarizing voltage can be supplied by an internal or external battery, or by an external power supply, or can be obtained from associated equipment already equipped with a power supply.

With no sound input to the microphone, the *potential*, the polarizing voltage across the condenser element, remains fixed. However, a sound pressure, supplied by voice or musical instrument, will cause the diaphragm to vibrate and, as it does so, the charging ability of the condenser element will vary accordingly. As the diaphragm moves in toward the backplate, the charge across the condenser will become larger, as it moves away it will become smaller.

Note what we have here. We have a voltage change, but more important is that this change will be in step with the incoming sound pressure level. We can take this voltage change and lead it into an amplifier, either a tube-type or solid-state. In either case, we can amplify the voltage change until it is strong enough to supply a corresponding varying current for recording on magnetic tape or for driving one or more speakers.

The very small signal voltage developed across the plates of the condenser microphone cannot be used directly. This output voltage could be "wiped out" simply by connecting a load, such as an output cable, across the condenser plates. An impedance converter is incorporated directly in the microphone to act as an interface between the required input impedance of the following component and the capacitor.

The output impedance of condenser microphones (Fig. 4-17) is extremely high and, to avoid the use of connecting cables, the amplifier is built right into the microphone. Even with miniature tubes, vacuum-tube amplifiers aren't noted for their small size, so early condenser mics were large and heavy.

When solid-state devices came along, two things became possible. The first was that transistor amplifiers could be highly miniaturized. They occupied a fraction of the space demanded by tubes. Tube amplifiers also required a substantially sized power supply. Transistor amplifiers, though, could work nicely from small batteries. The amplifier and its battery and the battery for supplying the polarizing voltage were all moved directly into the condenser mic housing.

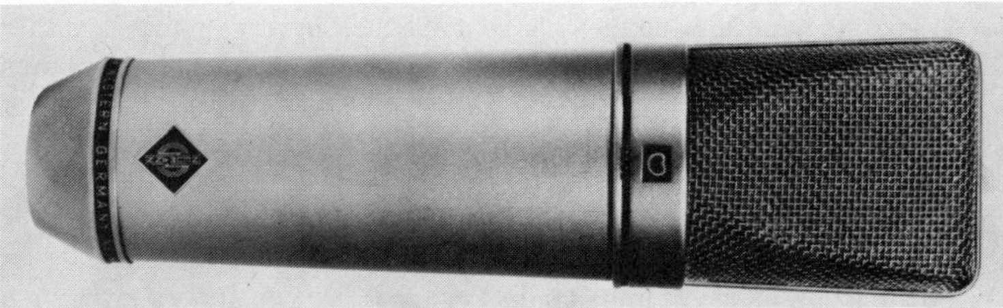

Fig. 4-17. Popular professional condenser microphones. The Neumann U-87, top, and the AKG C-414 EB with accessory H-17 suspension/windscreen.

Not only do condenser mics have excellent frequency response and low distortion, but because the mass of the diaphragm is extremely low, transient response is excellent.

From a technical viewpoint, the so-called "amplifier" in the condenser mic also functions more like an impedance changing device. A transformer is used in the carbon mic to convert the very low impedance of the mic to a somewhat higher impedance through the use of a step-up arrangement. The amplifier in the condenser mic does the same work—but in reverse. It takes the extremely high impedance of the capacitor and converts it to a much lower impedance. Without this impedance converter, the impedance of the condenser mic is in the order of 10,000,000 ohms (10 megohms).

At the output of the converter, the impedance is somewhere between 50 and 200 ohms.

MODULAR CONDENSER MIC SYSTEM

One of the most significant innovations in the line of mics for professional use is the Modular Condenser System. These are quick change mics utilizing screw-on capsules and pre-amplifier modules giving the microphone user the same flexibility as the photographer with interchangeable lenses and filters. As a result the modular mic is a complete system and can be used in a variety of applications: in the studio, for sound reinforcement, and in film and broadcast work.

TUBES VS TRANSISTORS

The amplifier in a condenser mic is as voltage amplifier and a such is directly related to similar amplifiers used in the early stages of an audio high-fidelity system. Most such amplifiers are now completely solid-state but there are some music enthusiasts who insist that vacuum tubes produce "Sweeter" sound than transistors, whatever "Sweeter" sound might be. The matter is purely subjective, and the claim to having "golden ears" is a private matter.

Although a solid-state amplifier for a condenser mic has numerous advantages, it is possible to obtain a condenser mic built around a vacuum tube, to satisfy those who claim that a tube's distinctive sound has never been equalled by any solid-state design. AKG created a new version of the classic vacuum tube microphone with its inherent susceptibility to electrical noise and hum induction, non-linear transfer characteristics, and its physical delicacy. Alternatively, solid-state offered new possibilities for circuit layouts, including miniaturization, convenient powering, lowest possible self-noise, and high overload capabilities along with ultra-low distortion characteristics.

The AKG microphone is a pressure gradient transducer having a double diaphragm. It has a variety of directional characteristics: omni, cardioid, figure-eight, and six intermediate positions remotely controlled from the microphone on the powering unit. The tube used in the mic amplifier is a type 6072, sometimes referred to as an "N" tube. The frequency range is 30 Hz to 20 kHz, plus-minus 2.5 dB. The electrical impedance is 200 ohms.

PHANTOM POWER

Until recently, condenser microphones were very expensive and cumbersome, due to the associated power supply needed. Modern condenser microphones can be powered with batteries, or they can use a "phantom powering system" with power taken from an associated amplifying or recording equipment.

Unlike other types of mics which are output devices only, the condenser mic, in addition to supplying a signal voltage, requires a dc input voltage. In early condenser mics the units were four terminal devices, with one pair for signal output and the other pair for dc voltage input, thus requiring two cables for each condenser mic.

This arrangement was considerably simplified when the signal output cable was used also to supply dc input to the mic. Because the signal output voltage doesn't "see" the dc input and operates as though it did not exist, the dc voltage is referred to as a "phantom."

Figure 4-18 shows a power supply that can be used with mics that depend on a remote source of power. The unit has one 3-pin balanced input connector from the associated mic and one 3-pin balanced output connector which can be joined to a mixer or console. The purpose of the LED indicator is the glow it supplies when the battery test pushbuttons are depressed.

While the circuit shows two batteries, only one is used at a time with the other held in reserve as a backup unit. This power supply can be used for any mic which operates from a 9-volt dc power source.

Figure 4-19A shows the input circuit of an amplifier capable of supplying dc power to a condenser mic. In this arrangement the cable used for carrying the signal from the microphone to the amplifier is also used for carrying dc power from the amplifier to the mic. The cable is an unbalanced type and the shield of the cable is the negative (minus) side of the connection. The plus terminal of the dc voltage is connected to the center tap of the amplifier's input transformer. The plus wire of the cable is connected to the plus terminal of the mic amplifier via the secondary winding of the mic's output transformer.

Figure 4-19B shows the connections for a separate, phantom power

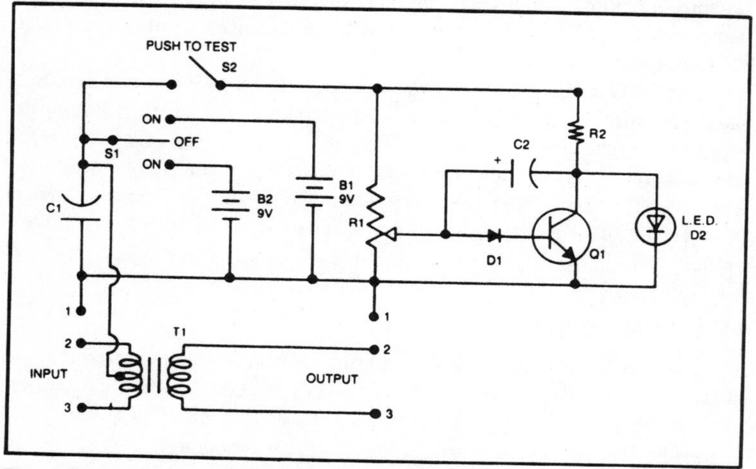

Fig. 4-18. Power supply for condenser mic.

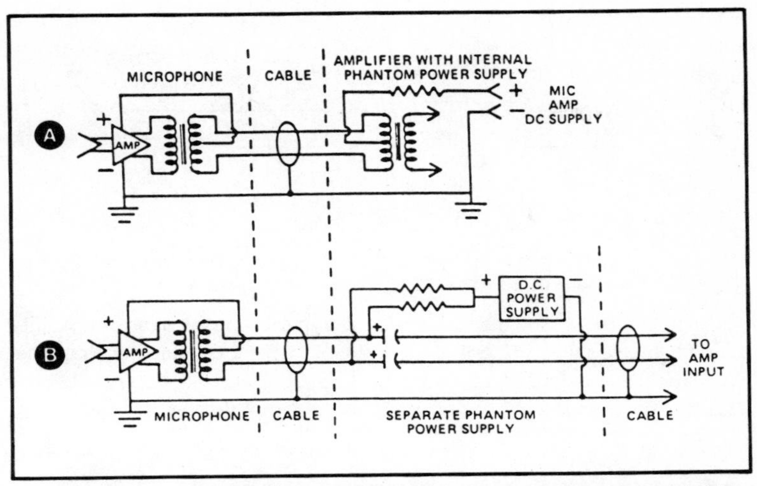

Fig. 4-19. Phantom powering for unbalanced cable (A) and for balanced cable (B).

supply. The minus side of the power supply connects to the negative terminal of the amplifier used in the mic via the connecting cable's ground lead. The two separate, independent conductors of the balanced cable both receive positive voltages from the phantom power supply.

THE ELECTRET MICROPHONE

There is no question that the condenser mic is a superior type, but its structure is complicated by the fact that it does require two voltages—a voltage supply for the self-contained transistor amplifier or impedance converter, and a polarizing voltage for the condenser element.

The *electret* microphone (Fig. 4-20) belongs to the condenser microphone family. The powering requirements of the electret microphone are handled by incorporating a self-polarized or electret capacitor element within the microphone. If you take an ordinary capacitor of good quality and put a voltage across it, the capacitor will become charged. If you then remove the voltage, the capacitor will retain its charge, quite often for a long time. The electret is a specially designed capacitor that will hold a charge indefinitely. This means, then, that the electret can be charged by the manufacturer during the process of constructing the microphone, a step which eliminates the polarizing voltage requirement of the condenser mic.

The voltage impressed on the electret at the time of its manufacture, known as a bias voltage, can be more than 100 volts. The changing capacitance of the electret mic, due to incoming sound, causes the voltage to vary a plus and minus amount.

The impedance converter inside the microphone case still demands its own power supply, but this can be handled by a small penlight type battery.

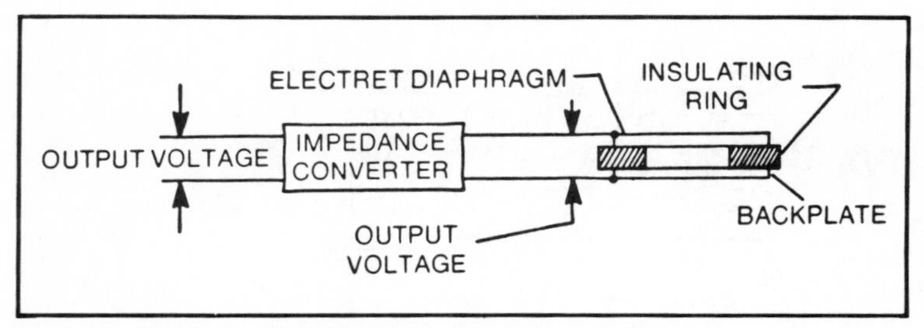

Fig. 4-20. Structure of an electret condenser microphone.

The electret technique has brought a higher performance capability to inexpensive mics, but the electret should not be considered a replacement for the standard condenser mic. The self-polarized electret element can only hold about 20% of the charge placed on it during manufacturing,

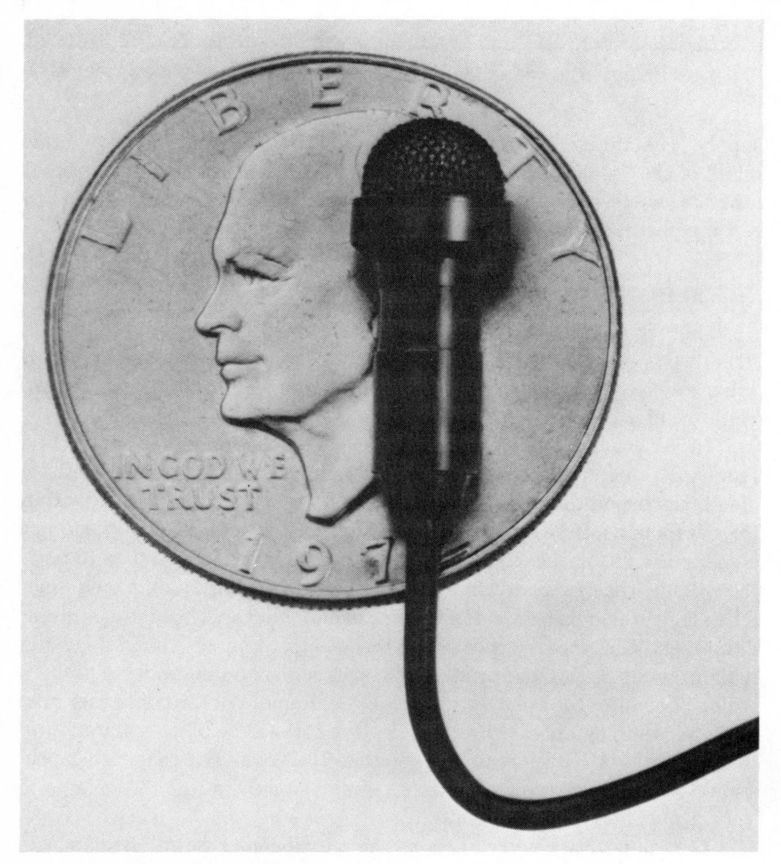

Fig. 4-21. Beyer MCE-5. World's smallest electret microphone.

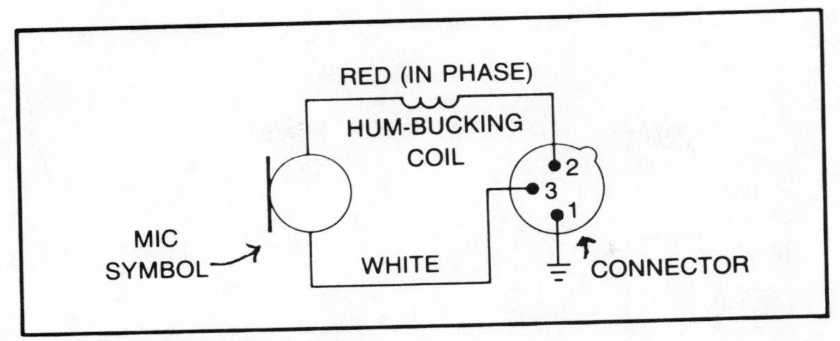

Fig. 4-22. Microphone symbol.

and as a consequence the plates of the capacitor must be closer together than in the standard condenser mic. The dynamic range and sensitivity of the electret are less than that of the standard condenser unit because of close spacing in the capacitor element. Early electret mics were subject to performance deterioration because temperature and humidity effects caused the diaphragm to sag, resulting in reduced output and lessened dynamic range. State-of-the-art gold-vapored diaphragms have eliminated these problems.

In the electret mic, the dielectric of the capacitor element is bonded to a perforated backplate and consists of a coating that is permanently charged electrically. A field effect transistor (FET) impedance converter circuit is usually included in the electret mic. It can be powered by a small battery positioned inside the mic housing, but some electret mics are "phantom" powered obtaining dc operating power remotely from the mic mixer via the mic cable.

Electret mics are suitable for high-quality recording since they have a uniform frequency response and have a good transient response capability. They can also be made extremely small as shown in the photo in Fig. 4-21.

THE MICROPHONE SYMBOL

Fairly standard symbols are used when drawing electronic circuits which may include a microphone. While there are several variations of mic symbols they are close enough to be easily recognizable. Figure 4-22 shows such a symbol. The symbol supplies no technical information about the mic nor does it indicate its type. Data, if required, is supplied separately.

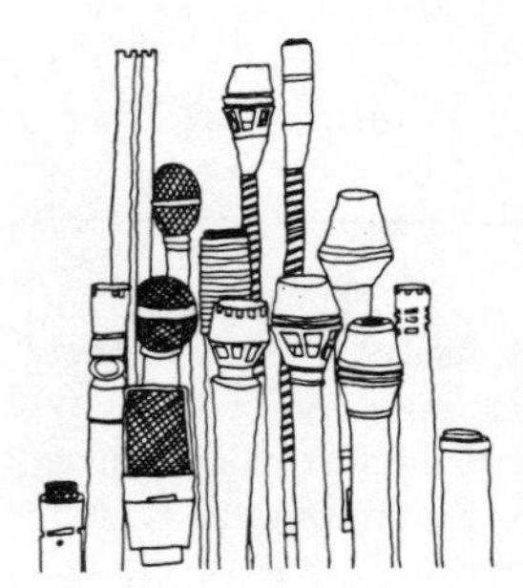

Chapter 5
The Making of a Mic

The making of a microphone is both an art and a science. It is an art in the sense that the assembly of a microphone demands considerable manual and machine skill, involving the assembly of components that have been manufactured to extremely fine tolerances; it is a science since microphone design is subject to mathematical analysis. It takes a combination of these two, art and science, to produce a top quality microphone, not just for a laboratory prototype, but also on a quantity manufacturing basis.

MICROPHONE SHAPE

The shape of a microphone supplies no clue as to its performance capabilities. You cannot equate size with quality, since extremely small size may be a recording necessity. Weight can be meaningless; it would be unthinkable to sell microphones by the pound. Comparing mic specs is helpful, but in the final analysis it is the performance of a mic in a specific application that counts. A mic isn't a good mic because it is a dynamic, ribbon or condenser type. There are good and lesser designs in all types.

For studio use, sound engineers prefer slim, slender mics of a rod shape. This is, in fact, what most condenser microphones are. This not only applies to single capsule mics, but to two-way units as well.

MICROPHONE CASE

The case of the microphone can be die cast zinc, zinc alloy or machined aluminum or steel, with the case finished in chrome, bronze anodized aluminum, or some non-reflecting or brass color. For electret types there may be a battery compartment made of aluminum.

Microphone designers use techniques employed by speaker engineers. Thus, dynamic mics are sometimes made with a housing that works as a bass reflex. A mechanical bass attenuation control may be included to govern the amount of bass response, done by rotating a ring located beneath the head of the microphone.

MICROPHONE DESIGN

Microphone manufacturers rely for their microphone designs on in-house research and development departments and so the machinery for manufacturing such microphones is also in-house designed and constructed. For these reasons, each microphone manufacturing plant is unique.

Designing and making a microphone is an incredibly complex task demanding the highest level of mechanical and technical precision. The task is further complicated by the fact that all the materials used in the construction of a mic have a significant bearing on its performance. The mic design engineer must know how these various materials interact and how they respond to changes in humidity and temperature.

PERFORMANCE CHARACTERISTICS

The performance characteristics of a mic are influenced by the housing used, by the volume, size, and position of the openings in that housing, and the kind of acoustical resistance around the capsule's rear sound entrance. As a result of these often conflicting requirements the development of a particular kind of microphone can be a slow and painstaking process.

Beginning in the 1960s, microphone engineers began to use computer simulation of various acoustical circuits for mics, leading to the development of some completely new concepts in transducer technology. This resulted in the manufacture of microphones that combined ruggedness with specific high performance characteristics.

THE PROTOTYPE

The ultimate result of microphone design leads to a prototype. This sample is tested both electrically and mechanically. Every aspect of its acoustic ability is researched and then the prototype is exposed to heat, moisture, and air pollution to determine if it can survive the transition from the laboratory to the real world.

In some instances, material weaknesses revealed during prototype testing leads to a demand for substances having specific desired characteristics, and this, in turn, leads to the creation of completely new synthetics. One example has been the pioneering efforts by AKG in the use of sintered bronze grilles in microphones, permitting a highly controllable acoustical resistance in a small package.

ASSEMBLY OF A DYNAMIC MIC

The basic dynamic mic is a coil of wire (thinner than that of a human hair) fastened to a diaphragm. The diaphragm is a synthetic material, heated and then formed. It must be of uniform thickness and stiffness, no easy task since its thickness is measured in microns. In some Electro-Voice mics, a polystyrene disc is laminated to the diaphragm to prevent diaphragm breakup at high audio frequencies. Breakup in this case means the segmentized movement of the diaphragm, preventing it from having a piston-like movement.

Positioning the voice coil on its non-metallic diaphragm requires the greatest precision. The diaphragm and its mounted voice coil are then centered in a material capable of being magnetized. After this assembly is placed in its capsule, a housing for the transducer, it is subjected to a high intensity magnetic field. This field converts the material surrounding the voice coil into a permanent magnet. In this way the resulting magnets are always properly polarized. Finally, the capsule is placed in its housing which will give the microphone its final performance characteristics.

HUM BUCKING COIL

A hum voltage can be induced in the moving coil of a dynamic mic by the heavy magnetic fields accompanying the strong ac currents needed for stage lighting or power transformers. A hum bucking coil is used to counteract the effects of hum induced voltages, decreasing hum pickup on the order of 25 dB.

TONE COMPENSATION

Some mics have built-in tone compensation filters that are switchable and which are used to change the performance characteristics of the unit, emphasizing bass, midrange or treble, as in AKG's D1200E.

THE CAPSULE

The capsule contains the elements of the acoustic transducer system and in AKG's D-300 series (D310, D320B, D321 and D330BT) is made of a thermoplastic material using an injection molding process. All parts of the magnetic system are precisely positioned and firmly encapsulated.

Suspension of the acoustic transducer presents directly conflicting requirements. The suspension must be robust enough to protect the mic in case it is dropped, particularly important in the case of hand-held mics.

On its outer surface in AKG's D310, D320B and D330BT, this element carries a number of tiny hemispheres or cone shaped protrusions. When the transducer responds to low sound levels it rests on these protrusions only. With higher sound amplitudes the protrusions are compressed, limiting the maximum possible amplitude. Thus, if the mic is dropped, there is a progressive decrease of elasticity, limiting the maximum possible amplitude. The damping action of this suspension also limits impact sound and handling noise.

POP SCREENING

The microphone grille (Fig. 5-1) sometimes called a windscreen or pop screen, is made of shock absorbing, heavy stainless rigid steel mesh or of a soft foam. Open cell polyurethane foam, sometimes called a pop filter or blast filter is positioned between the grille and a safety basket located above the capsule, not only to reduce popping, but breath and wind noises as well, and also to keep dirt and moisture out. The cellular material is acoustically transparent to normal sound pressures but stops sudden air blasts produced by wind, movement of the mic, and "p" and "t" pops. The purpose of the safety basket is to protect the transducer module from impact and also supports the grille, preventing its possible deformation. The grille assembly is threaded into the microphone housing but can be easily removed for replacement or cleaning. Windscreens are available in colors such as gray, red, blue, green and yellow. In some mics a colored foam windscreen encloses the capsule almost completely.

MAKING A HYPERCARDIOID DYNAMIC MICROPHONE

AKG's D-330BT hypercardioid dynamic mic is a member of its D-300 series. The mic is a dynamic pressure-gradient type and has a normal, unequalized frequency range of 50 Hz to 20 kHz. The low-frequency response characteristics are adjustable by a three-position bass rolloff switch accessible through a cutout in the mic's housing. To provide nine different equalization contours, the microphone's bass rolloff and presence rise switches are separately and/or simultaneously adjustable in nine possible combinations of settings.

The microphone consists of four special modular assemblies (Fig. 5-2), that are interrelated. These consist of a removable shock-absorbing three-layer windscreen/pop filter; a removable safety basket, internally reinforcing the windscreen/pop filter and protecting the front of the transducer system; a specially suspended and vibration/hum compensated plug-in module, containing the transducer system, plus the presence rise switch with its associated network; an impact resistant main housing, mating with the first three assemblies and containing the bass rolloff switch with its as-

Fig. 5-1. Microphone windscreen. (AKG Acoustics, Inc.)

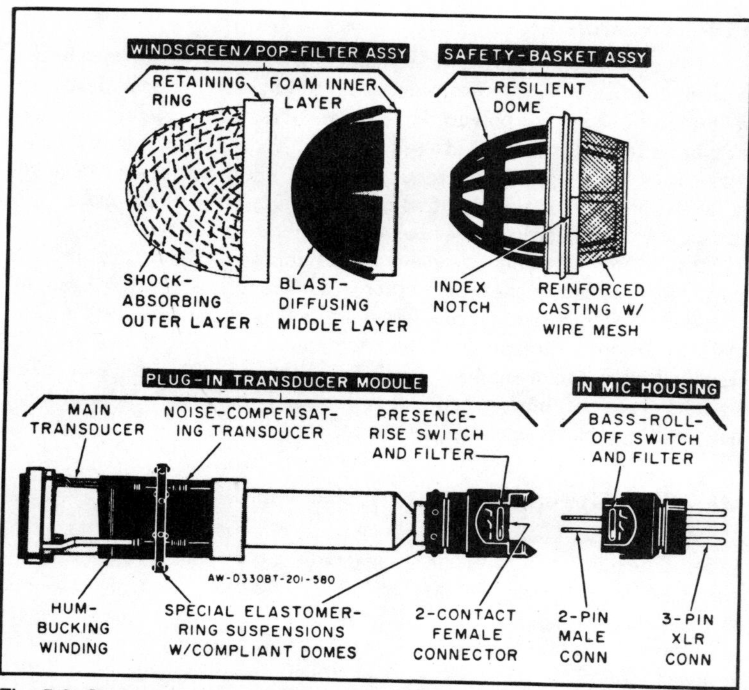

Fig. 5-2. Construction techniques of AKG's D-330 BT hypercardioid dynamic microphone.

sociated network plus the microphone's audio output connector.

The three-layer windscreen/pop filter assembly suppresses wind noise, breath pop and similar acoustic interference and also partially isolates the transducer module from the effects of head-on impact damage. The assembly consists of a shock-absorbing stainless steel wire mesh outer layer, a removable liner made up of a blast-diffusing fabric middle layer sandwiched with an open cell reticulated polyurethane foam inner layer, and an internally threaded retaining ring to mate the entire assembly securely with an external thread on the main housing.

The internal safety basket assembly is contoured to fit supportively just inside the windscreen/pop filter and protectively around the front of the transducer module, thus reinforcing the windscreen/pop filter against the effects of head-on impact damage. The safety basket consists of a resilient dome shaped ribbed cage joined to a reinforced truncated conical open frame casting fitted with a fine wire mesh screen to coincide with side ports in the main housing and having indexing notches to align the entire assembly with structural members in the main housing.

To further isolate the transducer module from the effects of impact damage, as well as to reduce the effects of handling noise and spurious vibrations, the entire module floats in all directions within the main housing. This isolation is achieved by two special ring-shaped elastomer suspensions, one fitted around the module near its front end and the other fitted to the module at its rear.

To further reduce the effects of mechanically and motionally induced handling noise by an additional 30 dB at 100 Hz, the transducer module has a special noise bucking arrangement. The microphone has two transducers: one serves as the main sound pickup device and the other transducer works as a nonacoustic noise compensating device acting strictly as a motion sensing accelerometer. The noise compensating transducer is positioned behind the main transducer and faces rearward. The two transducers are connected in parallel and electrically in phase. Thus, the inward direction of diaphragm movement in each, generates like polarity of output signal from each. However, because of their opposed orientation, the two transducers operate effectively out of phase so that any handling noise signal generated by the main transducer is nulled by an equal but opposite handling noise signal generated by the noise compensating transducer.

MAKING A CONDENSER MICROPHONE

The condenser mic transducer is made with a metal plate diaphragm which is tightly stretched but still capable of movement. The diaphragm is sometimes made of a polyester film, but since it is nonconductive is coated with a fine covering of gold.

The diaphragm with its conductive surface is one plate of the condenser with the plastic as the dielectric. The other condenser plate is a fixed position ceramic backplate which is also gold plated. The two plates are charged by a polarizing voltage from a dc voltage source.

The amount of charge on the two plates depends on the surface area of the plates, their separation and the type of dielectric used. When the microphone is not in use, the charge on the two plates remains constant. Vibration of the diaphragm because of sound pressure changes the charging ability of the capacitor and so the plate charge varies in accordance with the sound pressure level. This voltage change, varying at an audio rate, is fed into a solid-state preamplifier incorporated in the microphone. This amplifier positioning is essential because of the small signal voltage change produced by the plates and by the very high impedance output of this type of microphone.

Studio condenser microphones are hand made in air-conditioned rooms which are immaculately clean. The condenser diaphragm assembly is repeatedly tested with early tests evaluating its acoustical and electrical properties. It is then installed in a microphone test housing with the testing done in an anechoic chamber equipped with acoustical measuring equipment. It is in an anechoic chamber that the diaphragm assembly, in its test housing, is evaluated for performance prior to installation in its own casing. Final assembly is a hand operation with the completed microphone once again placed in the anechoic chamber for final testing and evaluation.

Some mics feature capsule interchangeability. Neumann makes a line of miniature condenser mics all having the same electronics and dimensionally identical, but using three different interchangeable screw-on capsules (the KM 83, an omni, and the KM 84 and KM 85, both cardioids). The KM 88 has three pattern switchability, cardioid, figure 8 and omni. The capsule's dual membranes are made of nickel.

MAKING AN ELECTRET

The electret is a member of the condenser microphone family and works in the same way. The flexible diaphragm of the electret consists of a metallized plastic foil with a permanent electrostatic charge. In some units the diaphragm is gold vapored Teflon. The advantage of the charge, set at the time of the manufacture of the mic, is that it eliminates the need for an external polarizing voltage, but like the condenser microphone the electret requires a built-in voltage amplifier. Either the condenser or electret units can be phantom powered by simplexing the dc supply from associated equipment via a standard shielded two-conductor audio cable.

In the construction of an electret mic, such as TEAC's PE-120, there is a heavy duty metal mesh windscreen with its interior lined with a colored foam type windscreen. The capsules consist of interchangeable polar pattern elements, such as an omni and cardioid. Below the capsule is a switchable attenuation pad followed by a built-in solid-state preamp. The preamp is supplied by a 8.4 to 9.0 volt battery housed in the mic handle. This handle contains a two position voice/equalization switch. Coming into the bottom of the mic is an XL type connector with 15′ of balanced cable. Some mics, such as Nakamichi's CM 300 electret, have a built in 10 dB attenuator, a switchable pad to avoid preamp overload when excessively high sound pressure levels are encountered.

MAKING A RIBBON MICROPHONE

The ribbon or velocity mic has a thin stretched ribbon made of duralumin or pure aluminum, from less than 1″ to 4″ long and 1/4″ wide. Often weighing as little as 0.0005 gram, or less, it is clamped at its ends but can vibrate between the poles of a permanent magnet. The ribbon works on the same principle as the coil of a dynamic since its motion in a magnetic field induces a voltage across it. At the same time it functions as the diaphragm of the microphone.

In their line of ribbon mics, Beyer uses a thin metal foil ribbon which measures 0.05″ in length and weighs only 0.00034 gram. This mass is only a fraction of that of the voice coil and diaphragm used in dynamic microphones.

DOUBLE RIBBON MICROPHONE

The Beyer M 160 N is a double ribbon microphone with the ribbons having a thickness of 0.002 millimeter, mounted one above the other within a distance of half of a millimeter. It does not use venting on its shaft to achieve its hypercardioid characteristic. In double element moving coil mics, also called two-way or coaxial mics, one element works as a tweeter or treble transducer; the other as a woofer for bass and midrange tones.

In the double ribbon mic the transducing elements are thinner and shorter than in most other ribbon mics. The shorter length makes the mic more immune to mechanical shock problems and also more responsive to transients. In Beyer's M 160 N there are two red dots on a ring around the head of the mic to mark the longitudinal axis of the two ribbons as an aid in determining optimum recording orientation.

DIAPHRAGM FORCES

When the transducing element of a microphone is exposed to a sound field, air pressure is exerted on both sides of the diaphragm, assuming complete exposure of that diaphragm. The amount of driving or moving force on the mic diaphragm depends on the distance of the front and rear sound ports from the diaphragm and also the frequency and the angles at which the sound pressures arrive at the diaphragm.

The resultant moving force exerted on the diaphragm, known as the pressure gradient, is the difference of the pressures in front and rear of the diaphragm. If both acoustic pressures are identical, the diaphragm will remain motionless since the acoustic forces are equal and opposite.

We can regard the acoustic pressures as a pair of vectors 180° out of phase. The magnitude of this vector is a function of the distance of the front and rear sound ports. This is a simplistic approach since frequency and angle of incidence are also involved. Consequently, the vectors represent the composition of a large number of small forces. In any event, we cannot assume that the vectors will be of equal amplitudes and phase angles nor will they all be normal (perpendicular) to the diaphragm. The composition of these forces acting on the membrane supplies a resultant and it is this resultant force that moves the diaphragm. The only time the vectors can be added directly is when they are in phase or out of phase by 180°. At all other times vector addition is required.

However, it is possible to make a mic in which only one side of the diaphragm is exposed to sound, with the other side sealed off against the acoustic environment. In this case the diaphragm will be moved by sound pressure, not by a sound pressure gradient. Instead of a vector quantity, we will have a scalar quantity only, that is, just a number representing the amount of pressure.

ACOUSTIC PHASE INVERTER

If we assume a microphone whose membrane is completely exposed to a sound field, it is possible to have a condition in which the pressure gradient will be zero. If this element receives sound at an angle of 90° off axis, for example, sound pressures in front of and behind the diaphragm will be equal, the pressure gradient will be zero and the membrane will be at rest.

With the existence of a larger sound pressure from the front, on axis or 0°, the transducing element will respond in an on-axis direction but at 90° off axis it will not respond, since there will be equal and opposite sound pressures at this point. The directional characteristic of such a mic will have a figure 8 pattern. Maximum response will be at 0° and 180°; zero response at 90° and 270°.

The directional characteristics of this mic can be changed by the use of an acoustic phase inverter element acting as an acoustic barrier between entry of sound from the rear and the rear side of the diaphragm. The mic will then begin to adopt unidirectional characteristics, producing a so-called unidirectional pattern or cardioid.

This assumes that all of the rear sound will travel via the acoustic barrier and that none of that sound will enter the mic via any other path. A more likely condition is that the sound will take two paths; one through the acoustic barrier and the other around the mic, making its entry via the front port. Depending on the physical geometry of the mic it is possible for the transit time of the sound to be equal for both sound paths, both on axis, to exert equal and opposite pressures on the diaphragm. Under these conditions there will be no output from the mic since there will be no diaphragm movement.

One of the earliest techniques for including an acoustic phase inverter for dynamic mics consisted of a group of holes or ports in the rear half of the mic housing with these holes covered by a substance having a high absorption coefficient. The holes weren't perforated at random but spaced at various equidistances from the diaphragm. The problem with this approach was that at low frequencies the front to back ratio was about 12 dB while the frequency range was about 1-1/2 octaves below the fundamental resonance frequency of the diaphragm.

An improvement was subsequently made by Gorike of AKG who used a single diaphragm loaded by an acoustic plug. In the low and medium frequency ranges the front-to-back ratio was between 15 and 25 dB and about 6 to 15 dB in the treble.

TESTING A MICROPHONE

Microphones are tested in anechoic chambers with a sound generator operating at some fixed frequency, such as 1 kHz and kept in a fixed position while the microphone is rotated circularly around it at a distance of one meter. The output of the microphone is then plotted on a polar diagram. The test is repeated for a number of frequencies, possible beginning at 125 Hz and spaced an octave apart, with the final test at 16 kHz.

FREQUENCY RESPONSE

Offhand, it would seem that all microphones should be designed and manufactured to have a flat frequency response over the mic's specified range. Yet from a practical viewpoint this could lead to problems in recording applications. A microphone having a flat response down to 20 Hz would be sensitive to pickup of hum and also rumble from mechanical devices such as air conditioners or other kinds of mechanical equipment. Some microphones are deliberately designed to have a low frequency rolloff or drop, possibly 5 dB, at some frequency such as 100 Hz. Proximity effect, bass emphasis when the microphone is used in close proximity to a sound source, may be preferred by some recording artists, yet is a characteristic that under some conditions can lead to muddy recordings or sound transmission. Along these same lines, no mic can be expected to overcome the problem of poor studio acoustics. In a hard, highly reflective room both sound and noise will be bright and "pingy" because of a high level of reverberant to dry sound.

In amplifiers, it is customary to spec the unit in terms of flatness of frequency response, and there are now such components with a flat response from dc (0 Hz) to several hundred kHz. It is tempting to apply such a standard to microphone frequency response, and there are some mics that have a flat response from 20 Hz to 20 kHz, with a plus/minus variation of no more than 1 dB. Theoretically this may be ideal but may not be desirable in a practical recording situation. A microphone not too responsive at the low or high ends may be useful in eliminating low frequency problems, in emphasizing voice projection, or in producing certain sound effects.

In general, the frequency range of microphones is by no means as great as the frequency range of components, such as audio amplifiers used in high-fidelity systems. Individual musical instruments have a rather limited fundamental range. Thus, the plucked double-bass supplies sounds from a low of about 30 Hz to 4200 Hz while the triangle begins at 2000 Hz and continues to 16,000 Hz. Very few musical instruments go below 50 Hz while most musical instruments have a fundamental frequency below 5 kHz.

Consequently, because of the variety of microphone applications, the frequency response of a mic and its range are usually shaped or tailored to some particular use. For instance, a musical instrument microphone is ideally "flat" across the range of that instrument, whereas a vocal microphone may have a "presence peak" in its voice frequency area, so that the vocalist stands out from the instruments. The frequency response of a communications microphone is carefully tailored to the voice-frequency spectrum to eliminate unwanted high- and low-frequency background noise.

Figure 5-3 shows a method of making a frequency response measurement. Using an anechoic chamber, an audio frequency generator produces sound waves over the frequency range being checked. These are picked up by the production mic with the response recorded on a graphic chart.

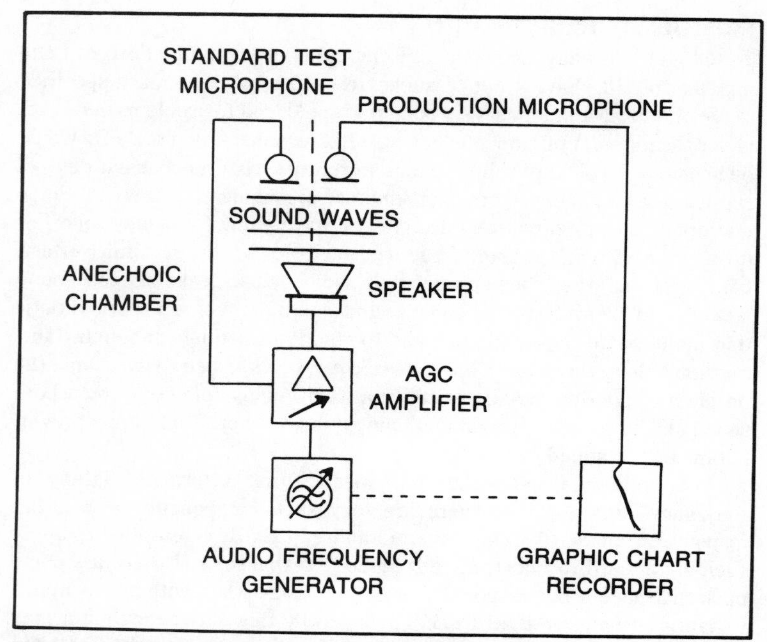

Fig. 5-3. Method of checking frequency response of a mic.

The output of the audio generator, hence the amount of sound, is controlled by an AGC (automatic gain control) amplifier governed by the standard measuring microphone.

When the frequency response of a mic is tested it is done so in an anechoic chamber. But the frequency response of a microphone can be changed the way it is used in practical applications. A lavalier mic can be boomy when worn next to the chest cavity, but will have a different response when hand held.

MOUNTING THE MIC

A mic can be hand held or, as in the case of lavalier types, clipped to some article of clothing, or can be mounted on a stand. Usually, the grooves cut into the top of a mic stand have a diameter of 5/8 inch and are 27 threads to the inch. In a spec sheet, you may find this written as 5/8-27. In some instances the mic is manufactured with the intent of having it stand mounted and so will have an accommodating thread. This does not mean only such mics can be stand mounted as the alternative is the use of a clamp. The clamp is generally made of plastic and is threaded to fit the stand.

Chapter 6 Microphone Response

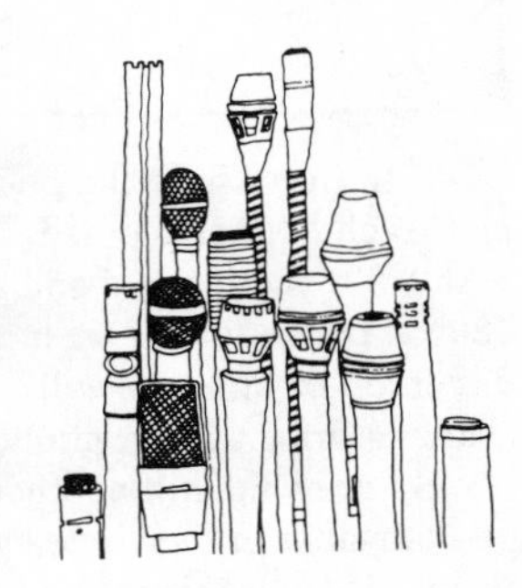

Depending on how a microphone is designed and constructed it will have "areas of sensitivity." These are the areas in which the microphone will be more or less sensitive to sound pressures. The word "area," though, could be regarded as technically inaccurate since it implies a two-dimensional surface. The air pressure that reaches microphone elements arrives at the microphone from all directions, above, below and from the sides. This does not mean a microphone will necessarily respond equally to all changes in air pressure produced by the voice and musical instruments. The microphone can be more receptive to sounds arriving from one direction than from another.

POLAR PATTERNS

While the response of a microphone is three-dimensional (Fig. 6-1) and while we could possibly sculpt the response in three-dimensional form using clay or some other medium, such an approach isn't practical. Instead we graph the response in two dimensions by drawing it on a flat sheet of paper.

A *polar diagram* is a type of graph that is particularly suitable for showing the directional response of a microphone. The polar pattern is a way of graphically indicating the sensitivity of a particular microphone to sounds which arrive from every possible direction. In its construction, as shown in Fig. 6-2, the diagram initially consists of a series of concentric circles, with the outermost marked 0 dB, the next circle, moving inward, as −5 dB, the next −10 dB, then −15 dB, and the final circle, the one closest to the centrally pictured microphone, −20 dB.

Every circle, no matter how large or small, can be equally divided into

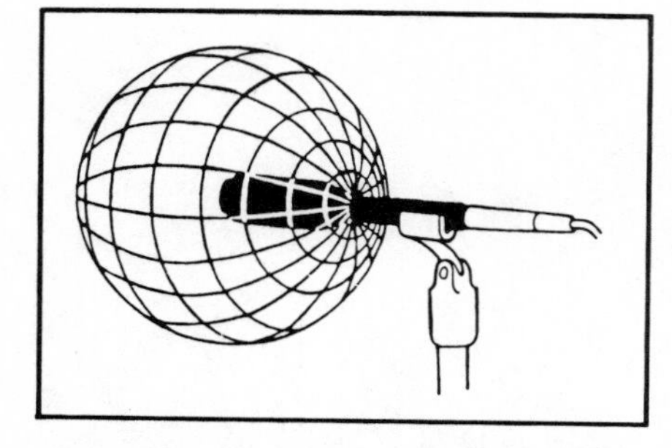

Fig. 6-1. Three-dimensional response of an omni mic.

360°. In the case of the polar diagram it is regarded as a pair of semicircles, each 180° and joined. Thus, in the polar diagram of Fig. 6-2, we start with 0° at the top and end along a vertical line drawn from that point to 180° at the bottom. If we move along the circumference, from the 0° point, either left or right, we will cover a total of 180° each way. At the 90° point a line is drawn horizontally through the center.

By drawing still another diagonal line halfway between 0° and 90°, on both sides of the circle, we finally have the circle divided into 45° segments. It is on this graph that we plot the directional response curve of a microphone.

Zero degrees on the diagram is a position directly in front of the microphone; 90° is a position at either side, while 180° faces the rear of the mic.

INTERPRETING THE POLAR PATTERN

At the start, a polar pattern is plotted on a polar diagram, such as the one shown in Fig. 6-2. A sound generator, working at a fixed frequency, such as 1 kHz, is kept in a fixed position while the microphone is moved in a complete circle around the sound source. The distance between the mic and sound source is usually 1 meter, with this distance remaining fixed. The result is a plot or graph of the microphone output. The circular movement of the microphone is a trip along the surface area of a sphere. This traverse can be horizontal to the sound source, vertical to it, or any angular displacement in between.

THE OMNI PATTERN

The pattern produced by the omnidirectional (often written as omni) microphone shows that it is more or less evenly receptive to sound coming at it from all directions. You will find omni mics in various styles, and shapes, including those that are held by the hand of the performer or mounted on a boom. Because the omni is non-directional you can use it where the possibility of pickup of extraneous noise, such as audience, doesn't exist or is minimum, or if you want to include sound other than "dry" or direct sound.

The omni is suitable for recording situations when feedback of audience

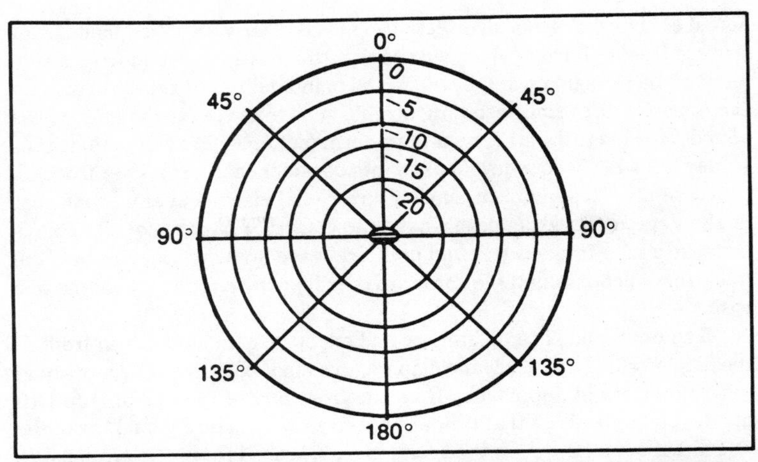

Fig. 6-2. Basic polar diagram. The small ellipse at the center is the microphone.

noise is no problem or even in cases when audience noise is wanted to add realism to a recording. It is almost useless in public address work but is widely used in broadcast when feedback control is not the primary problem.

As a general rule, it is correct to say that an omni mic will pick up

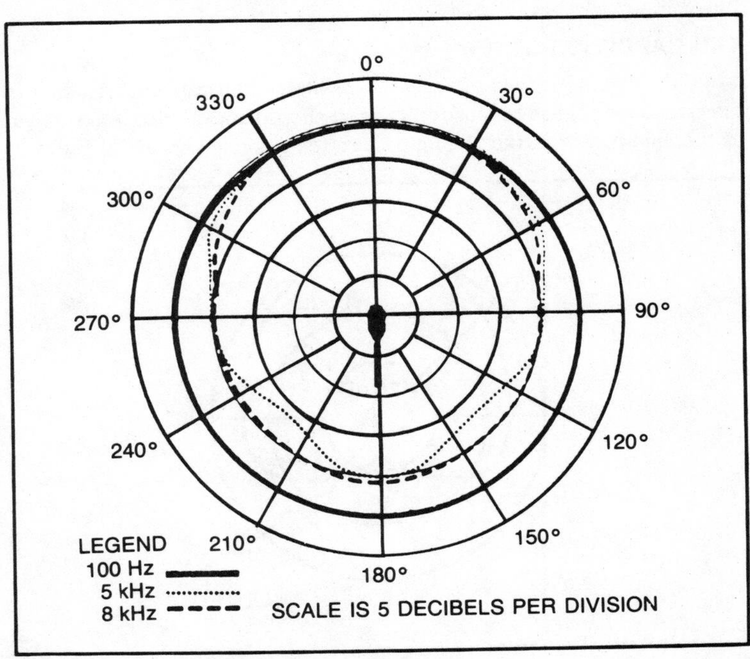

Fig. 6-3. Polar pattern of an omni mic.

sound equally well from all directions. This is clearly indicated by its polar pattern (Fig. 6-3) from which we see that the pickup of sound from a subject will be the same whether the mic is pointed directly at it or in exactly the opposite direction, assuming no change is made in the distance to the sound. However, the statement of uniform pickup from all directions fails somewhat when frequency is taken into consideration. At higher frequencies, even the best omnis assume a directional effect and any sound that is 180 degrees off-axis doesn't have the clarity of on-axis sound. This is because a high frequency sound pressure wave arriving at the rear of the mic cannot bend around the corner to the diaphragm at the end of the mic case.

The omni mic is completely sealed except for the opening in front of the diaphragm. This means the diaphragm cannot be affected by any sound unless that sound approaches from the front end of the diaphragm. Offhand, you might think that this sort of physical structure would make the mic directional. However, the sound in a room used for recording consists of a sound field—that is, there is sound throughout the room. As long as sound reaches the diaphragm of the omni, you will get electrical output from the mic. If you mount such a mic in a fixed position and then take a sound source having a constant pound pressure level output and move this source in a circle around the mic, you will get a constant output from the mic.

THE CARDIOID PATTERN

The omni is just one of several possible pickup patterns. Another is the cardioid, so-called because of its heart-shaped appearance. Also known as the unidirectional, the pickup pattern for this mic appears in Fig. 6-4.

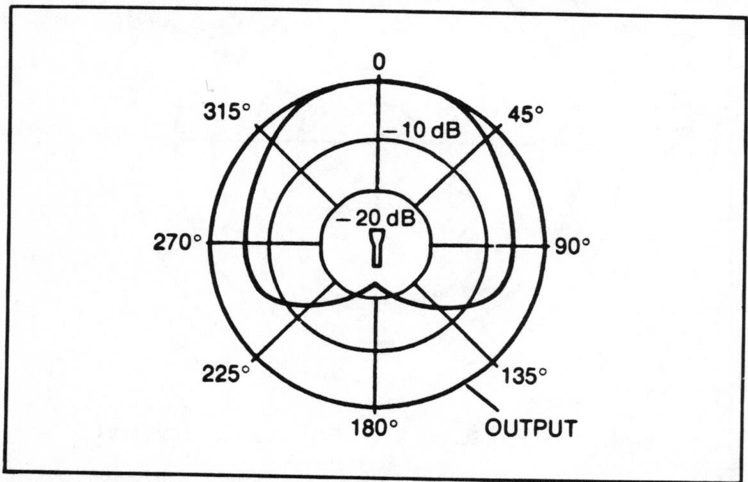

Fig. 6-4. Polar response of a cardioid microphone.

HOW MICROPHONE DIRECTIVITY IS ACCOMPLISHED

If only one side of the diaphragm of a mic is exposed to a sound field while the other side is sealed against the environment, the diaphragm will be moved by the sound pressure only. This sound pressure is a nondirectional (scalar) quantity, which implies that the mic is equally nondirectional. The pressure receiver has an omnidirectional characteristic. However, at higher frequencies, some directivity does occur through pressure boost by distortion of the sound field.

If both the front and the rear of the diaphragm are exposed to the sound field, the force moving the diaphragm is due to the difference between the sound pressures in front of and behind the diaphragm, a condition referred to as a pressure gradient. The magnitude of the driving force is a function of the distance of the front and rear sound ports, the frequency, and the angle of sound incidence. Thus, a directional quantity (a vector) is involved which can be utilized so as to produce a directional effect.

DIRECTIONAL CHARACTERISTICS: THE CARDIOID

The directional characteristic of a mic is a measure of the extent to which it will respond to sounds arriving off axis, keeping in mind that sounds are not confined to two dimensions, as implied by a polar graph. Note that the assumption is made here that sounds arriving on axis will always be satisfactory since the mic is generally pointed directly at the sound source.

This assumption, however, has its limitations. It is concerned with dry sound only, that is, sound produced by the source, and conveniently ignores reverberant sound which can reach the mic from all directions. It also does not take into consideration the fact that a single mic may be shared by a number of instruments, and so quite possibly only one of them, or even none of them, will be on axis.

Note that the cardioid does not have the circular response pattern of the omni, but is rather oblate. It is somewhat pushed in at the back. Its sensitivity is less in the rear than in the front. Physically, the cardioid microphone is open so that sound can reach the diaphragm of the mic from both front and rear. The cardioid case has holes or ports at the rear of the mic.

When the head of a cardioid microphone is pointed at the sound source, on-axis sound pressure drives the diaphragm directly. Sound waves arriving off-axis at the sides and rear of the microphone enter specially designed openings in the microphone body and are channeled through often complicated acoustic "plumbing" to the rear of the diaphragm.

As the angle of sound waves reaching the microphone increases from on-axis (0°), to the sides (90°), and to the cable end (180°), sound-pressure differences change between the front and rear of the diaphragm. In a certain sense you can consider the microphone as a kind of computer. It samples the difference in pressure on both sides of the diaphragm and decides which direction of sound reaching it will take precedence. Because of its construction, the cardioid microphone is programmed for maximum sensitivity in the forward direction and gradually reduces its sensitivity to sounds arriving from the sides and rear.

There are three reasons for using microphones having a so-called unidirectional pattern. The first is to get more pickup from an instrument, a vocalist, or a group. Another is to reduce either excessive reverberation or background noise (such as audience noise), and finally to minimize acoustic feedback when a sound reinforcement system is used, that is, when there are loudspeakers positioned in the same room with the microphones.

Technically, the word "unidirectional" is misleading. The fact is that cardioids do, in fact should, reproduce sounds directed toward them from the side. For effective reproduction of small groups, for instance, the mic should pick up not only sounds that reach it from the front, but from somewhat off-axis as well.

In general terms, then, the advantage of the cardioid is that it gives preference to sounds which are on axis, that is, which are incident to the front of the mic. This mic supplies less signal output from sounds which reach it from the sides, and even less from those which reach it from the rear. The cardioid, then, is less susceptible to feedback than the omni. However, a pattern generating microphone element doesn't mean an automatic improvement in technical performance. A cardioid pattern cannot be said to be "better" or "flatter" than an omni mic.

The difficulty with the cardioid polar pattern, as with all other polar patterns, is that (as mentioned earlier) it encourages us to think two-dimensionally. The cardioid pickup is *three* dimensional (Fig. 6-5). Visualize it as a balloon with one side pushed in to make an indentation. The indentation is located at the sound entrances in the microphone body. These sound entrances in the cardioid microphone may be concealed in the head, visible below the head, or slotted on the microphone body. One usual arrangement is found in an AKG microphone having a two-way double-element system with high-frequency slots around the head and bass-frequency entrances near the connector.

The word "cardioid," derived from the Greek, means heart shaped. Mathematically it has the general equation $p = a(1 + \cos \Theta)$ in polar coordinates. The response of the cardioid mic is frequency dependent with only quality units able to supply uniform frequency response at every angle on and off axis.

As indicated earlier, the directional characteristic of the cardioid is achieved by means of external openings and internal passages in the mic that allow sound to reach both sides of the mic's diaphragm. Sound that is 180 degrees off axis, which arrives at the rear of the mic, reaches the back of the diaphragm out of phase with the on-axis signal, causing some cancellation. The extent of this cancellation depends on the instantaneous amplitudes of the two signals and whether or not they are completely out of phase.

With the exception of the omni, then, all pattern generating methods involve phase cancellation to make the mic discriminate between sounds

Fig. 6-5. Three-dimensional response of a cardioid mic.

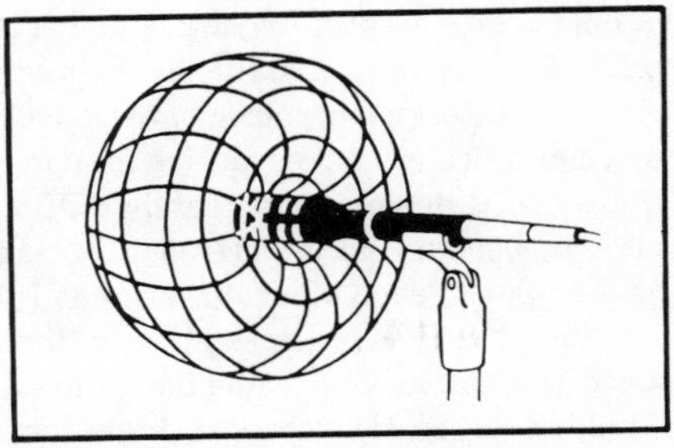

that arrive from the rear and sounds that arrive from the front. They only work predictably when the sources of sound are some distance away from the mic. If you place the mic one inch away, the pattern may cost you all the "punch" you need for mixdown.

UNIFORMITY OF CARDIOID REJECTION

It is desirable to maintain the essential uniformity of the cardioid rejection pattern at all frequencies. This is quite difficult in view of the many technical and acoustic problems involved. While slight variations at the two extremes of the frequency responses may be acceptable, several of the top brands of mics are noted for their polar uniformity at all frequencies.

The basic cardioid pickup pattern of a microphone can be pushed, pulled, and squeezed into a variety of modified, three-dimensional shapes, either by design or by accident.

Examine the cardioid polar pattern (Fig. 6-4) at the 90° line and you will see it is almost halfway between the 0 dB circle and the −10 dB circle. Theoretically, the cardioid is down an average of 2 dB at 45° and 6 dB at 90° off axis on both sides of the microphone. At 180° the output of the mic is down about 20 dB or more.

If you look at the top part of the cardioid polar pattern, you will see its output is similar to that of the omni for about 22° on either side of the vertical axis. The total angle in this example is 44°, so don't expect to be able to pick up a single instrument out of a group.

When examining the polar pattern for the cardioid, remember that each of the circles on which the pattern was constructed indicates a relative amount of signal attenuation. The quantity −10, for example, means an attenuation of 10 dB. While Fig. 6-4 does indeed indicate that each circle is a certain amount of decibels, sometimes patterns are supplied without the decibel designation.

Examining the vertical 180° line, you will see its interception is very close to −20 dB circle. Thus, this particular microphone has an attenuation of somewhat more than 20 dB to signals arriving at it from the rear.

Now consider the 135° diagonal. It intersects the pattern at about the −12 dB point. A signal arriving at the microphone from a position which forms an angle of 135° with the front of the microphone will have this amount of attenuation. According to this pattern, then, you could have signals coming from in front of the microphone to about 22° on either side of the 0° mark. But as you deviate from 22° on either side the microphone becomes less and less responsive to sound. Thus we see that the response of the mic is down approximately 6 dB at 90 degrees off axis. If we assume two sound sources are equidistant from the mic and that they are right angles to each other, that is, one is on axis with respect to the mic and the other is 90 degrees off axis, then the off axis sound source will sound just half as loud as the one in front, assuming, of course, that both supply equal sound volumes of single, identical frequencies. In short, a decrease of 6 dB is equivalent to a 50% reduction in volume.

AXIS OF A MICROPHONE

Figure 6-4 shows the head of the microphone pointing directly to the 0° point on the polar pattern. A vertical line drawn from this 0° point, passing through the center of the head of the microphone, is referred to as its *axis*. Sounds arriving head on, that is, from the 0° point, are referred to as being *on axis*. Sounds arriving at angular deviation from 0° are said to be *off axis*. Thus, sounds that are 90° off axis would be those arriving from 90° and also 270°. The 270° point is 90° from the 0° point.

There is no microphone made that receives sounds only on axis. However, it is the amount of off-axis attenuation that distinguishes one microphone polar pattern from another. Theoretically a perfect omni would have no signal attenuation at any off-axis point. The cardioid is different than the omni, since it is deliberately designed to have off-axis attenuation.

THE SOUND OF A MIC

Under identical operating conditions and identical positioning, it is possible for a pair of mics to produce different sound. If the sound recording is being made in a non-reverberant room, a pair of quality mics, differing only in their polar pattern (one an omni; the other a cardioid) will sound very much alike.

But put the same mics in a different recording environment, such as a church or a large hall. The result will be a difference in the sound for the omni will pick up the reverberation and the echoes as well, thus giving the sound its liveliness. For the cardioid, however, the resultant sound will not be much different from that made in the non-reverberant room.

Similar comparative tests can be made in a noisy environment. Here, with the cardioid pointed away from the noise, the resulting sound will be clearer, less noisy, than with the use of an omni.

SOUND REINFORCEMENT

Sound reinforcement consists of a microphone (one or more) supplying a signal to an amplifier which then operates one or more speakers. The amplified sound is an advantage to the performer who can then hear the sound as it is heard by the audience. This enables the vocalist, instrumentalist, or group to make an immediate evaluation of the sound.

In a sound reinforcement system, regenerative coupling of loudspeaker sound into the microphones can cause feedback squeal. Regenerative feedback doesn't necessarily remain constant but can change in amount. Regenerative feedback tends to make sound have a very sharp, edgy quality. As feedback increases, the sound is accompanied by squealing. This obviously calls for some action to be taken to eliminate the cause. Small amounts of regenerative feedback may go undetected since they may not be accompanied by howling or squealing, but it does give the music a ringing quality it should not have.

The reduced directional response of the cardioid to rear sound makes it especially valuable where sound reinforcement is used. Angling the speakers away from the microphones, as in Fig. 6-6, helps to reduce the intensity level of such sound and minimizes the possibility of feedback.

FEEDBACK

If you have ever used a two-piece telephone you may be familiar with feedback. With the telephone, it was obtained by putting the earpiece up against the mouthpiece. For telephones the problems of feedback were eliminated by making the phone into a one-piece unit.

In recording, feedback becomes possible when the amplified sound supplied by a speaker manages to reach a microphone whose polar pattern is such that the amplified sound is picked up and goes through the amplification process again. This is repeated over and over again, quite rapidly.

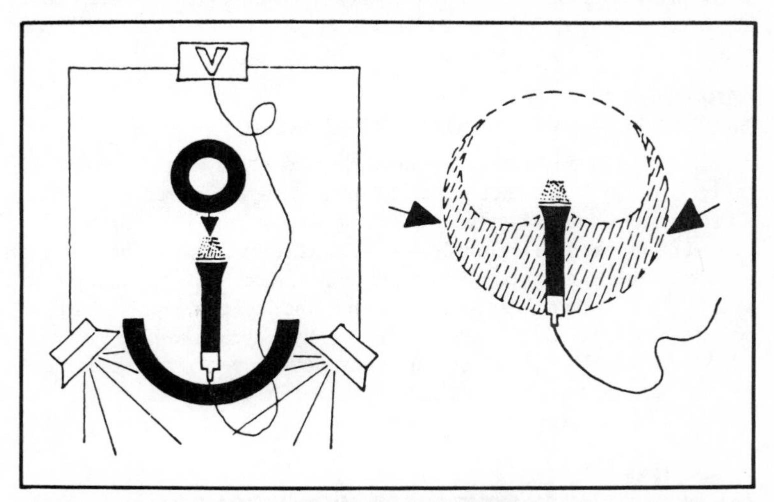

Fig. 6-6. The speakers are positioned so their reproduced sound does not reach the response area of the microphone.

The result is instability and howling. The sound may become sharp, there may be ringing, and the frequency response is narrowed.

Feedback can be minimized or eliminated in several ways: by using a cardioid instead of an omni; by adjusting the position of the mic so that it does not favor loudspeaker sound pickup; by moving the loudspeaker farther from the mic or by moving the mic farther from the speaker, or both; by lowering the sound level of the amplifying system. Sometimes turning the speakers a small distance can be helpful.

As a broad generalization, in a sound reinforcement system the mic, the loudspeaker system, and the room each play a part in the successful or unsuccessful performance of the reinforcement system.

The amount of feedback may be less than that necessary to produce oscillation. While there will be no howling, the sound will become much sharper. The system may also fall into and out of oscillation resulting in a condition of considerable instability. Thus, there may be no ringing or howling during soft sound passages to give evidence of the feedback that will occur when the sound level is increased.

No two systems will feedback at the same frequency. A room has its fundamental and harmonic resonances in addition to a certain amount of reverberation that produce their effects on the sound. A mic with a smooth frequency response on-axis and uniform cancellation off-axis can produce more amplification before feedback than can a mic with a peaky response and uneven cardioid rejection. As a general rule, the two-way cardioid mic with its smooth on-axis response, smooth 90-degree response and uniform back-rejection at both high and low frequencies (as compared to some cardioids which have good rejection at mid frequencies only) can provide up to 6 dB more amplification before feedback. Proximity effect, a peaky high frequency response, and uneven back rejection cause acoustic feedback in many reinforcement systems.

RESPONSE OF THE DOUBLE ELEMENT DYNAMIC MICROPHONE

Instead of having a single element, the two-way dynamic microphone has two, with each element designed to favor its respective frequency range. Further, each element can be individually cardioid-tuned to suppress side and rear sounds more uniformly with respect to frequency. The result is a more even frequency response at all angles, more uniform suppression of side and rear sounds at all frequencies, and less spitting, popping, or booming when the microphone is used properly. Cardioid response is maintained down to the lowest frequencies by slotted openings at the base of the microphone, a longer distance from the low-frequency diaphragm than in most critical mics.

RESPONSE OF THE DOUBLE-ELEMENT CONDENSER MICROPHONE

Although the cardioid pattern is widely used in studios, it is often necessary to change the characteristics to match the sound perceptivity to the acoustic conditions in the recording room or to the radiation characteristics of the sound source. Two principles have been established to select different polar patterns of microphones.

One principle uses switchable sound delay elements such as acoustic phase shifting networks. This technique is associated with some considerable disadvantages. The switchable acoustic elements, such as resonators or acoustic resistors, are usually of relatively large size and therefore disturb the sound field by an amount that isn't negligible. This results in a nonsymmetrical pattern (especially at high frequencies) and in an uneven frequency response. In addition, the need for movable mechanical parts can result in excessive wear and change of pattern due to decreasing tightness of the resonators with time.

The second principle is the double diaphragm system, described earlier in connection with dynamic microphones. In the condenser microphone the capsule consists of two condenser transducers positioned 180° apart from each other, that is, two capsules with cardioid patterns mounted back to back with the membranes pointing outward. Two backplates are used with a common volume in between. Acoustically, this volume is coupled to the transducers by separate and carefully matched and selected acoustic resistors. Both capsules must be selected for equal sensitivity and frequency response.

By electrically combining the outputs of the two capsules in phase, an omnidirectional pattern is achieved because the two 90° components which equal one half of the on-axis sensitivity are summed up to give the full sensitivity from the 90° direction. Of course the same full sensitivity is measured in the 0° and 180° direction.

A cardioid pattern can be obtained by using only one system and switching off the second. This can be done by switching off the polarization voltage of the second system. Connecting both systems out of phase will cancel the two 90° half sensitivity components while full sensitivity is gained from 0° and 180°.

Intermediate patterns, such as the hypercardioid, are possible by changing the sensitivity of the system pointing off axis (180° direction). AKG's Model C414EB-P48 (Fig. 6-7) allows the selection of four patterns: cardioid, figure eight, omni and hypercardioid, directly on the microphone by only switching the respective polarization voltages. For that reason, symmetrical patterns can be realized and all moving parts susceptible to wear are avoided.

ADVANTAGES OF THE CARDIOID

Cardioid microphones are the preferred choice in both recording and live performances where "house" and other ambient noise must be suppressed. They are used when instruments or voices must be emphasized and isolated from others in the group or orchestra, and to reduce feedback

Fig. 6-7. Double element condenser mic has four patterns: cardioid, figure eight, omni and hypercardioid.

when loudspeakers are used in the same room.

It is, in fact, the cardioid's inherent resistance to feedback, the squealing, howling sound that occurs when an amplified signal is picked again by the microphone and re-amplified, that makes this type of microphone a favorite among vocal performers, especially when used in less than ideal acoustic situations. Also, since with the cardioid the possibility of signal feedback is reduced, a higher level of amplification can be used, if desired.

If you are recording in front of an audience, the cardioid mic will not be responsive to a substantial amount of the rear sound and that includes audience noise. The degree of noise rejection depends on how far the sound is off axis. Compared to an omni, the useful reach of the cardioid pickup extends farther from the mic. This means that performers can work at a greater distance from the mic than is the case with a mic having an omni pattern. The cardioid also simplifies the problem of sound installation since sound from a loudspeaker system has less chance of signal pickup by the mic, decreasing the chance of feedback.

The narrower pickup pattern of the cardioid, though, means you may need to use more of them if you have a number of performers working along a stage. If not, using a single cardioid means that sound from performers near the edges will be weaker in relationship to sounds coming from closer to the center. If these performers on the outer edges of the stage use instruments whose output is normally weaker in comparison with others, then they will make little overall contribution to the sound.

It is also possible for sound waves to be reflected from the walls of the room used for sound recording. These sound reflections produce reverberation in rooms that aren't acoustically treated and which are normally used for purposes other than recording. Quite often the reverberation is rather strong at lower frequencies.

In the drawing of Fig. 6-8 part of the sound leaving the source travels to the left as shown by the straight line arrow, bounces off a wall, strikes the rear wall, bounces once again to an opposite side wall and is then reflected toward the microphone. The reflected wave ends up in the less responsive area. The drawing at the right shows, the reverberant sound entering the reduced response area of the cardioid mic is suppressed.

CARDIOID VS OMNI

One advantage of the cardioid is that you can move it further away from the performer, almost twice the distance, before you run into problems of picking up too much reverberant sound or too much background noise or running into feedback from a speaker to the microphone. If, as an example, you use an omni at a distance of 1 foot from the performer, you should be able to put the cardioid at a distance of about 2 feet. These are approximate figures and the amount of *working* distance—the distance between performer and microphone—depends on the amount of background noise and how you have your loudspeakers positioned (Fig. 6-9).

For vocalists in sound reinforcement systems, the cardioid does seem to be preferable. The performer need not worry about undue pickup of audience noise or the possibility of feedback howling occurs during the performance as compared with an omni.

It is also possible that the loudspeakers may be such a great distance from the microphones that there is no problem of feedback. Under such conditions, the omni may be preferable since omnis usually have a smoother frequency response than cardioids. Omnis are less afflicted by popping than

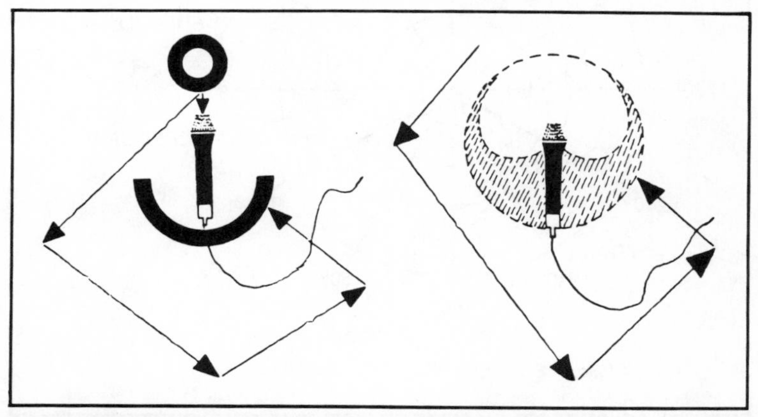

Fig. 6-8. A sound wave can reflect several times from various surfaces. In this example the reflected wave enters the less responsive area of the microphone.

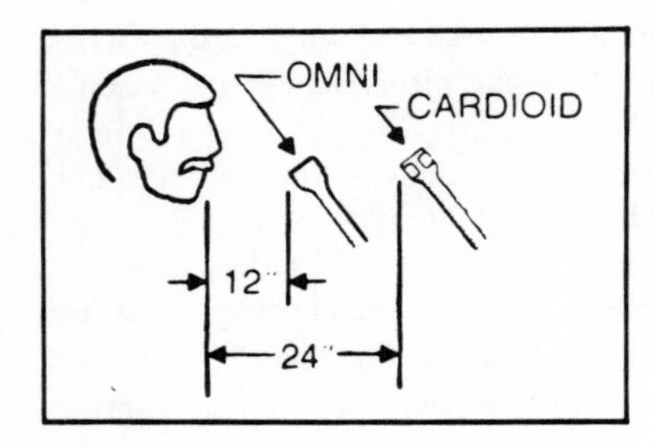

Fig. 6-9. Working distance of the cardioid vs the omni.

cardioids and may be less sensitive to mechanical shock. Sometimes, depending on the manufacturer, it is possible for an omni to be constructed more ruggedly than a cardioid.

SUPERCARDIOID

The supercardioid is somewhat similar to the cardioid in its response pattern. The pattern has two lobes—a front and a rear, with the rear much smaller than the front. Note, also, as shown in Fig. 6-10, the working distance is somewhat greater than the cardioid as the shape of the directional response in front is more elliptical. The null regions—regions of little or no pickup—are toward the left and right sides of the mic instead of at 180°.

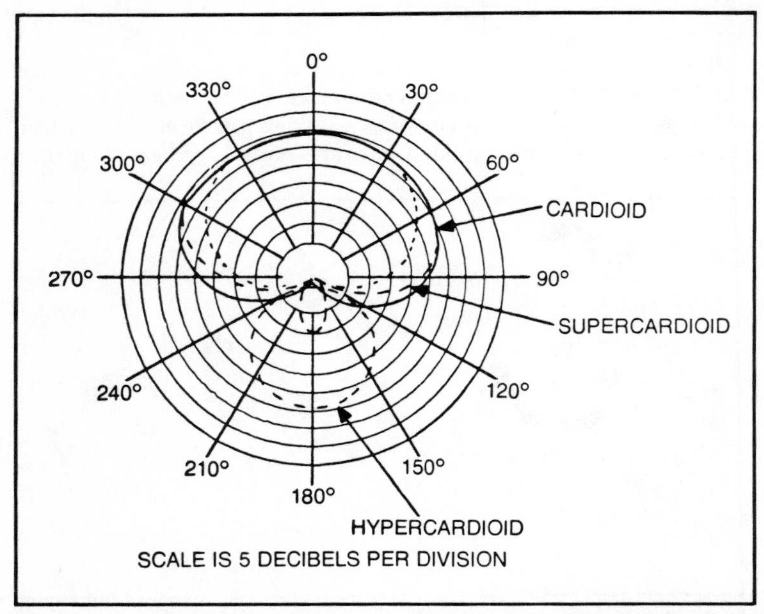

Fig. 6-10. Polar response of cardioid, supercardioid and hypercardioid microphones.

The supercardioid provides its greatest rejection at 150 degrees off axis. Compare this with the typical cardioid which has its minimum response at 180 degrees. This assures greatest rejection in the horizontal plane when the mic is tilted in its most natural position, that is, 30 degrees from the horizontal, as on a boom or floor stand. Figure 6-11A illustrates the polar response of the supercardioid while Fig. 6-11B is a wiring diagram.

While the supercardioid, like the cardioid, is more sensitive to on-axis sounds, sounds which arrive off axis, that is, from the sides, are even weaker than in the case of the cardioid. Further, the supercardioid is excellent in terms of not being susceptible to feedback.

The directional efficiency of the supercardioid means its reach can be greater than that of the cardioid. One question often asked concerns the

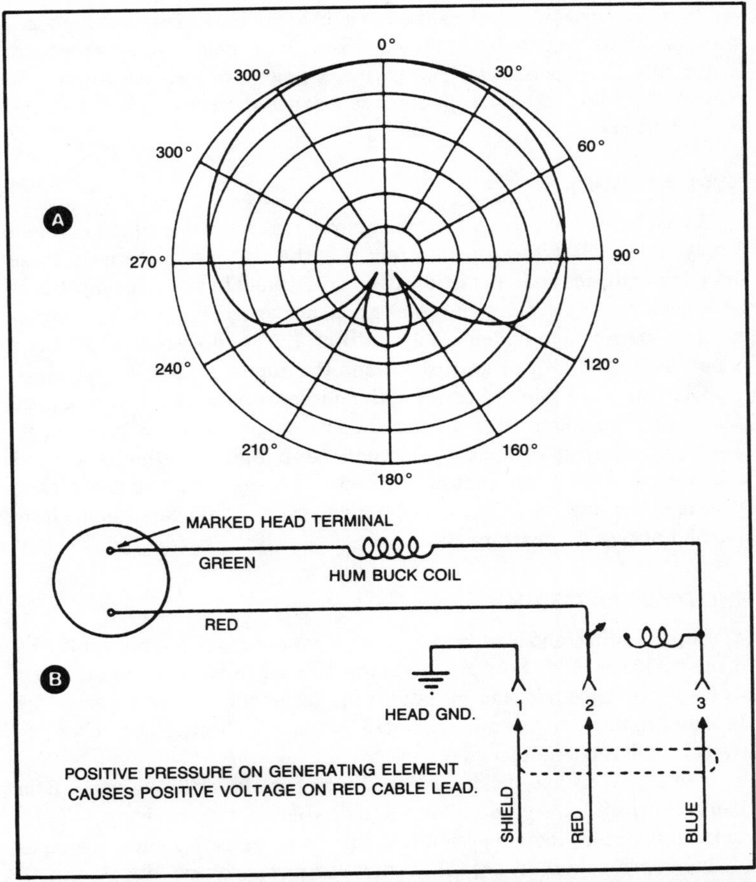

Fig. 6-11. Polar response of the supercardioid (A); wiring diagram (B). Each circle on the polar diagram indicates 5 dB.

extent of microphone reach. The same question could be applied to your ears. Over what distance can you hear? Given an adequate sound pressure of the original source, an absence of noise, and a pair of ears in reasonably good physical condition, the hearing distance could be miles. And even in the presence of noise, our hearing is capable of discriminating against the noise and focusing on the sound of interest. Nevertheless, we do put a practical limit on our hearing range. We do not expect to hear someone whispering several hundred yards away.

No more than your ears does a microphone hear all sounds equally well. When the ambient (unwanted) noise becomes disturbingly loud along with the voice or music in a recording, reach is limited. When recording in a quiet studio, distance pickup ends when the system electronic noise becomes apparent in the recording.

Both the cardioid and the supercardioid are unidirectional microphones with the differences shown by a comparison of the polar patterns for each. Essentially, the supercardioid has narrow segments of response toward the rear while its 90° off axis response is somewhat more attenuated than that of the cardioid.

HYPERCARDIOID

Like the supercardioid, the hypercardioid is a member of the cardioid family. To be able to see the differences in the response patterns of these three microphone types, cardioid, supercardioid and hypercardioid patterns have been plotted on the same polar graph shown in Fig. 6-10. The hypercardioid microphone is down only 6 dB at 180° and so this microphone is not used as often as the cardioid and the supercardioid.

Generally, we think of pointing the microphone so that it is in a position to pick up sound from a desired source. The converse is also true. We may, for example, point the microphone so it picks up the least sound from a specific direction. This not only applies to noise but to signal pickup as well, something you might consider if you wanted attenuated sound from a group of nearby instruments.

SHOTGUN PATTERN

Supercardioid and shotgun pattern microphones are simply variations of the basic cardioid. Forward coverage may be tighter, sound collected over a narrower angle, but variations in placement will be governed by the same application information noted for sound collection control of any event by microphone distance.

The supercardioid picks up less energy out of a diffused sound field than a cardioid. This effect allows higher volume without acoustic feedback. In addition, the supercardioid capsule gives better suppression of adjacent sound sources, e.g., from monitor loudspeakers. Such a microphone can be brought nearer to the loudspeaker before causing acoustic feedback.

There are also some microphones whose directional patterns are midway between those of the cardioid and hypercardioid. They tend to suppress hall and auditorium noises more effectively than standard cardioids. Feedback problems are minimized even under higher amplification levels.

BIDIRECTIONAL (FIGURE-EIGHT)

The bidirectional microphone response is also called a figure-eight because of its seeming resemblance to that number. As shown in Fig. 6-12, there are approximately equal front and rear lobes and so the front and rear on-axis sound pickup is approximately the same. The attenuation of the signal is maximum at 90° off axis.

A microphone of this kind is very useful when a pair of conversationalists face each other across a table and you want optimum pickup of both talkers and minimum sound pickup from the sides.

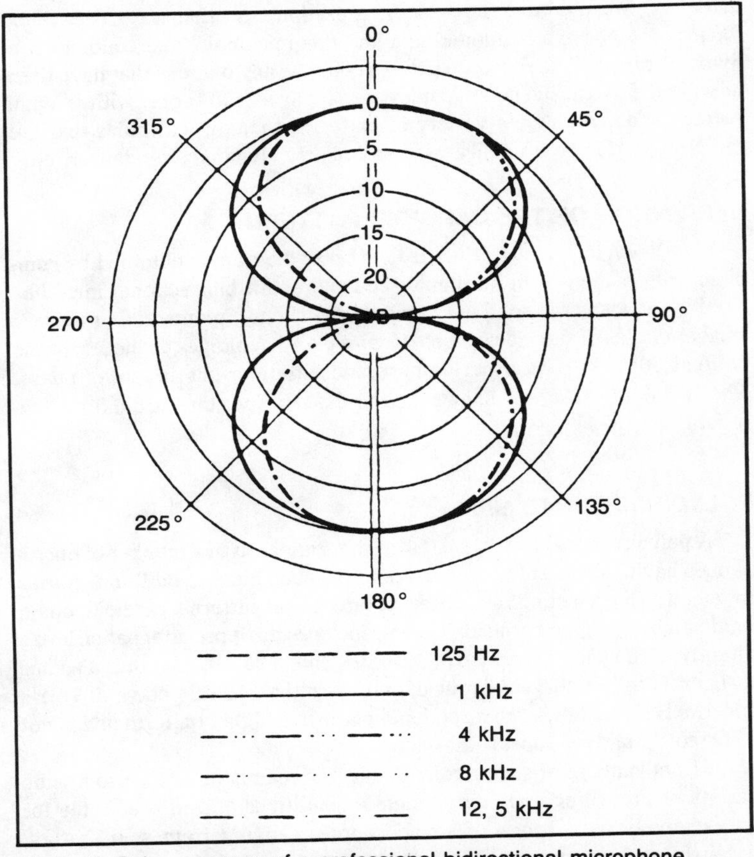

Fig. 6-12. Polar response of a professional bidirectional microphone.

The bidirectional mic is well suited for use when a pair of performers or groups of performers are to be positioned on opposite sides of a mic. However, it is well to keep in mind that the rear of the figure-eight mic is out of phase with its front. This characteristic can sometimes lead to some strange recording effects. Thus, if the rear of the figure-eight is positioned near the front of another mic, whether that other mic is another figure-eight, cardioid or omni and if both mics are connected to the same input, possibly through a mixer, there may be some signal cancellation when a sound source such as a voice or musical instrument is positioned between the two mics. The cure is simple. Try to keep the performer much closer to one or the other of the microphones.

The pattern depends on the interior physical structure of the microphone and not upon the type of reproducing element used. A bidirectional mic could be a moving coil type, a ribbon mic, a condenser, and so on. Further, the pattern does not depend on whether the mic is boom mounted, hand held, worn around the neck, or is used in any other way. You could, for example, have a hand-held mic and this mic could be a condenser or dynamic type and its response pattern could be any of those that have been described. Crystal and ceramic mics are usually available only with an omni pattern. You can have a variety of patterns, though, with mics that are more suited for high-fidelity use, such as the dynamic, condenser, etc.

DERIVATION OF THE CARDIOID PATTERN

As shown in Fig. 6-13 the cardioid polar diagram is obtained by combining the responses of an omnidirectional and a bidirectional mic. The omni is shown as the solid line circle at the right, circumscribing the two smaller circles of the bidirectional (figure-eight). Voltages to the left of the vertical 90°/–270° axis are in phase, those to the right are out of phase. The resultant voltage is shown by the dashed line, the cardioid characteristic.

POLYDIRECTIONAL MICS

A polydirectional mic is one that is the equivalent of a number of microphones having different polar patterns. One such mic has built-in facilities for electrically selecting four different directional patterns: cardioid; omni; figure-eight and hypercardioid. The mic includes three pre-attenuation levels directly at the transducer: 0 dB, –10 dB, and –20 dB. The unit also has three bass rolloff curves: flat at 0, and more than 12 dB/octave at 75 Hz and 150 Hz. Further, each directional pattern is highly uniform and is not afflicted by pattern lopsidedness.

The advantage of such a mic is that it is readily adaptable to a wide variety of prevailing and environmental conditions and so is suitable for an extremely wide range of sound sources ranging from solo orators, vocalists, and instrumentalists through rock, jazz, and chamber groups to

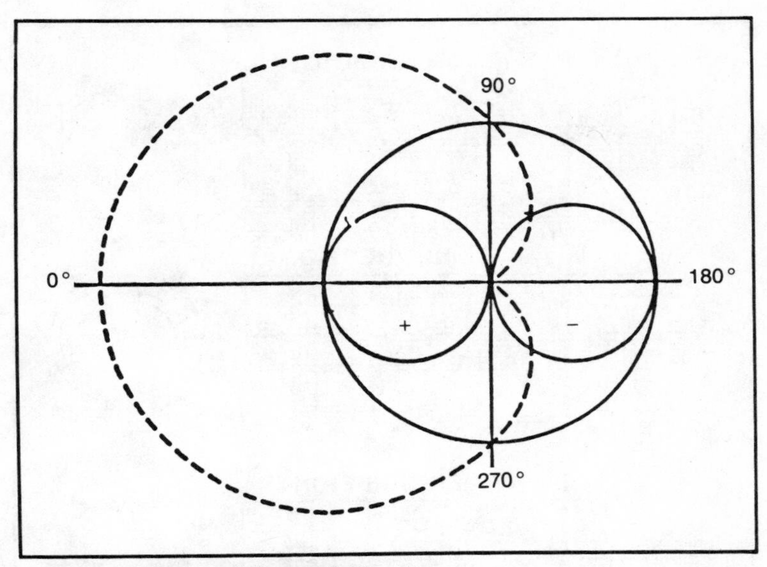

Fig. 6-13. Derivation of the cardioid pattern.

the largest choruses and orchestras.

The mic described here is a condenser type prewired for a 12-volt phantom power supply but can be easily modified for 48-volt phantom power instead. The mic is a low impedance balanced output type fitted with a standard 3-pin male XLR type connector.

Figure 6-14 shows the frequency and polar response of this mic for each of its four operating conditions.

POLAR PATTERNS AND FREQUENCY

A polar pattern should be referenced to some particular frequency. While the pattern without a frequency specification can give you an overall idea of the sort of pickup you can expect, the frequency will give you a more definite concept of what happens to directional response as frequency is changed.

Figure 6-15 shows polar diagrams taken at a distance of 1 meter (39.37 inches) from the source of sound. These diagrams are of an omni. The initial frequency is 125 Hz and is represented by the diagram at the upper left. The frequency is shifted in multiples of two— $125 \times 2 = 250$ Hz, $250 \times 2 = 500$ Hz, and so on, until a frequency of 16 kHz is reached. Note how similar the diagrams are to each other. However, the diagram of this microphone at 16 kHz shows a small change in on axis directional response. Other than that, the pickup appears to be quite uniform over the entire range—125 Hz to 16 kHz.

Figure 6-16 shows another set of polar diagrams, also taken 1 meter

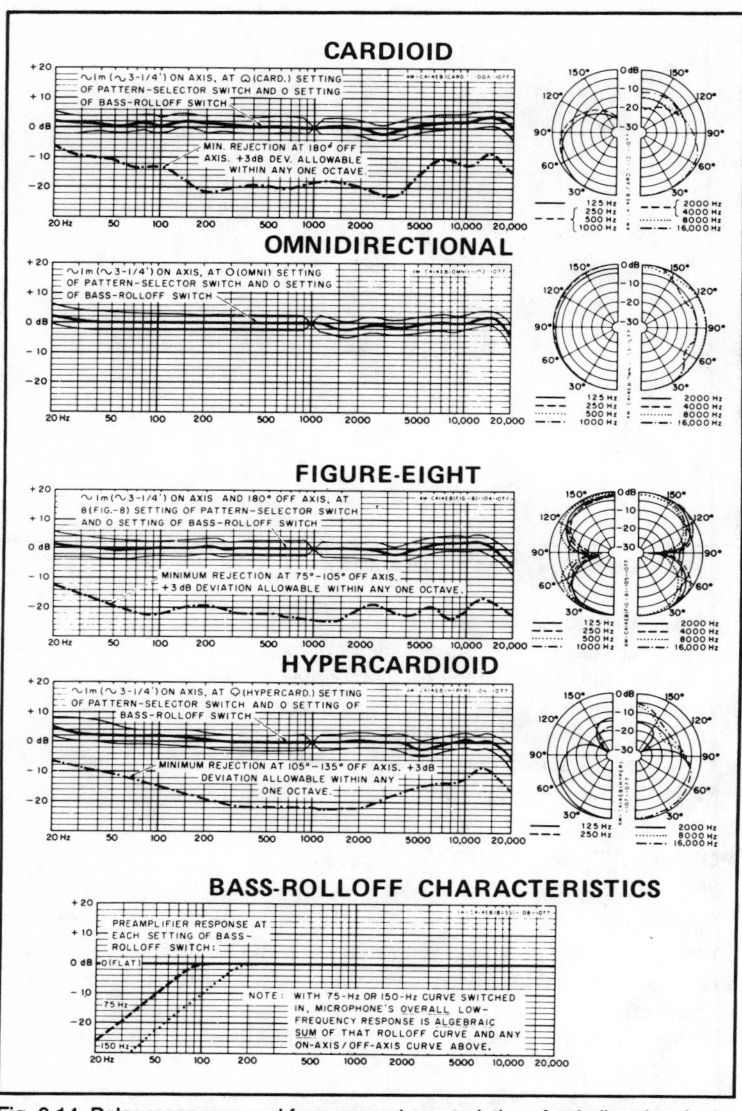

Fig. 6-14. Polar responses and frequency characteristics of polydirectional mic.

from the sound source. These are the patterns of a cardioid microphone. Note that beyond 4 kHz this particular cardioid shows more dramatic change in off-axis response than the omni. This doesn't necessarily mean it is better or worse than the omni, just that it is different. It all depends on what you want in the way of a recording.

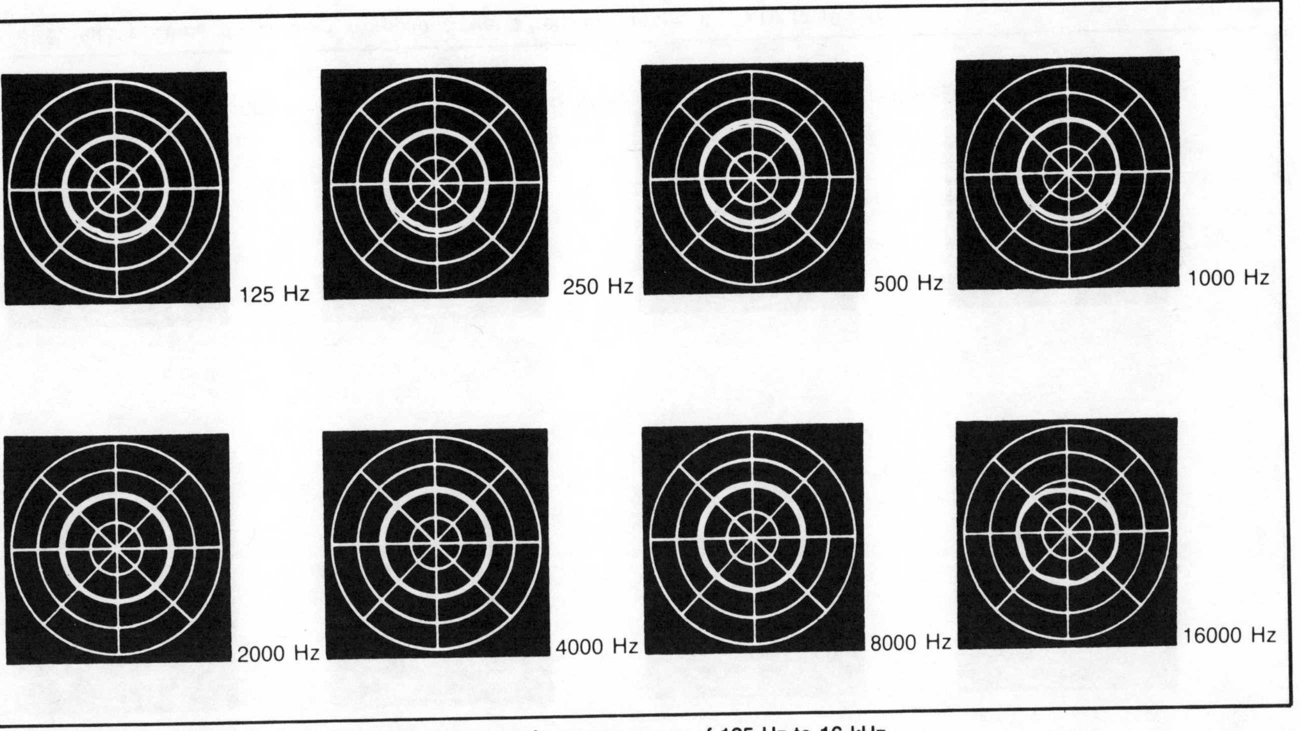

Fig. 6-15. Polar patterns of an omni microphone over a frequency range of 125 Hz to 16 kHz.

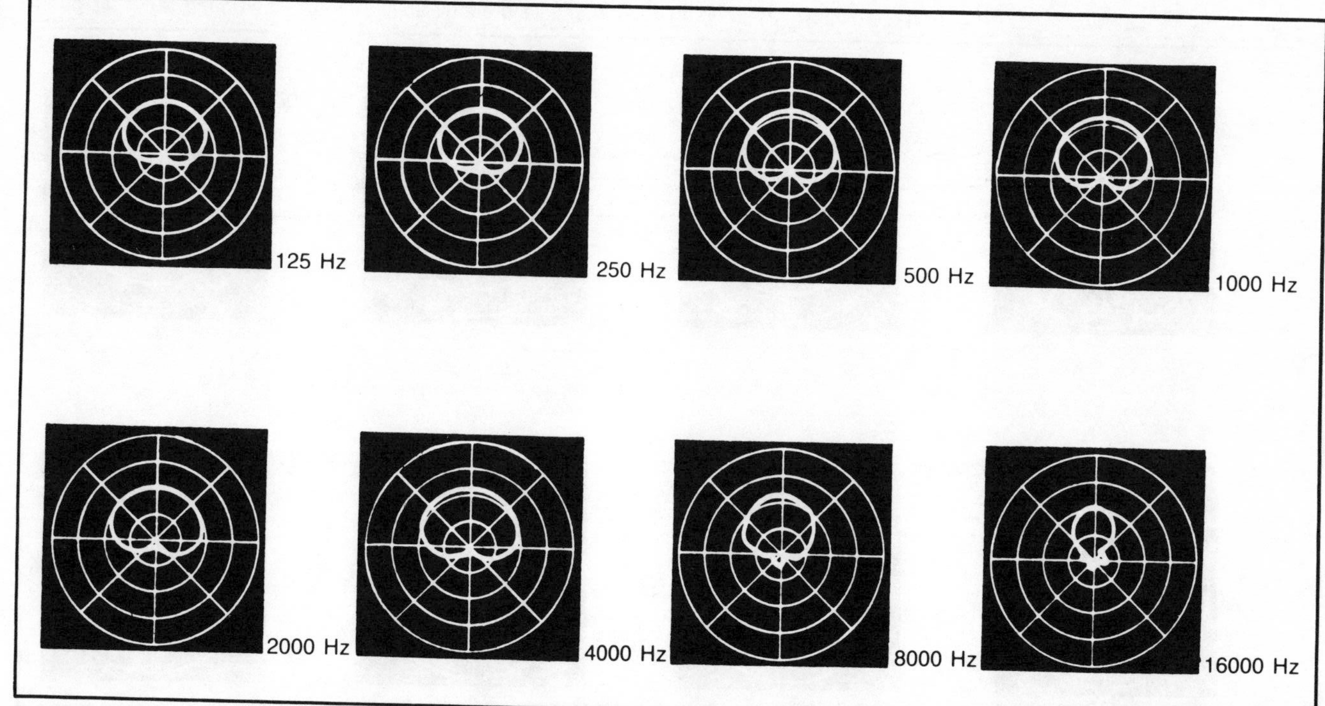

Fig. 6-16. Polar patterns of a cardioid microphone over a frequency range of 125 Hz to 16 kHz.

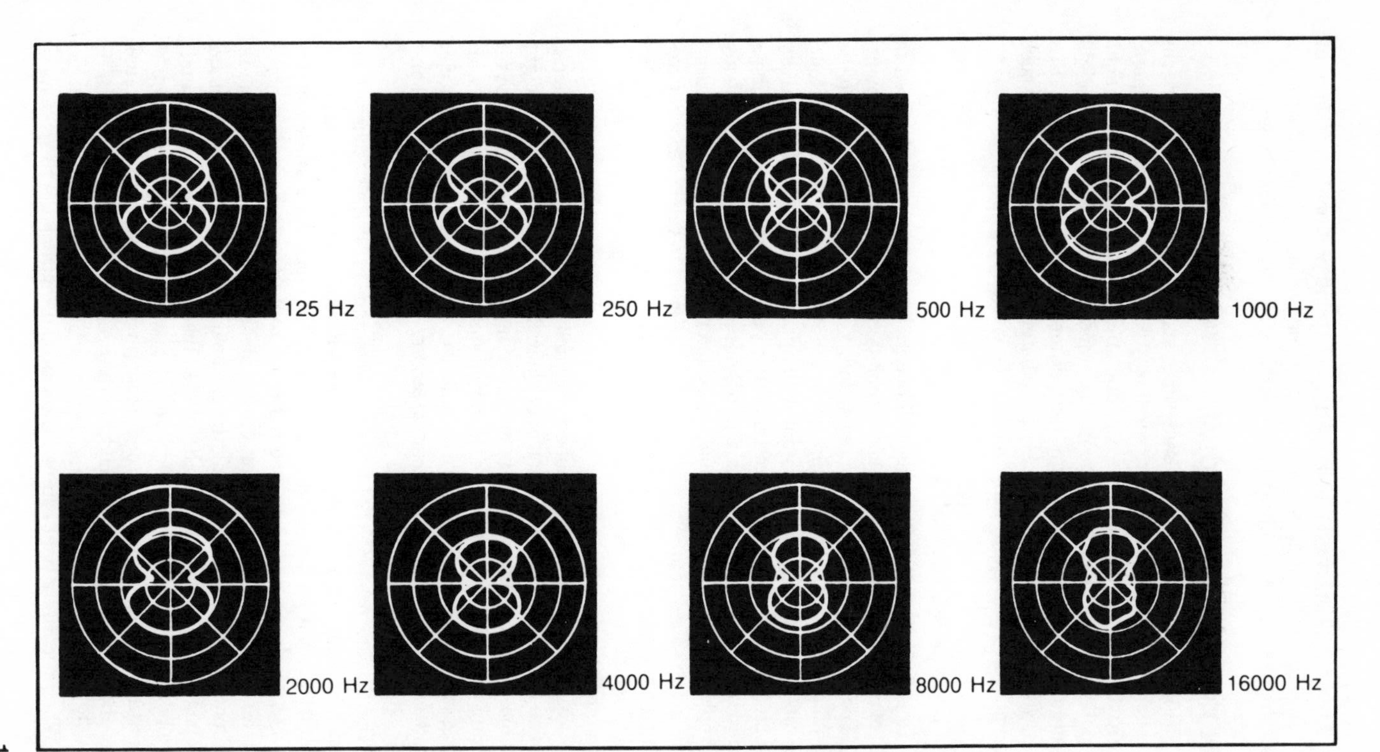

Fig. 6-17. Polar patterns of a bidirectional microphone over a frequency range of 125 Hz to 16 kHz.

Figure 6-17 shows the polar patterns of a bidirectional mic taken under the same test conditions as the omni and cardioid. This bidirectional also shows changes in directional response over the audio range.

MULTIPLE-FREQUENCY POLAR PATTERNS

While a manufacturer will sometimes supply a total of eight patterns for a particular microphone, more commonly you will see the patterns shown in Fig. 6-18. What you see here is a total of eight patterns drawn on similar polar graphs. The advantage is that you do not need to examine eight different graphs, looking for significant differences. With the setup in Fig. 6-18, you can see the differences immediately. The problem is that such a graph gets to be quite busy and can be confusing.

PATTERN APPLICATIONS

In practical applications, microphone pickup patterns can be used to help overcome the problems of a less-than-perfect recording environment, while concentrating only on wanted sound. You can use an omni when the possibility of sound feedback and ambient noise is relatively nonexistent. The 360° pickup angle of the omni with approximately equal sensitivity is useful for covering a group of talkers or performers or for recording a conference with the use of fewer microphones. Practically all *lavalier* body microphones are omni-directional. Such microphones are normally used close to the sound source and are usually designed to reject resonant frequencies of the chest cavity. Cardioids aren't used in such applications because reflections from the body would play havoc with cardioid characteristics.

While the lavalier microphone is normally suspended from the neck, you can also hang it from ceilings, over choirs for example. Its light weight also makes it desirable when a microphone must be hand held for several hours. Don't make the mistake of equating small physical dimensions with poor quality, however. Despite its small size, a well-designed lavalier can sound as good as or better than a larger, less well-made omni hand microphone.

Cardioid microphones are a distinct advantage in difficult acoustic conditions, both indoors and outdoors. When pointed at the sound source, unwanted sounds from the rear are suppressed. This includes audience noise, live and speaker sounds reflected to the microphone, air-movement noise from air-conditioning equipment, traffic noise, and any ambient sound.

Loudspeaker sounds entering the cardioid microphone are suppressed enabling more system loudness before feedback howl, squeal, or ringing. A well-designed cardioid will allow an increase in the order of 1.7 in working distance or reach for equal room pickup when substituted for an omni microphone. At the same distance from the desired sound source, the cardioid will reject unwanted side and rear sounds.

When recording, you can use the cardioid microphone to *spotlight* a

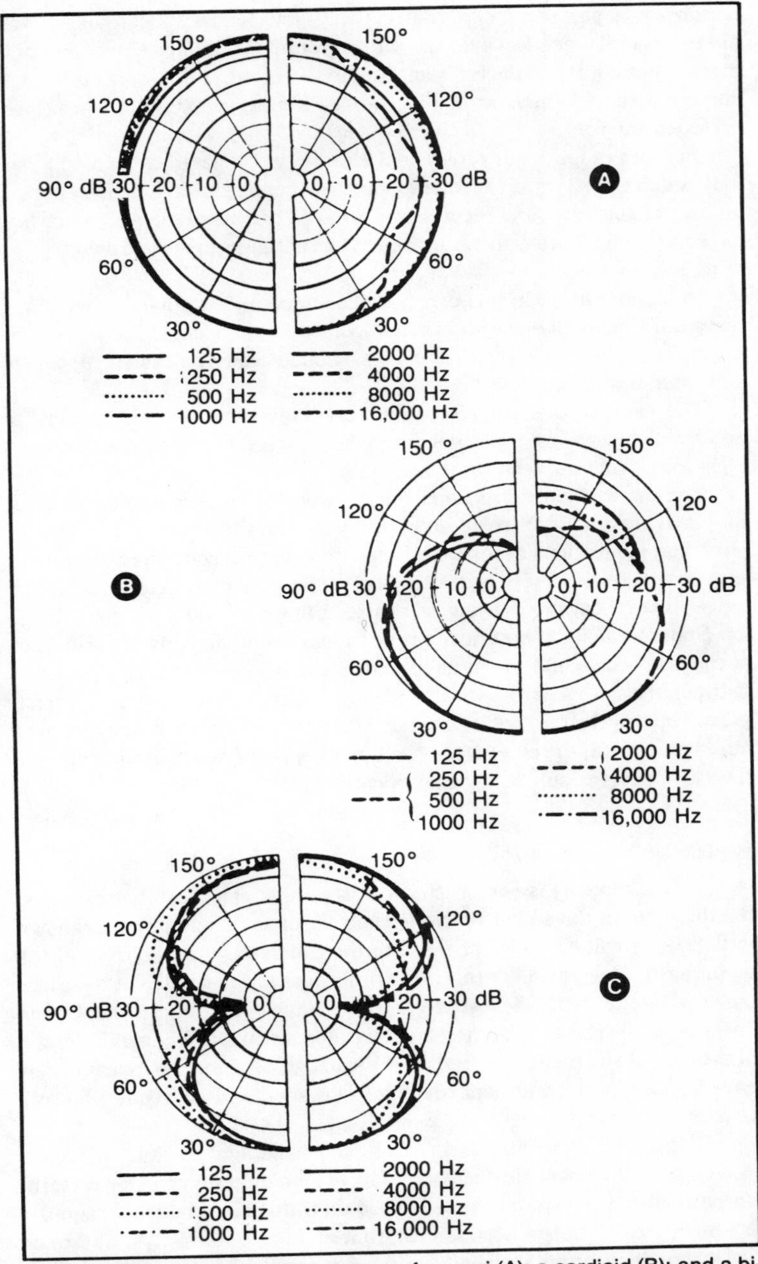

Fig. 6-18. Multifrequency polar patterns for omni (A); a cardioid (B); and a bidirectional microphone (C).

performer, or part of a group. Alternatively, you can use the cardioid microphone at a distance to cover an entire group, and by placement, permit more or less of the reflected room sound to enter the microphone. By including a suitable amount of reverberant sound you can add warmth to recorded music.

Because of its better directional efficiency compared to an omni, you may require more cardioid microphones in sound reinforcement systems, conference pickups, and recording. Cardioids can mean the difference between a practical, working arrangement and a useless, marginal or nonoperating system.

In broadcast work, you can use directional microphones to help control studio, announcement booth, and control-room background noises. In TV news applications, the cardioid can subdue crowd noises for reporting and interview pickups.

You can use the cardioid to pick up stage entertainment sounds at greater distances, yet the same microphone, when used in close vocal applications, will suppress stage sounds.

You can position bidirectional microphones between two sound sources for equal pickup, while rejecting sound from the sides, above, and below. However, bidirectionals aren't used that often, because one-direction pickup supplies more flexibility in controlling the environment, easily achieved by two cardioids, or by using one, placed low and pointing upward.

Supercardioids theoretically reject more sound arriving at 150° than at the 180° cable end. In actual practice, the best rejection can fluctuate between the sides and rear with frequency. Supercardioids also have a small pickup lobe at the tail, not evidenced with a pure cardioid. When mounted on a table, desk, or lectern, they will not suppress sound reflected from the close surface any better than a cardioid.

PRESSURE GRADIENT

If a microphone is constructed so that sound can reach both sides of the diaphragm, the sound reaching the rear of the diaphragm, as shown in Fig. 6-19, will be out of phase with the sound striking the front of the diaphragm. The reason for this is the difference in the lengths of the paths traveled by the sound. The strength of the force operating the diaphragm depends on the phase difference, and this depends on two factors: the length of the path D and the wavelength of the sound. The amount of phase difference because of the additional distance D is small at low frequencies and greater at higher frequencies.

That part of a microphone containing the diaphragm assembly is known as a capsule. In a capsule designed to have a unidirectional polar pattern, a front and rear sound entrance is used so as to drive the diaphragm by the pressure difference between its front and back, and is referred to as a pressure gradient microphone. The pressure gradient is the physical quantity moving the diaphragm in unidirectional microphones. Before the sound

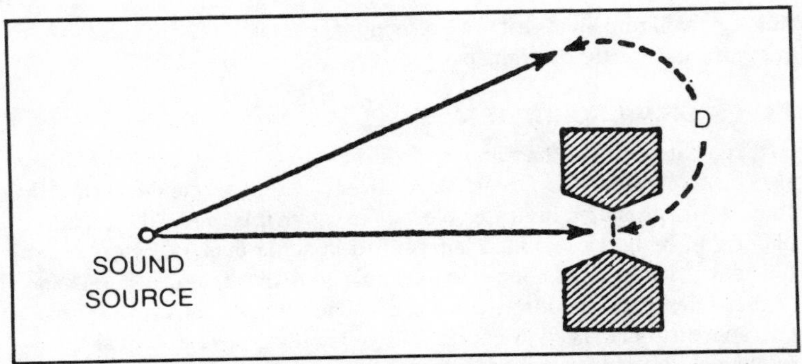

Fig. 6-19. Time difference in arrival of sound on both sides of a microphone diaphragm.

waves reach the back of the diaphragm, they have to pass an acoustic phase shifting network. In case of the cardioid, this element delays the sound waves from behind by exactly the same time they need to travel around the transducer system and reach the front side of the diaphragm. Therefore, the front and back pressure components are equal and the diaphragm remains steady (Fig. 6-20).

By different adjustments of the acoustic phase shifting network and the sound path detour, it is possible to get a hypercardioid pattern. In all cases the length of the sound path detour must be as short as possible, or at least shorter than the shortest wavelength of the received sound wave. This is necessary to maintain the cardioid or hypercardioid shape even at the highest frequencies.

The simplest embodiment of a pressure gradiant receiver is a diaphragm with both sides directly exposed to the sound field. If such a system receives sound at an angle of 90 degrees, the pressures in front of and behind the diaphragm will be equal. The pressure differential will be zero and the di-

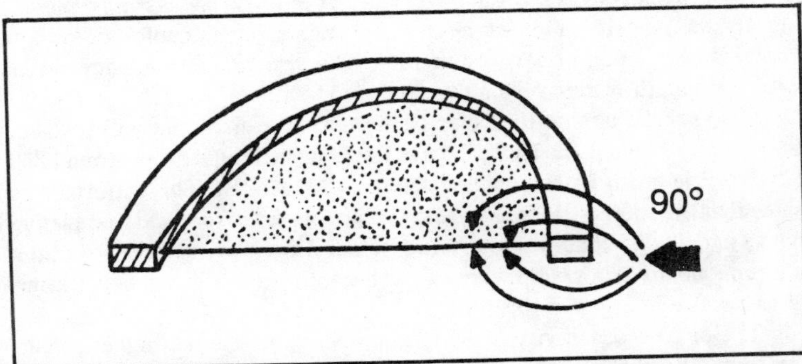

Fig. 6-20. When the front and rear sound waves have equal pressures and when they both form an angle of 90° with the diaphragm, the pressure differential is zero and the diaphragm does not move.

aphragm will remain at rest. The directional characteristic of such a structure will have a figure-eight pattern.

TYPES OF MICROPHONES

You can specify a microphone in a number of ways—by its *polar response*, its *physical structure* (such as dynamic, ribbon, condenser), or by some designation that indicates the way in which it is used. Thus, a microphone can be boom mounted, suspended in some manner, worn around the neck, hand held, positioned in a flexible goose neck, or in a stand. None of these designations describe a microphone totally. Thus, to say that a microphone is supported in a stand tells you nothing of its electrical characteristics. Similarly, to say that a microphone is a condenser type gives you no information about the way the microphone is to be supported physically. Some microphones are general-purpose types; others are highly specialized for particular applications.

Shotgun Microphone

Figure 6-21 shows the patterns of the *shotgun* microphone. The shotgun is a highly directional microphone mounted at one end of a long tube. It has reception patterns that vary between modified cardioid shapes at low frequencies to elongated pears at high frequencies, with first a *doughnut* and then a *tear drop* attached to the stem. Rejection of side sounds is much greater than with a standard cardioid. The directional efficiency is even greater than the supercardioid and is narrower. The rear of the pattern sometimes shows a number of small lobes.

Shotgun microphones provide concentrated sound pickup only over a narrow acceptance angle. The microphones are usually a combination of interference tube and cardioid designs for reducing sound pickup from the sides and the rear.

This microphone isn't used in high-fidelity recording unless it is needed to obtain special effects. It finds applications in the broadcasting of sporting events, question and answer microphones at press conferences, and boom use in film and television studios. The directional efficiency varies with the length of microphone tube.

If you will examine the polar patterns in Fig. 6-21 you will see that the shotgun microphone isn't all that directional at frequencies around 250 Hz (A). It is much better starting around 1000 Hz (B). The patterns also show that the shotgun does have rear pickup, but with front and rear pickup not as evenly distributed as the bidirectional. Useful areas for the shotgun microphone are in a studio where reverberant sound is controlled, or outdoors where it may not exist.

The on-axis sensitivity of a shotgun microphone is not higher in output than a hand microphone with the same type element. The usable degree of pattern off-axis without coloration is simply narrower than cardioids and similar, in addition, to higher directional efficiency.

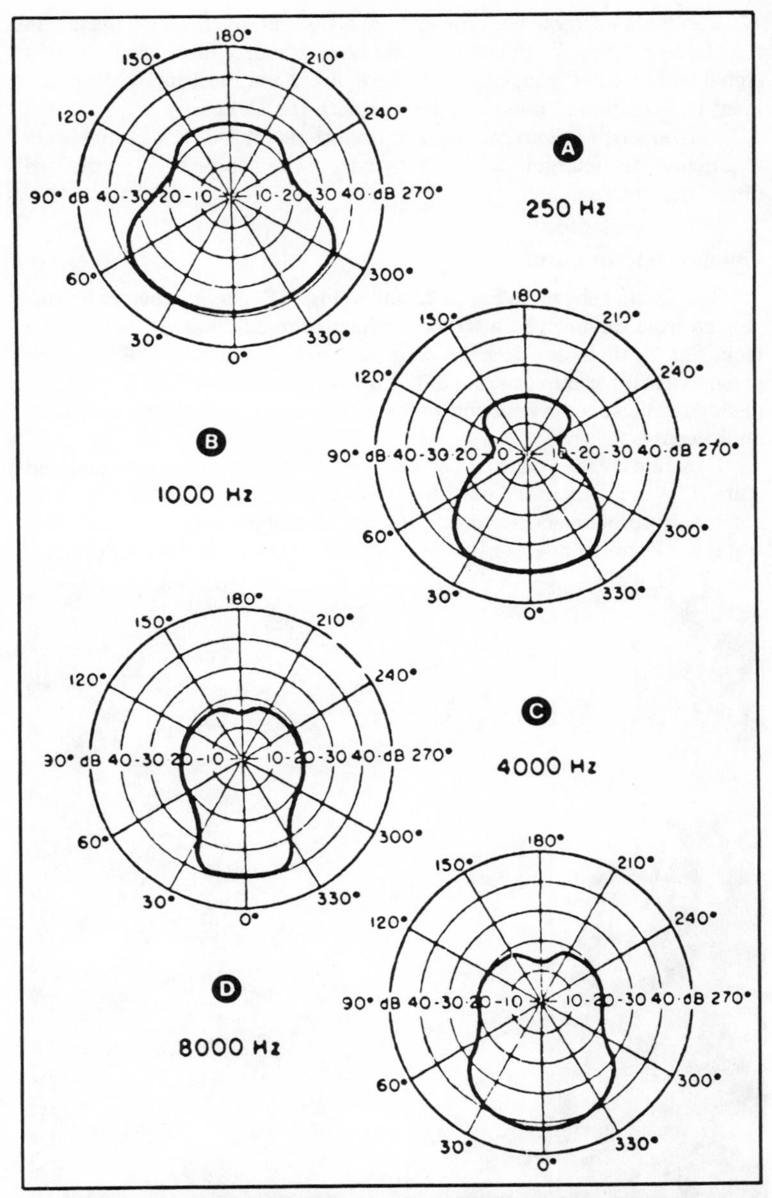

Fig. 6-21. Polar patterns of a shotgun microphone taken at different frequencies: 250 Hz (A); 1000 Hz (B); 4000 Hz (C); and 8000 Hz (D). Note that the polar patterns have been rotated 180°. The 0° point is at the bottom. Both polar drawing techniques are used, with 0° either at the top, as shown previously, or at the bottom, as shown here.

The higher directional efficiency of a shotgun microphone means, in the end result, that for the same microphone to source distance, the direct signal will be equally loud as a standard directional microphone, but ambient or background noise will be less with the shotgun.

Conversely, for equal ambient or background noise ratio to direct on-axis sound, the shotgun can be farther from the source than a standard directional microphone.

Lavalier Microphone

The *lavalier* shown in Fig. 6-22 and in Fig. 6-23, is designed to be suspended from the neck by a string or attached to clothing via a clip or tie tack. The lavalier has an omni pickup pattern and is most useful when the person wearing it moves around, but it is desirable that the microphone-to-subject distance be kept constant for optimum sound pickup and minimum feedback.

Dynamic lavalier units can be supplied as standard or shock-mounted units. The shock-mounted units minimize cable and clothing noise but do not supply absolute shockproofing if the microphone is dropped on a hard surface. Electret-condenser lavalier microphones provide the smallest size.

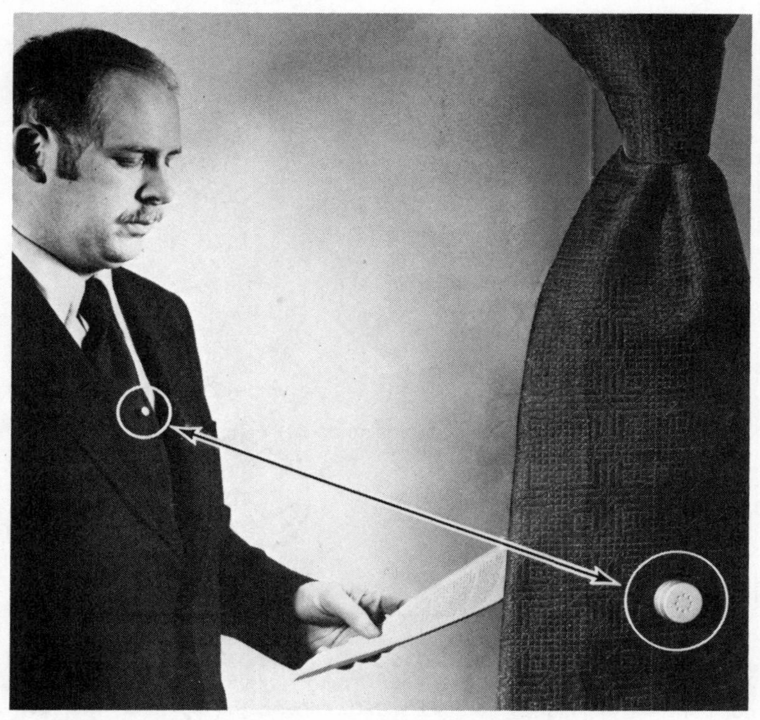

Fig. 6-22. Lavalier mic can be clipped to necktie. (Electro-Voice)

Fig. 6-23. Lavalier microphone.

This is useful in film and television applications where the microphone must be inconspicuous.

In case of AKG's Model D-109 lavalier, a sliding shield boosts the upper-range frequency response. The unit can be freely suspended around the neck or held in position with a tie or clothing clip.

The lavalier mic has specific problems of its own, suffering from "chest dips," off-axis depreciation, clothing noise, cable noise and hollow sound. Like mics that are mounted directly on or in musical instruments, the lavalier is positioned right on the source. (Fig. 6-24).

Differential or Noise-Cancelling Microphone

The differential or noise cancelling microphone (Fig. 6-25) is a special purpose type which works on the principle that if the sound source (a person's mouth) is very close, such as a few inches from the front of the microphone diaphragm, the higher sound pressure level on the front of the diaphragm will produce electrical output while equal pressures on both sides of the diaphragm will cancel output. This type of microphone is most useful for speech communications in noisy environments, such as factories,

Fig. 6-24. A lavalier microphone maintains a relatively constant working distance.

athletic fields and stores, and recording-studio control rooms. The noise-cancelling mic is a special case of directionality because, this type of mic, in addition to being insensitive to sounds from its sides, also discriminates against distant sounds in favor of in-close sound sources. An example of this type of mic is the special, screw-on mouthpiece for telephones which are used in noisy areas.

Fig. 6-25. Differential or noise-cancelling microphone is useful in noisy environments. These microphones are suitable for paging applications and factory base-station communications. The noise cancelling feature minimizes interference from background noise.

Stereo Condenser Microphone System

Ordinarily, separate microphones are required for various pickup patterns, such as omni, cardioid, or figure-eight. However, a condenser microphone, such as AKG's C-422, contains two twin-diaphragm condenser-microphone capsules mounted within the microphone body permitting *stereo* pickup with a single mic. The spacing is only 1 1/2 inches, making any difference in time between the two outputs negligible. The upper microphone system can be rotated 180° to provide any offset angle desired. Nine different directional patterns can be selected for each of the two twin-diaphragm capsules. These patterns are identical as to their phase relationship and sensitivity and maintain their polar characteristics independent of frequency. The microphone can supply three basic patterns: omni, cardioid, and figure-eight, plus six intermediary positions.

The Two-Way Cardioid

The *two-way cardioid* dynamic microphone has a total response range that is divided between a pair of transducers, one for high frequencies and the other for low, analogous to the two-way system used in speakers. In the two-way cardioid, the two systems are connected by a crossover network, with the crossover frequency at 500 Hz. The advantage of this arrangement is a complete absence of proximity effect, a wider flat-frequency response over the entire audible range, and linear off-axis response. This latter characteristic means that sounds reaching the microphone from off axis, such as 90°, are reproduced naturally.

In its construction the two dynamic transducers are mounted coaxially, with the high-frequency unit placed closer to the front grille and facing forward and the low-frequency transducer placed behind the first and facing rearward. The low-frequency unit incorporates a hum-bucking winding (Fig. 6-26) to cancel the effects of stray magnetic fields. Both transducers are coupled to a 500 Hz inductive-capacitive-resistive crossover network that is phase corrected.

The treble system in the two-way mic is very small, has a thin and light diaphragm and an extremely short sound path detour, while a larger diameter bass system, with an especially long distance between the front and rear sound port is used.

The small dimensions of the treble unit make sure that no impact pressure occurs, even at higher frequencies. The polar pattern is uniform because the distance between the front and rear sound entrances is much smaller than the shortest wavelength of the sound waves.

The long sound port distance of the bass system results in a total lack of proximity effect because the entire length of the mic body is used as a sound path detour. The result is a flat frequency response even at the highest audio frequencies, a frequency-independent cardioid pattern and a transient response which approaches that of condenser mics.

The two-way cardioid dynamic mic has a wide range of on-axis fre-

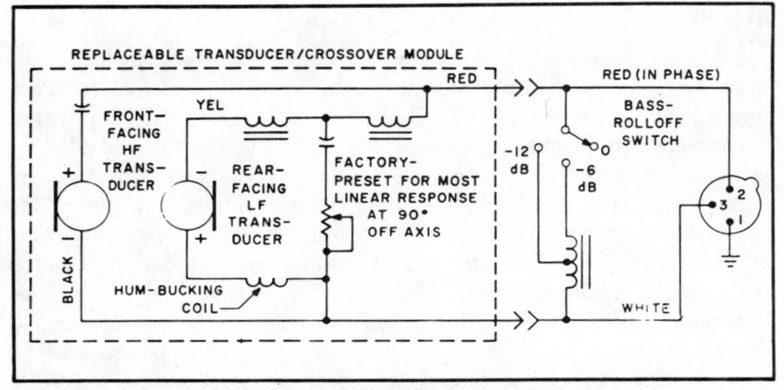

Fig. 6-26. Circuit arrangement of the two-way cardioid.

quency response, resembling that of a condenser mic at frequencies up to 14 kHz.

Pressure Zone Microphone

With a conventional mic it is possible to have sound cancellation caused by the combination of direct and reflected sound, supplying an effect in the output known as comb filtering (Fig. 6-27). This effect is practically eliminated in the pressure zone microphone (PZM) made by Crown (Fig. 6-28). The microphone has a boundary plate, 5″ × 6″ × 1/8″ thick. A thin rubber pad in each corner lifts the plate off its supporting surface by about 1/32″.

The PZM transducer is an electret, placed approximately in the center of the plate with a gap of less than 1/32″ between the transducer cover and the plate. Since the electret transducer needs a power supply there is an XLR output connector at the rear of the plate housing for connection to the power supply. The power supply can be a phantom type or a pair of alkaline cells.

The PZM takes advantage of the fact that at the primary boundary direct and reflected waves are so close together in time that they act as one reinforced signal. Time displacement effects disappear at the boundary. The panels of the PZM microphones function as the primary boundary in many applications, particularly in the area of speech reinforcement. However, for applications where frequencies below normal speech range, that is, below 350 Hz, will be encountered, a larger boundary panel is needed to prevent the fall off of lower frequencies whose wavelengths are longer than the dimensions of the PZM plate. This can be done by mounting the PZM on a large flat surface such as a floor, table top or square of masonite or clear plastic whose dimensions are about 4′ × 4′. Where necessary, a smaller panel down to 2′ square will work well with only minimal drop in the lowest frequencies.

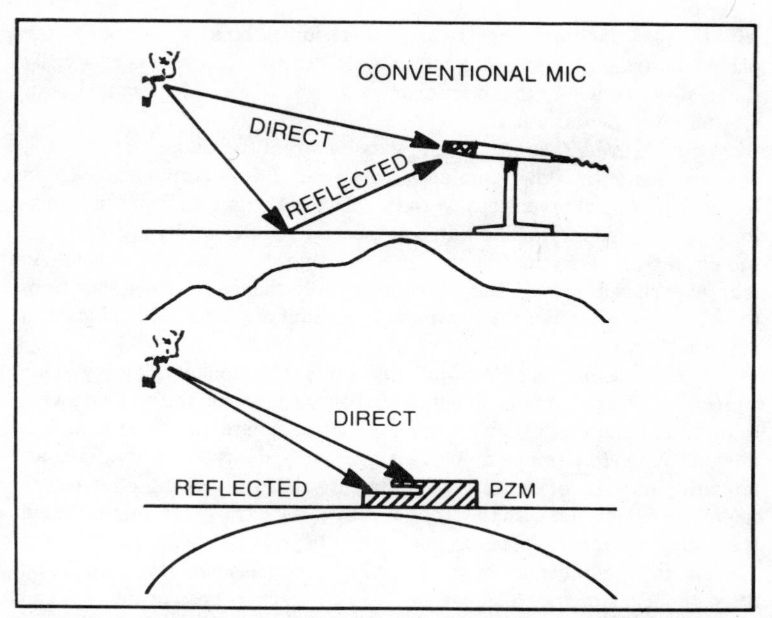

Fig. 6-27. Pickup of direct and reflected sound waves by conventional mic (above). With the PZM the direct and reflected waves are practically superimposed.

One of the problems involved in the use of these microphones is that they are at variance with long accepted rules of microphone technique. These microphones react to sound almost as if they were the ears in a human head and their positions can be predicted by thinking of them as such. All you need to do is to "walk around the sound" until you find the spot where you think it sounds best and then place the microphones at that spot

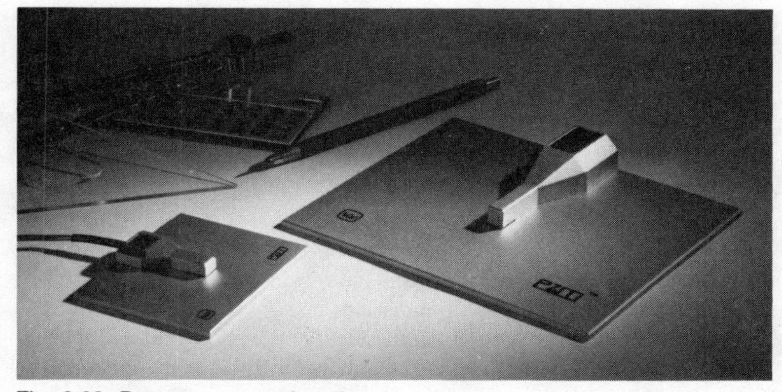

Fig. 6-28. Pressure zone microphones.

with the plate facing directly to the sound source. This will usually be the place where the monitors also sound best, although you may occasionally find a situation where the microphone needs to be moved a few inches or feet one way or the other.

You can use a floor as a common boundary if the floor is bare. A rug will attenuate the low frequency response. PZ microphones have a hemispherical pickup pattern, with the plate of the mic defining the equator of the hemisphere. The mic will pick up equally well anywhere in a full circle as long as the sound source is above the plane of the plate. As long as performers or speakers do not radically change their distance from the mic they can move freely around it without any noticeable change in tonal quality.

If, when using the PZM, you need some directionality, simply place a piece of carpet with the nap out, folded over and facing the area you wish to block, as you might in a speaking situation where you don't want audience sounds to be picked up. This gives you about 20 dB to 25 dB attenuation in the blocked direction. Be sure that the leading edge of the folded carpet is even with the front of the PZM cartridge. It can be behind it as much as 1/4" but should not extend beyond the edge.

The PZM family of microphones will reproduce sound levels up to 150 dB without distorting. You can put a PZM inside a bass drum, you can clip it to the F-hole of a string bass, you can put it inside a piano, you can record a brass quintet with all five instruments aimed directly at the mic.

The Wireless Microphone

There are many recording situations where it isn't possible to use a cable connected microphone. If the performer is on a revolving stage, for example, or is mobile in a car, float or other traveling display, a cable connected microphone is impractical. In place of the usual microphone/cable setup, a miniature battery operated radio transmitter and matching receiver are used. The advantage of the wireless (Fig. 6-29) mic is that it frees the artist from the microphone cable. This permits unrestricted movement on the stage and also allows the performer to move down into the audience.

The transmitter can be in the form of a small pocket-sized case or built right into the microphone handle. The sound, picked up by the microphone, is used to modulate an FM carrier wave, which is then transmitted and picked up by a remote FM receiver (Fig. 6-30). The receiver can be powered from the ac lines, if convenient, or may be portable and powered by a battery pack. Such systems may require an FCC station license if operated within the United States or its possessions. Because licensing depends upon the user's application, it is his responsibility to apply for a license from the Federal Communications Commission.

The receiver that is used, like all FM receivers, can pick up only one transmission at a given time, hence the receiver is tuned to the corresponding transmitter frequency. Think of the receiver as the end of your micro-

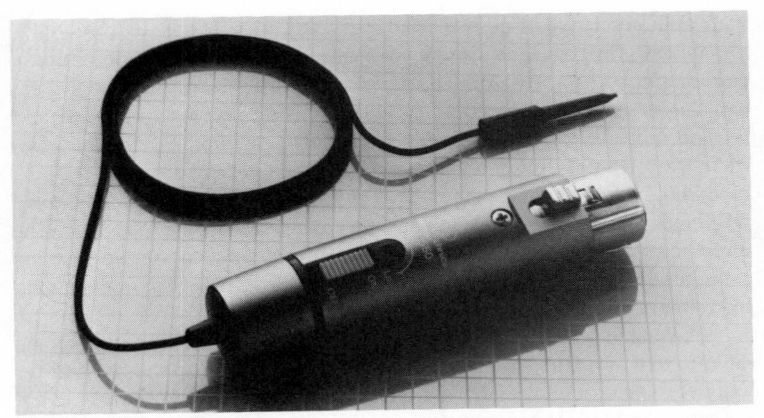

Fig. 6-29. Wireless microphone transmitter. (Audio-Technica)

phone cord with the connection to feed your mixer input. At the connector end of a given microphone you expect to have signals coming out from one microphone only. And this is also the case with a wireless receiver output. You can only have audio output from one microphone.

You can use several wireless microphones simultaneously, but never on the same frequency. Each transmitter/receiver combination must be at a different frequency from the other systems. The wireless microphone isn't more sensitive in one particular direction. The transmitter antenna radiates equally well in all directions, but this omnidirectional radiation pattern can be altered. Thus, the signal can be attenuated by the body of the performer, by walls and other objects. The antenna used at the receiver is also equally sensitive in all directions.

The receiver in the wireless microphone system can pick up other transmissions, such as business radios and other unrelated transmissions while the transmitter of the wireless microphone is turned off. As a precaution, if you are not using a given wireless microphone, turn the gain down on

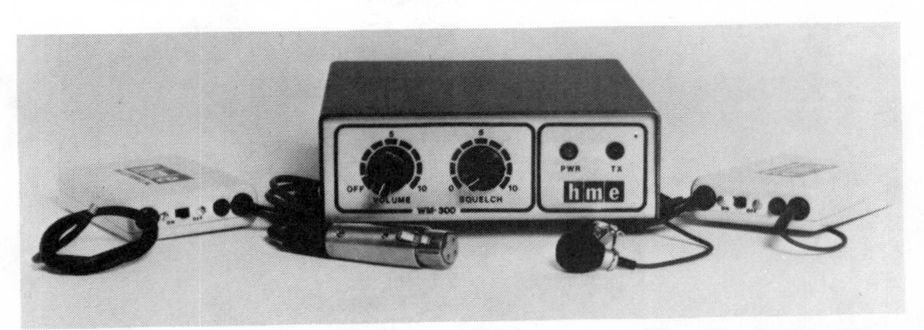

Fig. 6-30. VHF hi-band commercial grade systems with 60 dB dynamic range. System 42M comes with a lapel mic attached to the transmitter. System 42C has a standard 3-pin audio connector which can be used with any low impedance dynamic mic or self-powered electret mic. (HM Electronics, Inc.)

your audio mixer, just as you would a wired microphone.

Wireless microphone systems can suffer from drop outs and/or phase cancellation. The waves transmitted in the wireless microphone system can be absorbed or reflected. If a solid body comes between the transmitter and receiver it can absorb the radio wave, causing a signal loss. If a radio wave from the microphone transmitter bounces off a surface (usually metallic), it can arrive at the receiver out of phase with the primary wave which goes directly from the transmitter to the receiver and cancel it. The effect is a "whooshing" or "buzzing" noise. If dropouts occur without any associated noise, either the squelch on the receiver is set too high or a cable is making an intermittent connection.

The distance between the receiver antenna and the transmitter should be kept to a minimum. Ideally, it is good practice to keep the transmitter and receiver in line of sight and within 200 feet of each other. It is not necessary to place the receiver next to the mixer console and you can run a long audio cable if required. If overhead speakers are located near the microphone, either turn them off or move the speaker to avoid feedback. Wireless microphones are just as susceptible to feedback problems as cabled microphones.

Dynamic Expansion in the Wireless Microphone System

By using dynamic expansion, (Fig. 6-31) it is possible to reproduce the input signal linearly up to 95 dB at the balanced audio output. The monitor output has a slightly higher noise floor. In other words, a dynamic expansion wireless mic system sounds exactly like a cable microphone. There are no audio gain adjustments required on the transmitter. The only major

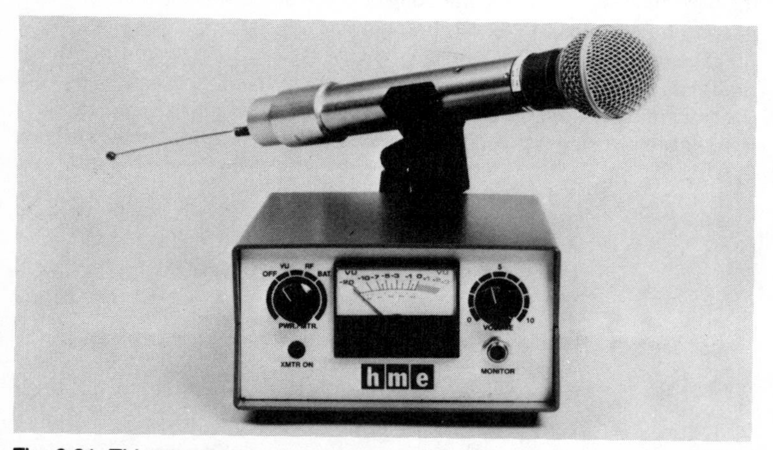

Fig. 6-31. This system has dynamic expansion for those who need maximum dynamic range and signal-to-noise ratio. Can be used by singers with soft voices or high level rock vocalists. Designed for professional use by television and stage entertainers. (HM Electronics, Inc.)

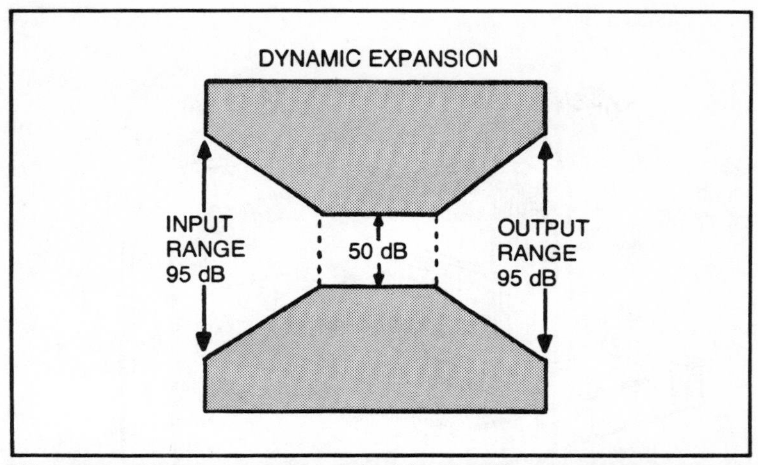

Fig. 6-32. In dynamic expansion method the signal is processed in the transmitter, transmitted over 200 feet, and then the signal is decoded in the receiver.

differences in using a wireless microphone with dynamic expansion (Fig. 6-32) and a standard wireless microphone are that in dynamic expansion systems the mic gain adjustment is in the receiver, as opposed to standard systems where the adjustment is made in the transmitter. As there is no limiter the mixer operator must ride gain just as he would with a cabled microphone and it isn't possible to overload the front end of the mixer, as with a "hot" cabled microphone.

Wireless Microphone Diversity System

The purpose of a diversity system (Fig. 6-33) is to eliminate dropouts or dead spots due to cancellation of radio waves at the receiver's antenna.

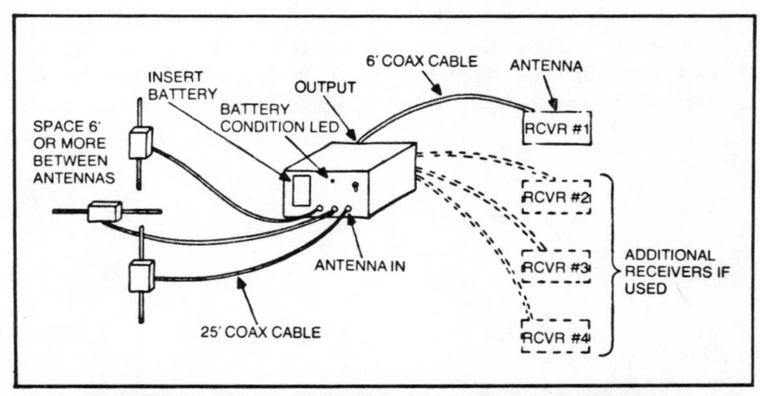

Fig. 6-33. Arrangement of diversity antenna system. (HM Electronics, Inc.)

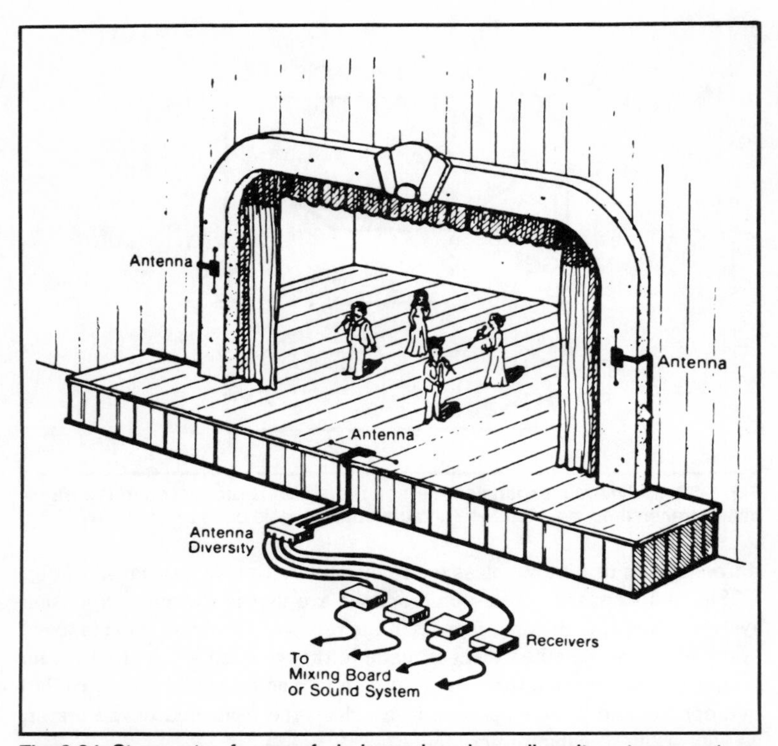

Fig. 6-34. Stage setup for use of wireless microphone diversity antenna system.

It will not, however, extend the distance between the transmitter and receiver. In the diversity system three dipole antennas are used with two of these installed in a vertical position and the remaining antenna in a horizontal position. The antennas should be at least six feet apart. They should not be mounted on a metal frame and should be kept at least a foot away from any large metal object. Three coaxial cables are required, one for each of the dipoles and these cables can be up to 25 feet long. If more than one receiver is used, position them at least one foot apart (Fig. 6-34).

Chapter 7 Microphone Characteristics

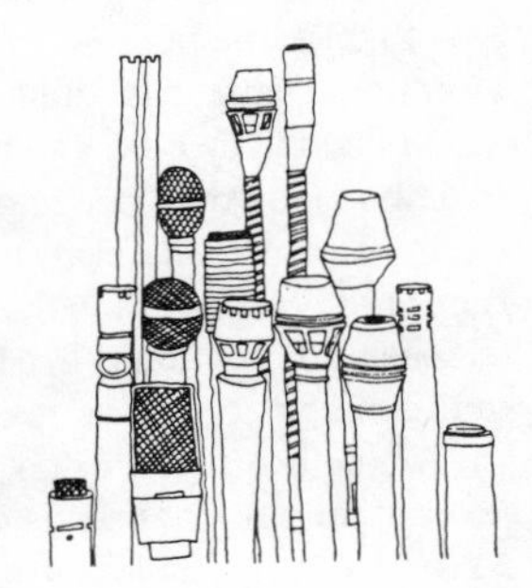

To say that a microphone is an *omni*, a *cardioid*, or a *bidirectional* is simply to put it within a general classification. Not all omnis are alike; neither are all cardioids, nor all bidirectionals. Within each group, microphones can have different characteristics.

There are three main design factors used to describe microphones: the *type of transducer* (moving coil, condenser, ribbon, etc); the *pickup pattern* (omni, cardioid, bidirectional); and the *impedance* (low or high). This does not mean these are the *only* mechanical and electrical features, for there are others, such as transient response, frequency response, efficiency, output, and so on.

MICROPHONE REQUIREMENTS

As a collector of sound a microphone must fulfill three basic requirements: to provide an electrical signal well above the microphone's self-noise level; to provide undistorted output over a wide dynamic range, and, when used with associated equipment, to respond equally well to all frequencies produced by the sound source.

All reproduction of sound, whether live or recorded, must begin with one or more microphones. The quality of reproduced sound can be no better than the quality of the microphones used. Consider the microphone as an electronic ear, a substitute for your own. A good-quality tape deck, a superb amplifier, and the best speakers you can buy can do no better than the microphone that precedes these components. It is the microphone that sets the quality level of the performance.

Compared with the cost of all the other components in a system for recording and reproducing sound, microphones are the least expensive.

There is a common error made in buying microphones, which is buying solely on the basis of price: either too little or too much. A microphone may be bought at a very low, so-called "bargain" price in an effort to economize somewhere along the line in the purchase of a total system. Or the most expensive microphone may be bought with the thought that the most expensive is automatically the highest quality and will therefore lift the overall quality of a recording system.

There are two criteria in selecting a microphone—*quality* and the *specific use* to which the microphone will be put. There is one guideline to use when buying microphones and that is to buy the best quality, regardless of the kind of sound system you have. If your sound system isn't everything you had hoped it would be, a good-quality microphone will supply reduced distortion and more natural sound reproduction. The microphone will not correct faults in your tape deck, amplifier, or speakers, but at least you can choose a microphone that will not add to your problems. A mediocre system can sometimes be greatly improved by changing the microphone.

While the finish or color of a microphone has no effect on quality or level of output, for television broadcast work the microphone should look attractive, yet should not produce reflection glare when working with the high intensity lighting systems used in video broadcasting.

Theoretically, the best microphone should cost more, but it is a fallacy to use price as the sole factor in influencing your buying decision. Engineering and manufacturing skills using similar materials do not always produce equivalent results.

MICROPHONE SPECIFICATIONS

Microphone specifications or *specs* are technical descriptions of the way a microphone behaves under certain test conditions. But it does not automatically follow that a microphone which performs well under laboratory examination will sound well. Microphones are tested in *anechoic chambers*—soundproof rooms—but a soundproof room is precisely opposite that of rooms used for recording. Such rooms permit sound to "leak-in" from the outside or produce noise from inside, or both. A recording room is a reflective environment in which sound bounces from the walls, floor, and ceiling. Such reflections do not exist in the anechoic chamber where microphone performance is recorded by laboratory test instruments. Your ears are the test instruments when the mics are finally put to work.

Quite often, microphones do not recreate the sound you hear. This creates a problem because we have been conditioned to expect high quality in reproduction at all times. You expect pictorial quality in the movies you see, you expect it in your television receiver, and the search for sound quality in a high-fidelity system is a constant effort.

You can identify a microphone by considering its physical structure or by its electrical characteristics, but usually the best approach is a combination of both. The heart of the microphone is the *transducer*, the part of the mic that is involved in converting sound energy to its electrical equivalent.

In any microphone, a diaphragm is that part which moves in response to sound input, that is, to variations in air pressure presented to the microphone. In turn, the diaphragm moves that part of the microphone which produces the output signal voltage, hence it is known as a *generating element*. In some microphones, the diaphragm and the generating or voltage-producing element are separate components that work together, as in the case of the dynamic (moving coil) microphone. In others, such as the condenser microphone, the diaphragm is an integral part of the generating element.

Any energy conversion, including the conversion of physical sound energy into electrical energy, involves the generation of undesired noise. To keep such noise from being audible, its noise electrical level compared to the electrical level produced by the useful sound should be as low as possible. This means that at a given acoustic sound pressure, a microphone to be used for high-fidelity purposes should produce as high a desired sound electric level as possible.

THE SIGNIFICANCE OF MIC SPECS

Mic specifications are useless unless they can be interpreted into meaningful guidelines for the mic user. Most low impedance mics are in the 150-ohm classification, but the actual impedance is a range, not a single specific value. This range is from 100 to 300 ohms and is specified with 94 dB SPL (sound pressure level) applied. When a mic output voltage is indicated as −50 dBm output that mic is 10 dB more sensitive than one rated at −60 dBm. Since conversational speech at a 1 foot distance is about 65 dB SPL, the 94 dB SPL used for a mic rating is typical of a pop vocalist or instrumentalist. However, some vocalists practically put the mic inside the mouth and extremely high sound levels are produced. As an example, a professional condenser mic with a specified output of −40 dBm produced a measured output of +5 dBm from a vocalist at mouth touching range. Such a high output signal made a line level input on a recorder appropriate for that application.

SENSITIVITY

Sensitivity is basically a straightforward characteristic and can best be described by a simple example. Two different microphones, A and B, are placed next to each other, both pointed directly at the same sound source and both exactly the same distance from that source. At any loudness level produced by the sound source, microphone A generates a larger electrical output signal than microphone B. Therefore, microphone A is said to be more sensitive than microphone B.

Within the frequency range of any particular microphone, the sound input should create an output voltage that is a direct function of the sound pressure input. Naturally, the farther away the sound source input is from the microphone, the lower the signal output. *Sensitivity*, then, refers to a signal input of fixed strength at a fixed distance from the microphone. Sensitivity is defined as the voltage or output level of the microphone, in *millivolts*, with the microphone placed within a sound of 1 microbar (a *microbar* is a unit of sound pressure). The measurement is made within a free sound field with the microphone *open-circuited*, that is, not terminated by a load resistor. The ratio of the two—the signal *output* to the sound *input*—is known as either the *response coefficient* or the *sensitivity*.

The sensitivity of a microphone does not refer to a fixed frequency input but, rather, to the entire frequency range specified by the manufacturer of the microphone. If the specs indicate a frequency range of 50 Hz to 15 kHz, the sensitivity refers to this entire range.

The upper illustration in Fig. 7-1 shows a comparison between sound pressure *input* to the microphone and voltage *output*. While the drawings show the sound pressure and output voltage as having the same amplitudes, this isn't at all significant, since sound pressure and output signal voltage are measured in completely different units. What is important is that the

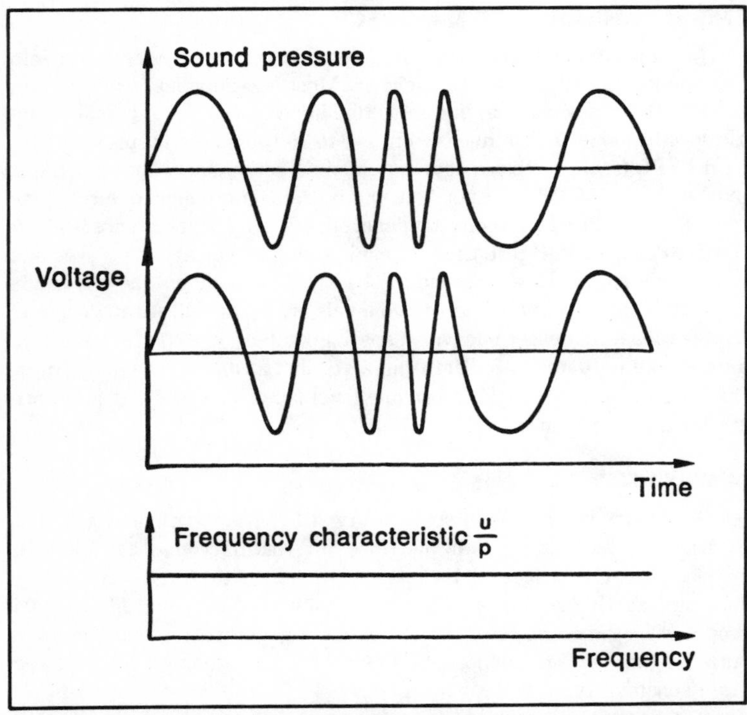

Fig. 7-1. Sound pressure input vs signal voltage output. The lower drawing shows an idealized frequency characteristic.

two waveforms be as identical as possible.

When plotting the logarithm of the magnitude of the response coefficient μ over the logarithm of the frequency (P) in a coordinate system, the result is a so-called *frequency characteristic.* Ideally, this should be a straight line, as shown in the lower drawing of Fig. 7-1.

A microphone's relative sensitivity can be expressed numerically, and all reputable manufacturers publish such figures for their microphones. But, be cautious. For sensitivity comparisons to be meaningful, keep these factors in mind:

☐ There are several systems of sensitivity notation—open-circuit output, maximum output (sometimes just called "sensitivity"), EIA G_m output—to name a few. Always compare the sensitivity figures for one microphone against those for another microphone *using the same system of notation*.

☐ Most of these systems of notation express a microphone's sensitivity in some value of dBm's or dBV's preceded by a minus sign. The microphone with the smaller number of dBm's or dBV's is therefore the *more sensitive* microphone. As an example, a microphone might be rated at −42 dBm while another less sensitive unit might be rated −48 dBm. This simply means that the first microphone is 6 dB more sensitive than the other and will generate twice as much output voltage for a given sound-source loudness.

The question is sometimes raised as to why a manufacturer will rate mic sensitivity in terms of minus decibels. This is because the reference level is well above the output level. Consequently, a mic having a sensitivity rating of −50 dB supplies more signal voltage across its output terminals than one rated at −55 dB.

A spec sheet may indicate mic sensitivity in three different ways using one or more methods. The mic can be rated with reference to 1 mW/10 microbars; EIA Sensitivity or Open Circuit Sensitivity with the output in millivolts but with no load on the mic.

☐ The sensitivity of a microphone is also often expressed as output voltage per units of sound pressure at the microphone diaphragm. Values may be given in mV/μb, mV/Pa or as mentioned above, in dBV. These values are also related to microphone impedance, e.g. the higher the impedance of the microphone, the higher its sensitivity figure. Both extremes of either too high or too low sensitivity may cause problems when the microphone is connected to other equipment like sound mixers, tape recorders, pre-amplifiers, etc. Either overloading or excessive amplifier noise would be the result when using an unusually sensitive microphone.

The sensitivity of a mic is a measurement, and like all other measurements, has significance only if compared to a reference. A generalized statement to the effect that a particular mic is highly sensitive is meaningless.

Consider, for example, that a mic may be set up at some specific distance from a sound source. The mic will then produce a measurable output voltage. Under exactly the same test conditions, a different mic will also produce an output voltage, presenting three varying possibilities. Its output voltage will be the same, less than or more than that supplied by the first test. What we have here then are relative sensitivities. The EIA sensitivity rating supplies a standard figure and precise testing conditions so that the sensitivity of a microphone can be compared to a standard.

One microphone isn't necessarily better than another simply because it is more sensitive. For one thing, the manufacturer of the less sensitive microphone may have intentionally "traded off" a certain amount of sensitivity for other, equally important, design considerations. For another thing, "best" sensitivity like other microphone characteristics depends largely on the microphone's intended application. A microphone that must often be placed relatively far from a soft sound source (a film or videotape dialog-recording microphone that must be kept out of camera range on a fishpole or boom) would benefit from relatively high sensitivity. Under these conditions it is imperative that the microphone generate enough signal to overcome the inherent hum or noise of a mixer's or tape recorder's input circuits. On the other hand, a rock-vocalist's or instrumentalist's microphone may not require as much sensitivity because it's rarely more than a few inches from the performer and almost always in the presence of high sound-pressure levels.

When considering the operating characteristics of a mic it is always well to remember that it may be necessary to trade off one for another. Thus, a mic which is directional can work with an amplifier having a higher output level with less possibility of acoustic feedback. But if a solo singer uses such a mic, the output sound becomes fuller and deeper, as the mic is brought closer to the lips. This effect may or may not be wanted. In that case, an omni can be substituted, but the tradeoff here is that the possibility of acoustic feedback is greater.

FREQUENCY RESPONSE

Frequency is the engineering term used to describe pitch. A tuning fork oscillating (vibrating) at 440 times per second will reproduce audibly the musical note "A above middle C"—the international tone for tuning any musical instrument. The number of vibrations in a time period of one second necessary to reproduce a given pitch is referred to as cycles per second (cps) or, more commonly, Hertz (Hz). Thus, the frequency of the musical note "A above middle C" is said to be 440 Hz.

The frequency response of a microphone is either indicated by a graph over the full audio spectrum or expressed by two frequency limits in cycles per second or Hertz, within which the microphone responds uniformly and with little variation of sensitivity.

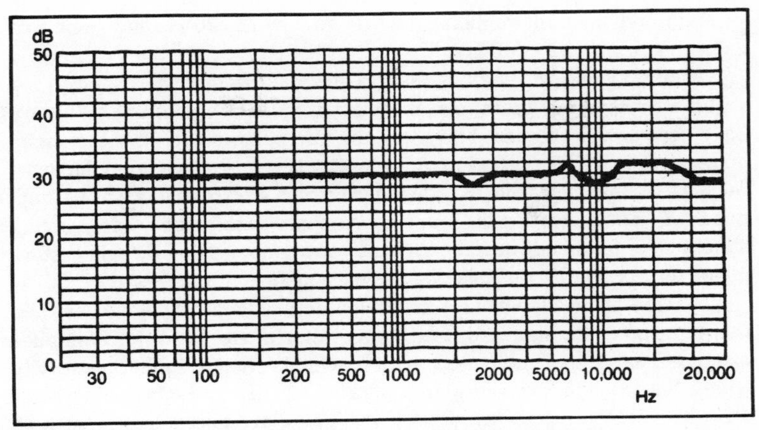

Fig. 7-2. Partially flat frequency response of a microphone.

Flat Response

The performance of the microphone whose frequency response is shown in Fig. 7-2 is approximately flat. This means that the microphone responds equally to all frequencies across the full spectrum, from 20 Hz to 20 kHz.

Boosts and Rolloffs

At first look it would seem desirable for all microphones to have flat response but in practical applications variations in frequency response are designed into a microphone to compensate for varying acoustical conditions, environmental problems or creative needs. Such variations are referred to as boosts and rolloffs and the degree of variation is expressed in decibels. An example of such a frequency-response is shown in Fig. 7-3. This is often done because a microphone with extended bass response would tend to emphasize low-frequency rumble from air conditioners and other mechanical equipment.

You can produce a tone as low as 30 Hz by plucking a double bass. Cymbals have sounds at about 7.5 kHz to above 10 kHz. And the harmonics of certain musical tones go right on up to 15 kHz or more.

What is produced in the way of sound and what you can hear are two entirely different matters. What you can hear depends on your sex, age, physical condition, and musical training. We like to think of human hearing as extending from 20 Hz to 20 kHz. These are outer limits, though, and most of us hear well within that frequency range. Actually, most natural sounds have hardly any low frequencies and quite often what you will find down at the lower end of the frequency scale is noise. This is important, for it establishes a practical set of margins for microphones. A practical response for a microphone could be from 50 Hz to 15,000 Hz, but you will find some microphones that do go down to 20 Hz and some that extend to beyond 20,000 Hz.

A microphone will not only have a specified frequency response, but within its range may prefer certain sound frequencies to others. What we have here now is the possibility that the microphone will alter the sound you pour into it. The variation in frequency response is given in decibels. Such variations range from as little as plus/minus 2 dB to as much as plus/minus 6 dB. A specification of frequency response without an accompanying statement of frequency deviation means you are not getting the whole story about the frequency response of the microphone.

In microphones we can set up a hypothetical ideal which we can then use as a reference standard. *Ideal* is an unfortunate word, for it implies *best*, but ideal is not synonymous with practical. The ideal microphone will have a flat-frequency response from 20 Hz to 20 kHz, possibly with a deviation of no more than plus/minus 1 dB. There are microphones, such as AKG's C414EB-P48 and C460B, which are as close to the ideal as you can get. But you should not arbitrarily rule against a microphone response which is limited or which has a boost in its midrange, or which is weak at the low- or high-frequency ends. You may find such a response quite useful in eliminating low-frequency problems, in emphasizing voice projection, or if you want to deliberately color sound to produce certain sound effects or to minimize noise pickup, or to decrease the pickup of reverberant sound. Consider also that the frequency response of a microphone can be altered by the way in which it is used. A lavalier-type microphone when suspended around the neck may produce a certain booming resonance because of its proximity to the human chest cavity. And the same microphone can have a different response when hand held.

As indicated earlier, a microphone is a device for converting sound energy to electrical energy. However, we usually refer to the output of a

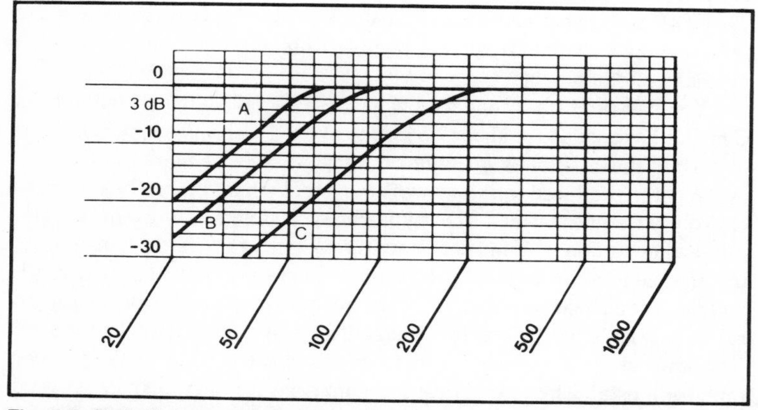

Fig. 7-3. Rolloff characteristics. (A) bass cut at 50 Hz; (B) bass cut at 70 Hz and (C) bass cut at 150 Hz.

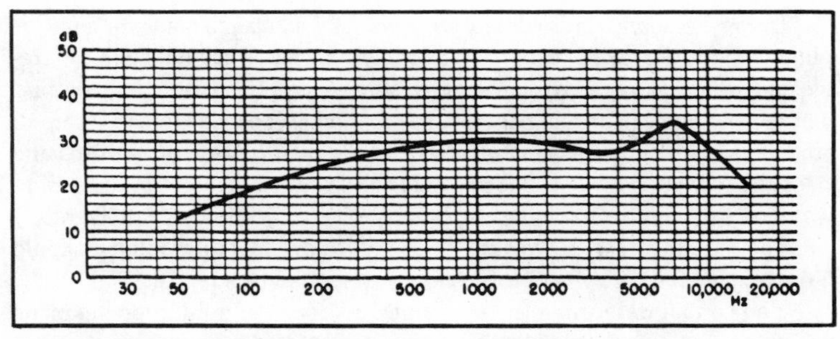

Fig. 7-4. A microphone response curve may have dips and peaks.

microphone in terms of *volts*. We put sound in, we get voltage out. If the sound input level remains constant, the output voltage will also remain constant. If we vary the frequency of that sound, the output voltage will still remain constant and will be constant over the entire frequency range. However, the voltage output of the microphone will increase or decrease when its frequency response isn't uniform. This produces the dips and rises or peaks as shown by the curve in Fig. 7-4. The graph indicates that at certain frequencies the microphone is unable to convert a constant sound pressure into equal voltages as the input frequency is varied. Of course, if the input sound level is increased, the output voltage will also increase—or it should. Conversely, when the input sound decreases, the output voltage of the microphone will also decrease.

Frequency response is just as important in microphones as it is in preamps, power amps, tape decks, and speakers. A wide-band, smooth-frequency response always gives cleaner, more distortion-free sound within the range that is actually used (Fig. 7-5). You can equalize—increase or decrease bass, midrange, and treble—to compensate for the acoustic deficiencies of your recording room, but you should not be forced to use equalization to compensate for the frequency response deficiencies of a microphone. That is not the function of equalization.

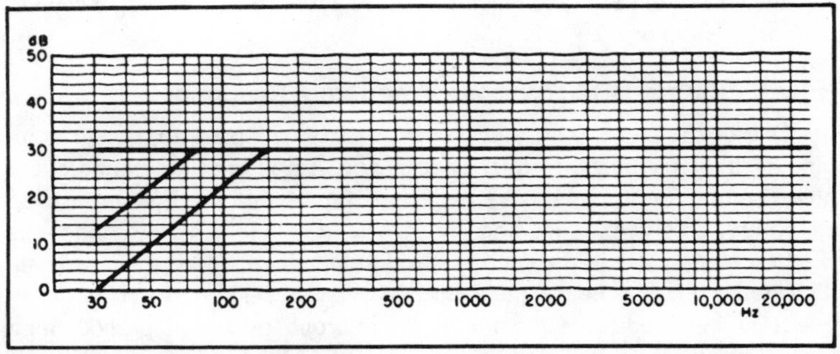

Fig. 7-5. Frequency response curve of a condenser microphone showing positions of bass attenuation.

In some manufacturers' literature, you will find claims made for a microphone having a peaked response as supplying greater brilliance in the treble end, or a peak in the response in the bass as supplying more bass output. There is no point in arguing against microphone response which is limited, stacked in the midrange, misses lows and highs, in peaky, or has dips and rises. As indicated earlier, it is possible to use such characteristics to eliminate low-frequency problems, emphasize voice projection, and deliberately color sound for effect. However, the standard or ideal microphone should have a ruler-flat, wide-range frequency response to sound.

You cannot determine the frequency response of a microphone or its possible response by its physical shape. A pair of mics may appear absolutely identical in physical appearance, but the fact that they look like twins is no assurance they will have identical frequency responses. And this is also applicable to all the other electrical characteristics of microphones as well.

How Low Should Microphone Frequency Response Be?

Preamps and power-amps in modern high-fidelity systems have enormous frequency ranges—20 Hz to 20 kHz is quite common. There are some that go down as low as 5 Hz (or lower) and beyond 50 kHz (or higher). Five Hertz is practically dc and 50 kHz is far beyond the upper limits of human hearing.

Shouldn't a quality microphone, then, also go down to 5 Hz? But before microphone engineers work on this problem, isn't it logical for them to ask why? As the frequency response of a microphone is lowered to much less than 50 Hz, it becomes more sensitive to the pickup of hum and rumble. Further, there is very little music information below 50 Hz. And, finally, very few loudspeakers have a clean bass response under 50 Hz. There are very few musical instruments that can go below 50 Hz, such as the bass viol. But this is the very bottom end of the bass viol and such a tone is seldom, if ever, used. *Bottom C* on the piano is around 16 Hz, but this tone is a rarity. The human voice can go down to around 82 Hz, but it takes a trained basso to be able to reach it. Tenors, altos, and sopranos, of course, all start well above 100 Hz.

How Important Is Microphone Frequency Response?

One of the more important characteristics of a preamp, power amp, tape deck, or speaker is frequency response. Amplifiers, for example, are now manufactured which have responses far beyond the human hearing capability, both at the low- and high-frequency ends. The microphone is the first component in the sound reproduction chain, and if a microphone distorts, or if it does not respond equally well to sound pressures at all frequencies, any modification made by the microphone will be passed along to the next component in the recording setup. The deviation, whatever it may be, will not only be passed along, but will be amplified as well, by the preamp and by the power amp. So the frequency response of a microphone is not only just as important as the frequency response of following components, but more so.

There are microphones which are close to the ideal in frequency response. A good condenser microphone design can be very close to the perfect microphone in all respects, though some condenser types aren't essentially better than excellent dynamics. It is interesting to note that manufacturers making dynamic microphones use only a condenser mic as their laboratory standard reference for measuring their dynamic products. Condenser mics are the first choice by professional recording, sound-reinforcement, and broadcast engineers for quality sound and maximum suppression of noise, acoustically and electrically.

Which Microphone for You?

From all this it is easy—too easy— to rush to the premature conclusion that the microphones to buy must all be condenser types. However, there are a number of other factors to consider, including the type of use, whether in or outdoors, the musical results you want to achieve, whether in fixed or mobile work, and, of course, the cost. Thus, aside from condenser and double-element dynamic microphones, the single-element dynamic microphone is most capable of extended frequency response and the smoothest performance for quality sound at a reasonable price. Most microphones used today are dynamics.

How Microphones Are Tested for Frequency Response

The frequency response of a microphone is measured in an anechoic room or chamber, a walk-in type enclosure which is carefully lined throughout with highly sound absorbent material. If you were to walk into such a room you would hear any number of sounds normally masked or hidden by noise—sounds such as your heart beat, your breathing, and possibly the sound of your clothing rubbing against your skin. To some the experience is an uncomfortable one, particularly when alone. The presence of others in the chamber contributes some bit of customary, extraneous noise. The entire anechoic chamber arrangement is designed to enable a microphone to hear only the direct sound from a high-quality loudspeaker.

The loudspeaker reproduces sound electrical signals supplied by an audio-frequency signal generator and automatic-gain control (AGC) amplifier. Mounted next to the microphone being measured is a calibrated condenser microphone. It feeds the loudspeaker signal back to the AGC amplifier, which corrects any variations in loudspeaker output.

The microphone under test is then pointed at the loudspeaker and its frequency response curve is recorded. The microphone is rotated at various angles to the loudspeaker to measure the response at different angles, from direct (0°) to side (90°) and rear (180°). Sometimes angles in between are also recorded. The voltage output in proportion to the frequency and angle of reception are shown as polar patterns. However, these are nothing more than a two-dimensional slice taken out of the three-dimensional pickup pattern surrounding the microphone.

The voltage output of the microphone will increase or decrease when its frequency response isn't uniform. This results in the dips and peaks revealed by the curves indicating the ability of the microphone to convert sound pressures into equivalent variations in voltage output.

The off-axis angle measurements also indicate how well a directional microphone suppresses side and rear sound, since these angles should produce less voltage. The angle measurements will be lower on the polar graph, since output reduces with direction. The spacing of the rear measurement below the direct-to-head curve is measured in decibels. This becomes the front-to-back specification of a cardioid microphone.

Rear rejection of sound by a cardioid isn't uniform with frequency and varies all over the entire range. When the microphone tail is pointed at the loudspeaker test source, sound waves must pass over the microphone body to reach the diaphragm. This creates a turbulence, like water passing around a rock, and disturbs the sound field, producing additional variations in sound pressure at the diaphragm. This is why the rear response of a cardioid looks uneven compared to front response.

Cardioid front-to-back ratios in decibels are meaningless for comparison unless the frequency of measurement is specified. Many cardioid designs achieve maximum suppression of rear sounds up to 25 dB weaker at 1000 Hz, but may only suppress rear sounds 2 dB at 8000 Hz. This is a state-of-the-art problem, and final performance of the microphone is best judged under actual-use conditions.

It is also entirely possible that a cardioid with a larger front-to-back ratio number will not work as well as a cardioid with a smaller ratio, since uniformity of suppression is more important than maximum suppression at any one frequency.

Misconception of Frequency Response

A common misconception about frequency response is that the response should be uniform for wanted sound only. Thus, there is an idea that a cardioid need be uniformly frequency-responsive only to sounds within its plotted response pattern. Not so. The ideal microphone should also have a flat, wide-range frequency response to sound arriving from any direction, including sounds which it is suppressing and which are weaker than the direct pickup. The microphone must have this overall-sound flat-frequency response if the total sound is to be natural and acceptable.

But what if the microphone frequency response isn't identical for sounds reaching it from all directions? As talkers move off axis, their voices will change quality. Identical instruments on and off axis will not sound the same. The overall result will be that the total sound is "colored," the side-sound pickup will be dull, and the effect will be that the *total sound* will seem unnatural.

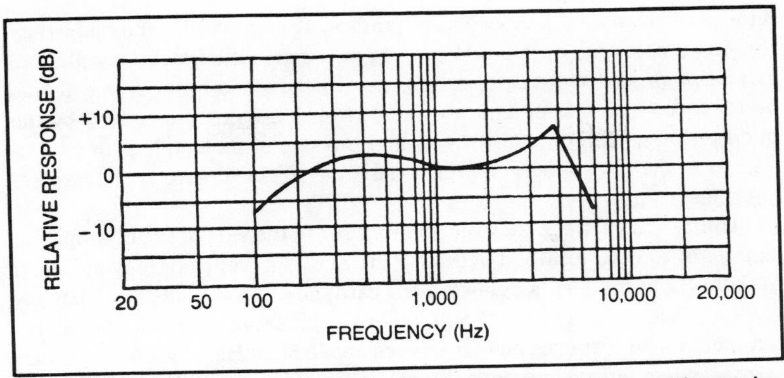

Fig. 7-6. A microphone for commercial applications with high ambient noise levels has a relatively narrow frequency response and is peaked for maximum sound intelligibility.

The frequency response of a microphone is plotted in *decibels* (dB). The decibel scale is a logarithmic scale for measuring acoustic quantities, such as sound level, microphone output, etc. It is related to the physiology of hearing. The sensing device of the human ear is sensitive to low-level sound and much less sensitive to high-level sounds.

A slight rise at the higher frequencies is referred to as *presence rise*. This is desirable at times to add "brilliance" to recordings made in acoustically dull surroundings. Furthermore, presence rise is desirable when recording over a distance so as to capture the quicker decaying high-frequency sound waves.

In industrial uses, in airports, garages, factories, machine shops, offices, restaurants or other noisy environments, the prime requirement of the microphone is in supplying voice transmissions that are distinct and understandable. Voice intelligibility—not voice quality—is the prime requirement. Figure 7-6 shows the frequency response curve of a microphone in such applications. The curve, starting at 100 Hz has a gradually rising characteristic, peaks at 4,000 Hz, and then has a sharp rolloff.

ON-AXIS VS OFF-AXIS RESPONSE

While a polar response pattern does supply some information about microphone response, it is two dimensional, as discussed earlier. However, the polar response as shown in the graph is but a single continuous line on the surface of a spherical type of three-dimensional object, such as a ball or an apple. To get a truer picture we should examine a large number of polar responses, each forming a sort of circle around the centrally located microphone. Further, we should do this for the entire frequency range of the microphone, selecting spot frequencies. However, such an approach is quite impractical.

In simple fact, the single-element dynamic microphone does not have the same frequency response at all angles. It is generally the frequency response measured on axis. However, an examination of the single, on-axis response curve is a clue, since generally the smoothest curve will give better results on side pickup. Response losses off axis are caused by the microphone getting in its own way and the size of the diaphragm in relation to frequency, which creates a narrower pickup pattern at high frequencies.

Uniform high-frequency response at 90° to the side of the microphone can begin to deteriorate above 1000 Hz and is never good beyond 5000 to 6000 Hz in the best single-element cardioids. Good high-frequency response at side pickup to 90° will generally preserve voice character. Instruments with response and harmonics above 5000 Hz will sound dull and colored when out of line with direct pickup. However, with the double-element type dynamic microphone, sound reaching the microphone 90° off-axis is reproduced naturally.

The frequency response of an omni mic is not affected by changing the distance from the event. Any difference in omni fidelity is basically the amount of room acoustics introduced by varying the location of the mic.

The majority of cardioid mics do change fidelity when placed closer than approximately five feet to certain sources. Cardioid designs are easy to identify. They have extra holes in the cardioid mic body for directional control of sound collection. These holes are very close to the head. The end result is an increase in bass response when such mics are placed progressively closer to small sound sources such as the mouth or f holes in the acoustic guitar, cello and string bass.

HOW TO MAKE YOUR OWN CARDIOID RESPONSE TEST

You can make your own evaluation of the off-axis response of a microphone. Speak, sing, or play an instrument with the microphone pointed at the source. Position the microphone at a distance of one to two feet. Set the recording level of your tape deck and record. As a voice test, keep repeating the number 7, selected since it contains both low and high frequencies. Of course, you can also speak, sing, or play an instrument.

Now rotate the microphone off axis. Maintain the same distance from the microphone and use the same sound source you did originally. You will need to raise the recording level to the same volume previously indicated in the first test. This will compensate for the drop in level, up to about 6 dB, due to cardioid suppression at 90° pickup. Now record again.

Rotate the microphone again and point the cable end at the sound source. Remember to keep the same distance from the sound source to the microphone. Raise the volume level once more to compensate for the up to 20 dB average drop in level from cardioid rear rejection of sound. Record again.

If you've done this experiment correctly, the playback level from the three microphone angles should should equally loud. You will hear the character of sound change with each microphone angle. The microphone which maintains basically the same sound quality over the widest off-axis angle is the better microphone, regardless of price.

You can also make yourself more conscious of possible changes in sound quality when attending a reinforced public sound stage or performing arts stage installation. Close your eyes and listen carefully as the speaker, vocalist, or instrumentalist moves to the left or right of the microphone. With a little practice you will soon begin to note changes in sound quality.

CLOSE-SPEAKING EFFECT

With some microphones the pitch of sound changes in vocal applications depending on the distance the mic is from the mouth. Bass tones are emphasized when the mic is worked in close with such tones becoming weaker as the mic is moved away. The *timbre*, the quality given to a sound by its overtones, also changes. A performer, familiar with this characteristic of a particular mic, can use it to advantage. However, there is always the possibility that it could also make the sound of the voice seem unnatural.

PROXIMITY EFFECT

Except in the case of omnidirectional microphones, the frequency response of a microphone can become distorted when it is used too close to the mouth. The result of such microphone positioning is rasping, spitting, popping, booming, or exaggeration of the low frequencies.

The mouth is a rather unusual sound source, for it generates spherical sound waves with very high impact pressures. However, as the distance from the mouth to the mic increases, the spherical sound waves tend to flatten, becoming plane waves, so distortion diminishes with distance. You can hear a comparable distortion effect when someone talks or sings within an inch of your ear.

Known as *proximity effect*, it is the increase in low-frequency response produced in most cardioid and bidirectional microphones when the distance from the sound source is decreased, with the effect most noticeable at a distance of less than two feet (Fig. 7-7). Proximity effect can easily introduce up to 16 dB bass boost when the microphone is positioned about one inch from the mouth. Depending on the design and low-frequency efficiency of the microphone, the voltage output can jump many times the catalog sensitivity rating. As an example, a microphone catalog rated at −55 dB can put out a signal of −19 dB.

Cause of Proximity Effect

Proximity effect is due to microphone design. For certain applications some quality microphones are intentionally designed to provide proximity effect. In the simplest and, usually, the least expensive cardioid mics, the rear sound entrances for side and rear sound cancellation are close to the diaphragm. Because these rear sound entrances are near the diaphragm, the result is an exaggerated, boomy bass boost in voice—an effect that becomes more noticeable as the working distance of the microphone is reduced from about two feet to where the mouth practically touches the mic. Proximity effect is a byproduct of low-cost design, and you will not find it in the response curves shown in microphone catalogs. Microphones with rear sound entrances on the body more remote from the head have less proximity bass boost and will sound more natural when close-talked. The most objective natural sound is achieved when the low-frequency sound entrances are as far from the head as possible. Omnis aren't subject to proximity effect because sound pressures cannot act on the rear of the diaphragm.

How Proximity Effect Is Minimized by Design

Microphone designers are continually working on techniques to minimize proximity effect for those uses where increased bass at short working distances isn't wanted. One technique is the inclusion of a variable distance slot in the handle of the microphone. A tapered length of absorbent material is mounted inside the microphone handle directly beneath the slot, providing variable frequency absorption, depending on the thickness of the absorbent material. The thickest part is placed farthest away from the microphone diaphragm, so high frequencies have the shortest path to the diaphragm, low frequencies the longest path. This provides a more uniform cardioid pattern at all frequencies and minimizes proximity effect. A proximity effect of 4 dB at a distance of 2 inches at 100 Hz has been measured on microphones of this design.

There is a disadvantage, though. A performer can place a hand over part or all of the variable distance slot during a performance. This creates a double-humped peak in mid-frequency response and an increase in proximity effect, depending on the amount of the slot which is covered.

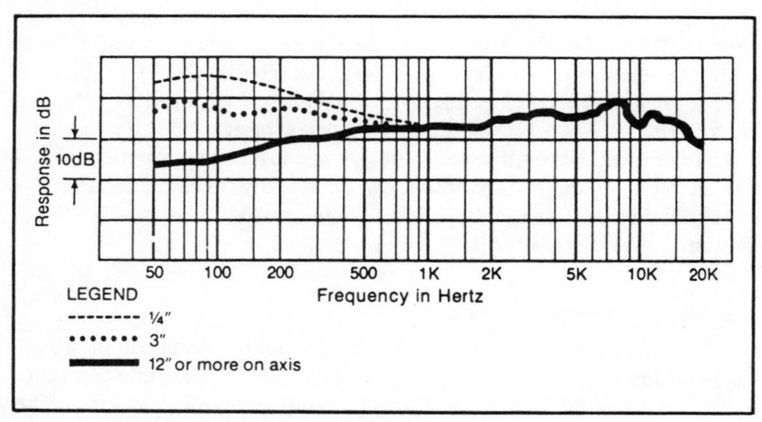

Fig. 7-7. Proximity effect increases bass output.

Another proximity-effect reduction has been developed by AKG based on the *two-way* or *coaxial* microphone, described earlier. Proximity effect is reduced because the high-frequency element of the microphone prevents the performer's mouth from getting up to the low-frequency diaphragm. The measured proximity effect of the two-way microphone is 2 dB at 2 inches at 100 Hz, an improvement over a variable-distance slot technique. A side effect of the two-way microphone system is that a midrange proximity effect is produced at 500 Hz, but this is limited to a 1.5 dB rise at a distance of 1/2 inch from the source. This is well within the response specifications of the microphone.

Utilization of Proximity Effect

You cannot characterize proximity effect by calling it either good or bad. For every microphone user who doesn't want it, there will be another who will depend on it. The best technique is to be aware of proximity effect—how to avoid it if that is what you want, and how to use it if that is what you want.

The cardioid proximity effect of artificial bass is subjectively preferred by many recordists and listeners despite its lack of objectivity and naturalness. It is possible to obtain a cardioid mic with a three-position bass shelving switch to subdue the proximity effect if not wanted when close micing. Normal response is easily restored for distance pickup, or the control can be used to subdue undesirable room effects when lower frequency content is not desirable in favor of clarity in the recording. However, the proximity effect of the usual cardioid sometimes keeps recordists from working in as closely as they would wish. In that case an omni can be substituted, allowing the mic to be moved in closer with no ill effect.

The most pronounced area in the sound spectrum for proximity effect is below 100 Hz. If the microphone has an extended bass response, such as a condenser mic, low-frequency boost will increase as the frequency is lowered. In hand-held microphones where the low-frequency response below 150 Hz is attenuated to minimize handling noise, proximity effect will produce a noticeable hump around 100 Hz.

There are applications where a low-frequency boost may be desirable. Announcers and vocalists, for example, may want to add fullness to the voice. Vocalists often like proximity effect because the increased bass energy produces a high signal to ambient noise ratio. This supplies the vocalist with greater isolation from accompanying instruments, supplying greater vocal penetration to rock groups and lessened acoustic feedback in some applications. Sometimes proximity effect appeals to vocal performers because it lets them "shade" their voices. But this same bass emphasis under other conditions could lead to "muddy" recording and/or sound transmission.

Proximity effect can also be troublesome. Thus, for a radio announcer who must move around to cue records, start various machines, or is in a situation where the distance from the microphone is constantly changing, proximity effect can cause the announcer's *on-the-air* voice to change character as he moves about. In this application it is better to select a microphone having minimum proximity effect.

Solo performers who are aware of proximity effect sometimes use it to produce a desired result. The performer may work the microphone close anyway, and not only does he keep stage sounds out of the mic, but the voice gets a deeper bass. The sound isn't "natural," it isn't high-fidelity, but the increased bass and exclusion of competing instruments is what the performer wants.

Proximity effect doesn't help public speakers. The exaggerated low-frequency boost produces echoes in the auditorium or hall and these destroy intelligibility. It makes no difference whether the microphone is hand held or mounted. The public speaker isn't able to maintain a constant distance to the microphone. There are sound dropouts whenever the speaker moves left, right, or backward. In some cases the result is that the speaker ends up hunching closer to the microphone. In this example, proximity effect can destroy the effectiveness of the speaker. The use of an operator to control sound-signal gain is no solution, for there is no way in which the operator can compensate for changes in the working distance between the lips and the microphone.

No other single microphone design has set the sound taste of the public or standards for microphone and user performance than the close-ported, single-element cardioid. Properly used, it is versatile. Performers can generate an effect that helps thin voices. The proximity cardioid will reach over distance in public sound and recording when the talker cannot get close to the mic. Improperly used, proximity effect is the greatest offender in obtaining unacceptable sound.

Microphones having a proximity effect characteristic have conditioned users of such microphones to expect close-talking bass boost, and the effect generated is mistakenly assumed to represent a standard for good performance. The effect can be tolerated with a performer or announcer whose thin voice needs enhancement. In sound-reinforcement systems, however, the bass-boost proximity effect is unnatural, and sound equalizers are used to cut its effects. Some microphones do have restricted low-frequency response to compensate for proximity bass boost, but such microphones produce thin sound in normal working-distance applications.

TRANSIENT RESPONSE

Like so many other things, musical notes have a beginning, followed by a sustained portion, and then an end as described earlier, in Chapter 1. If you were to listen to just the sustained part and not hear the beginning, you would probably have difficulty in distinguishing one musical instrument from another. If, for example, a violin and flute both played the same tones and you were not given an opportunity to hear the beginning of these tones, you might not be absolutely certain which instrument was the violin and which was the flute. If you could hear the beginning and the ending of the tones, instrument recognition would be much easier and more definite.

However, the beginning or ending of a tone has a much shorter time duration than the sustained portion. The tone starts at *zero* sound pressure level and reaches its peak value quickly. This is similar to the conclusion of a tone. It drops from some high sound pressure level to zero rapidly. These sudden starts and stops of a tone are called *transients*, possibly because they are so fleeting. This does not mean that every tone must have a sustained portion. You could have a transient consisting of a sudden start and an equally sudden finish, with no sustained section. It is these transients that give each tone its personality, its particular characteristic.

The transient nature of tones makes matters a bit difficult for microphones. When a sound pressure level rises almost instantaneously from zero to some peak, it means that the diaphragm of the microphone must move equally rapidly. While molecules of air are extremely light and can rearrange themselves into regions of compression and rarefaction without difficulty, the microphone diaphragm with its much larger mass is more difficult to get into motion and is equally more difficult to stop.

Not all instruments, of course, have tones which have sudden starts and stops. But the percussives do. The piano is a notable example and the ability of a microphone to reproduce the tones of a piano is a good test of the transient response of that microphone.

From a microphone design standpoint, this means that the diaphragm in the mic—the element that responds to changes in sound pressure level—must be lightweight because it must be able to get into motion as quickly as possible and because it must equally be able to stop quickly. One way of reducing the weight of the diaphragm is to make it smaller. And the trend over the years has been toward microphones whose size has been coming down. But if you reduce the size of the diaphragm and of the moving coil, the output voltage of the microphone decreases. Also, a microphone, like a speaker, is contained within an enclosure, and enclosures have *physical resonance* problems. This means they tend to boost sound for certain favored frequencies.

While manufacturers of mics issue spec sheets which cover characteristics such as frequency response, sensitivity and polar patterns, specs do not include transient response except in terms of generalities. No standards have been set up, so the only way in which to make a comparison is through a direct A/B test under identical conditions.

Quite often, the ribbon mic (the velocity mic) is regarded as being inherently superior to the moving coil dynamic mic simply on the basis of diaphragm weight. The diaphragm assembly of the moving coil mic has a weight, measured in milligrams, that is in the range of 30 to 40 times that of the diaphragm of a ribbon mic, with the ribbon mic weighing in at 2 mg or less and that of the moving coil at about 70 mg, or possibly a bit more.

But, this is only part of the story. The numbers given here overlook the acoustic impedance on both sides of the diaphragm.

EFFICIENCY

Some loudspeakers, such as the acoustic suspension types, have a very low *efficiency*, which means they require a much higher audio power input to produce speaker sound, and so such speakers are associated with higher-powered amplifiers. A much more efficient speaker can produce the same amount of sound output with a smaller amount of audio power input.

However, microphones do not have the advantage of being able to use power amplifiers. They must always work directly with the original sound source. Further, the signal-voltage generating element in the microphone must always be small and so microphones always develop weak output signals. These output signals are always in competition with noise, and anything that can be done to improve microphone output is always a step in favor of a better signal-to-noise ratio. With a high signal-voltage output microphone, residual amplifier noise and tape hiss are reduced in proportion to the loudness of reinforced voice or recorded music.

There is still another advantage of using a microphone with a high output. The high-output microphone can be used at a greater distance from the sound source before the noise level produced by the system becomes intrusive. But high microphone output is just one consideration among many. Other wanted microphone characteristics may be such as to override the desire for high output.

OUTPUT LEVEL

Microphone *output level* is an important specification. It is the amount of voltage generated by sound pressure driving the microphone diaphragm. Various standards are used, but all relate to a specific loudness with the microphone pointed, on axis, at the sound source, generally at a test frequency of 1000 Hz.

The applied sound pressure is based on practical use conditions. This is the 74 dB to 80 dB average loudness of a speaking voice 3 feet from the microphone, or the approximately 94 dB sound pressure of a speaker 1 foot from the microphone. Both measurements are related because microphone output voltage will double and increase 6 dB as the distance to the sound source is halved. The voltage will drop to one half, or 6 dB, as microphone distance is doubled.

The output level of a microphone depends on a number of factors: the working distance, the sound level of the performer, the sensitivity of the microphone, the amount of pickup of reverberant signal, and the noise level, including external noise and the self-generated noise of the mic. The actual overall signal voltage available at the mic's output terminals can range from as little as a number of microvolts (millionths of a volt), to a number of millivolts (thousandths of a volt) to more than one volt.

The output level, or sensitivity, of a microphone is an expression of the voltage or power output for a given sound pressure. The open circuit voltage is an "unloaded" figure, that is, there is no voltage drop due to the measuring instrument. The output is specified in both volts and decibels for convenience. A typical open circuit voltage for a low-impedance microphone could read: −80 dB re 1V/microbar, or −80 dBV. This means that for a sound pressure of 1 microbar, (74 dB SPL, equivalent to the pressure produced by a normal speaking voice two or three feet away) the unloaded output voltage would be −80 dB with 0 dB equal to 1 volt. A less sensitive microphone would have a larger negative dB number (for example, −82 dB) and a more sensitive microphone would have a smaller negative number (for example, −78 dB).

In general, the open circuit voltage of high impedance microphones is about 10 times (20 dB) greater than that of low-impedance microphones, and the impedance is about 100 times greater. The significantly lower impedance of low-impedance mics enables the use of long cables without signal loss or change in frequency response.

The power level is specified with a matched load, for instance, an actual 200-ohm microphone matched to an actual 200-ohm amplifier input impedance. A power level for this microphone might be: −60 dB re 1 mW/10 microbars. This means that the maximum power delivered is −60 dB with 0 dB equal to 1 milliwatt for a 10-microbar sound pressure (94 dB SPL). Note that the power output for a microphone with either low or high impedance would be about the same.

INPUT OVERLOAD

The microphone does not live in a world of its own. It is a signal generator and as such supplies a signal voltage to some following component such as a mixer or tape amplifier. These units have definite signal input requirements and the mic must meet those requirements. The problem is usually one of excessive signal supplied by the mic and this can result in distortion. This can happen by putting the mic too close to a high level sound source.

The mic is generally not responsible for overload distortion. It can be cured by using a greater working distance from the mic or by putting an attenuator pad between the mic and the component to which it is supplying a signal. Various types of pads are commercially available capable of reducing mic output by as much as 10 or 20 dB.

MIC PADS

Figure 7-8 shows a simple microphone input section. Since the gain

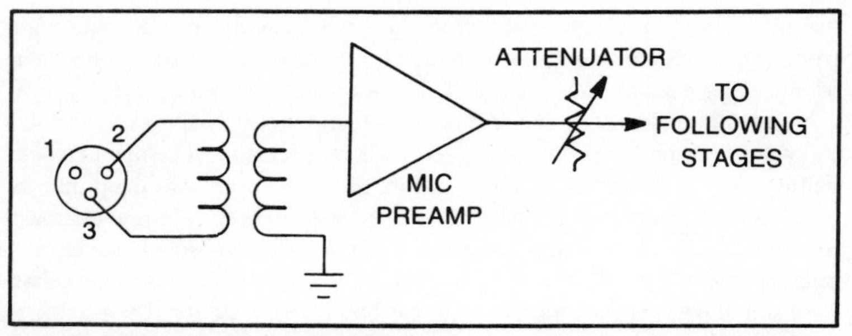

Fig. 7-8. With this arrangement the attenuator cannot prevent overloading of the first mic preamp stage.

control follows the first stage of amplification it can only prevent overload of following stages but not the first stage. Figure 7-9 shows an attenuator switch placed ahead of the first stage. This attenuator can be a simple resistive pad or a feedback loop around the amplification stage. In either case, it is able to reduce the signal so it does not overload the first stage.

Another potential overload point is the input transformer. Good, high level input transformers are fairly large, heavy, and not at all inexpensive. As a result many economical mixers simply do not have such a transformer.

Transformer overload can be avoided by placing the attenuator switch ahead of the transformer, or by using an external attenuator. The latter method should be used with caution in cases of phantom powering as certain in-line attenuator designs can short or imbalance the power feed. In that case it is desirable to have a mic with an internal attenuator. Figure 7-10 shows the right and wrong ways of using an attenuator with phantom powering arrangements.

Any situation in which a specific input gain control can be opened up only a "hair" before the signal drives the internal meter (or other level indicator) into the "red" may indicate that input attenuation is needed, especially if the master gain control is also opened up only a "hair."

Unnecessary input attenuation will result in a poor signal-to-noise ratio

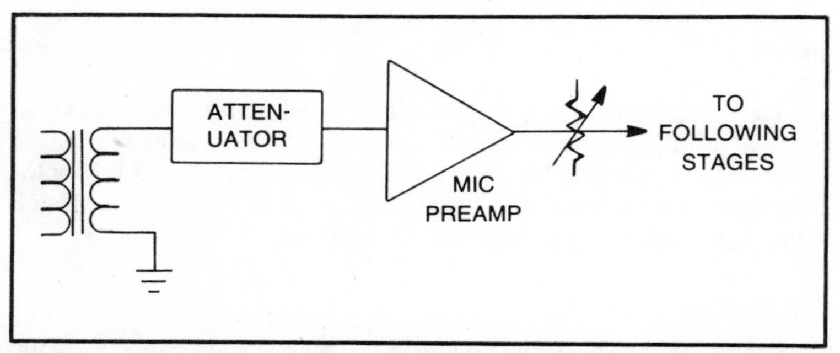

Fig. 7-9. Correct positioning of attenuator to prevent mic preamp overload.

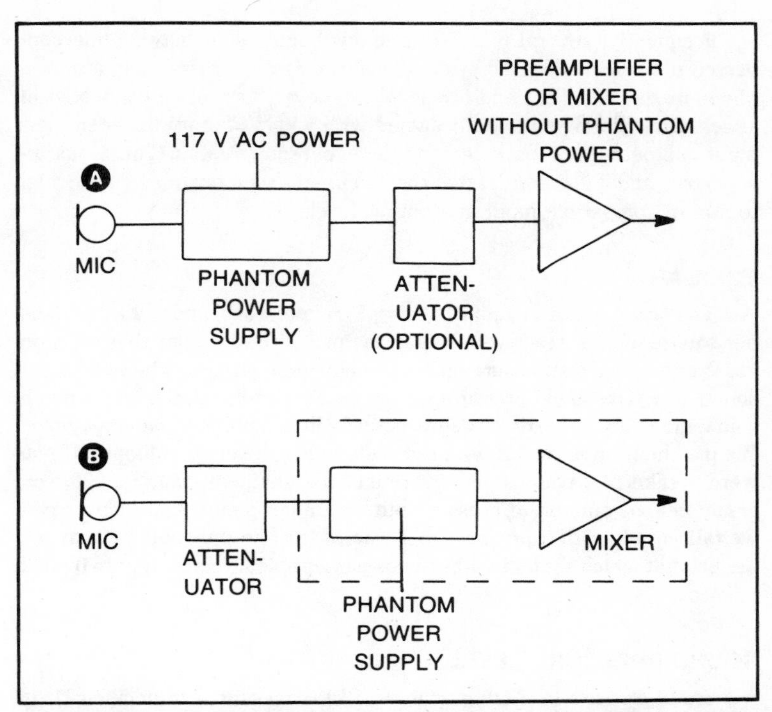

Fig. 7-10. Correct (A) and incorrect technique (B) for using an attenuator pad with phantom powering.

from the first stage. This is because the gain control will be turned up high and, as a result, will allow more noise from the first stage to be passed on to following stages.

It isn't always easy to detect input signal overload. If the condition is severe, the sound may appear to be "rough." This is due to uninterrupted overloading caused by high level signals. But if the signal has a wide dynamic range, the preamp may fall in and out of its overload condition and so may be difficult to detect aurally. Some preamps are equipped with LEDs (light emitting diodes) which function as input overload indicators. The correct setting of the attenuator control is the one which does not permit LED flashing for the strongest signal.

LINE LEVEL CONNECTIONS

The output of a typical microphone mixer usually provides a signal level on the order of one volt, which is considered to be "line level." If such equipment has VU meters (volume unit meters) they are often calibrated so that zero VU reading corresponds to an output level of +4 dBm (decibels relative to one milliwatt).

Frequently, several pieces of line level equipment may be interconnected to yield the desired system performances. For example, a microphone mixer, equalizer and low level crossover may all be connected in series. Often these units are designed with a gain of unity between their input and output and have compatible level requirements. This is not always true and it is essential that the equipment specifications be checked to assure compatible input and output levels.

THE BAR

The *bar* is the basic unit of atmospheric pressure, and is 14.7 pounds per square inch at sea level. This is the pressure of the air that rests on us. We exert an air pressure on a microphone diaphragm when we direct sound at it. The sound pressure we put on a microphone, though, is much smaller than atmospheric pressure, actually in millionths of *one atmosphere*. We use the term *microbar*, with the prefix *micro* meaning millionth. If you were to take 14.7 pounds per square inch and divide it by 1,000,000, you would get the amount of pressure put on a microphone diaphragm when we talk into it with a pressure of 1 microbar. The threshold of hearing, the point at which sound first becomes perceptible, is equivalent to 0.0002 microbar.

SOUND PRESSURE LEVEL

Sound pressure level (abbreviated SPL) is measured in decibels. It is an alternative form to the microbar of sound pressure measurement. The threshold of hearing is 0 dB SPL, corresponding also to 0.0002 microbar.

SPL is sometimes preferred as a unit of measurement in place of the microbar, since SPL is in decibels. Decibel ratios, like percentages, are convenient to use. Sound pressure levels, microphone, amplifier, and speaker frequency response curves, system gain, tone control, and similar audio functions are frequently measured in decibels. As a unit of measurement, SPL offers a convenient starting point for sound systems in which all measurements are in decibels.

What is not generally realized is that it requires just an extremely small change in normal air pressure to produce sound. Sound that is barely audible is caused by an increase or decrease in normal air pressure of about one-millionth of one percent. A variation of normal air pressure of about one-tenth of one percent puts sound at the threshold of pain.

THE DECIBEL

The *decibel* (abbreviated dB) is a *ratio*, a comparison of two voltages, two currents, or two electrical powers. It is also convenient as a relative measure of sound level, or sound-pressure level (SPL). The chart in Fig. 7-11 is a scale that supplies a comparison of the relative levels of familiar sounds.

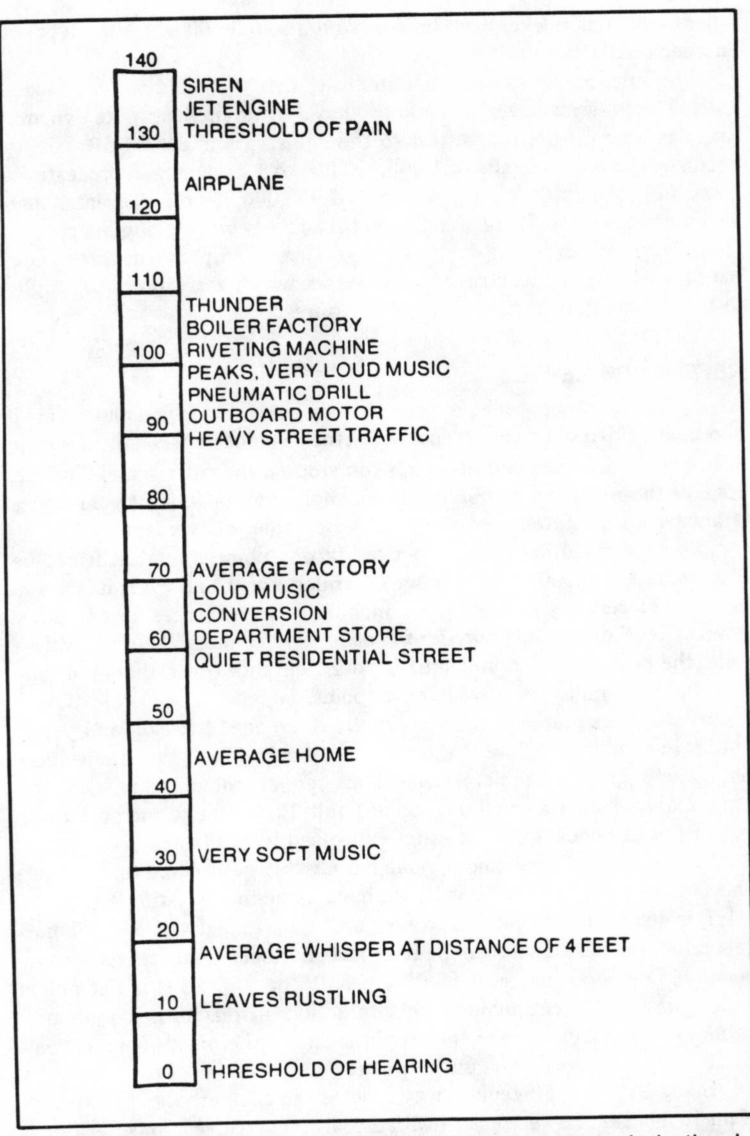

Fig. 7-11. Relative sound levels of ordinary sounds, in decibels. At the threshold of hearing (0 dB) we are barely able to hear sound.

Unlike other units of measurement, such as the inch, the yard, or the meter, the decibel isn't linear. Instead, changes on a decibel scale occur logarithmically. For example, one sound which is twice as strong as another sound has a level 3 dB higher. And a sound 100 times the level of another is 20 dB higher.

The smallest change in sound intensity that you could notice is about 1 dB. The pressure level of sound produced by an orchestra, its dynamic range of sound from the softest to the loudest, is about 90 dB. Ninety decibels, though, is a ratio of 1,000,000,000 to 1, so when an orchestra is going full blast, the SPL it produces is 1,000,000,000 times greater than its softest passages. Rock orchestras, particularly when using instrument amplifiers, can reach this dynamic range. The microphone must be capable of producing an electrical signal output which corresponds faithfully to this tremendous range of sound pressures.

REFERENCE LEVELS

A measurement is always a comparison, or a ratio. If you make a measurement with a ruler, the comparison is the distance between the left edge of the ruler and the point at which you stop on the ruler scale. The left edge of the ruler is a *reference*. For example, the location of your house is always with reference to some corner or other convenient spot.

A *reference* can be any number or starting point you wish. It can be 0, 5, 10, or 1,000,000. The left edge of a ruler is *zero*, and while it is a convenient reference, you could use 1 inch, 3 inches, 6 inches, or any other number on the ruler as your *starting point or reference*. When you walk a mile, the point at which you start is your reference, even though it may be a hundred miles from some other point.

Microphone output voltage is usually referenced to 1 volt and power output to 1 milliwatt. Note that a reference can be larger than the voltage or power supplied by a microphone. Thus, a microphone can have an output voltage which is a small fraction of 1 volt. Both voltage and power output of microphones are most often expressed in decibel ratios.

When the measured microphone output voltage is equal to the reference, either 1 volt or 1 milliwatt, we have a one-to-one ratio. Written as 1:1, it is designated as 0 dB. However, since low-impedance microphones generate a weaker signal than the reference, decibel output ratings become negative numbers, such as −50 dB, −55 dB, or −60 dB etc. The minus sign as used here does *not* indicate subtraction. Instead, it is just an indication of how much weaker the output signal of the microphone is compared to the reference voltage or power.

Using SPL is a convenient way in which to compare the sensitivities of microphones. The microphone with the largest negative number is lowest in output. A microphone rated at −60 dB has less output signal than one rated at −50 dB. Conversely, a −50 dB microphone generates more output voltage than one that is specified at −60 dB.

THE DYNE

The microbar and SPL are two different ways of describing the same thing—the pressure exerted on a microphone diaphragm. There is still another, known as the *dyne*, which is also a unit of force. The *dyne* is that force which will produce a velocity of 1 centimeter per second when acting on a mass of 1 gram. In other words, if you had an object that had a mass of 1 gram and if you were to push it so it moved 1 centimeter per second, you would be exerting a force of 1 dyne.

The dyne per square centimeter, written as dyne/cm^2 is equivalent to 1 microbar or 74 dB SPL—10 dynes/cm^2 is equivalent to 10 microbars or 94 dB SPL.

The output voltage of a microphone, then, can be given in millivolts or negative decibels less than a 0 dB reference (abbreviated re) of 1 volt or 1 milliwatt. The microbar is equivalent to the dyne/cm^2. Thus, a manufacturer might supply the sensitivity of a microphone as:

$$-74 \text{ dB re } 1\text{V}/\mu\text{bar}$$

Translated, this means that the microphone has an output of 74 dB SPL. The reference is 1 volt per microbar.

Another microphone might be specified as:

$$-56 \text{ dB re } 1 \text{ mW}/10 \text{ dynes/cm}^2$$

The output of this microphone is -56 dB, equivalent to 10 dynes/cm^2 with a reference of 1 milliwatt. The output is 1 milliwatt per 10 microbars.

The disadvantage of having a number of different ways to indicate microphone sensitivity is that it becomes extremely difficult to make comparisons among microphones. As an example, consider this listing:

(1) -74 dB re 1 V/μbar
(2) -58 dB re 1 mW/10 μbar
(3) -58 dB re 1 mW/10 dynes/cm^2

where

Item 1 is -74 dB referenced to 1 volt per microbar,
Item 2 is -58 dB referenced to 1 milliwatt per 10 microbars, and
Item 3 is -58 dB referenced to 1 milliwatt per 10 dynes per centimeter squared.

But all of these microphones have an identical output. The use of a higher reference level makes one microphone appear to have more output than another, as in a comparison of item 1 with 2 or 3. An examination of these negative decibel numbers will make one microphone appear more sensitive than another when, as a matter of fact, the voltage outputs of the microphones are just about the same.

Further, the microphone output level is meaningless unless the reference level is given along with the sound pressure loudness, the impedance, and the frequency response. To do otherwise would be like comparing the engine performance of two cars on the basis of gas-mileage only.

THE PASCAL

The pascal is a unit used in measuring the open circuit sensitivity of a microphone, generally at 1 kHz. Open circuit sensitivity is in mV/Pa (pascal) or in dBV. One Pa (pascal) is equal to 10 microbars or 10 dynes/cm^2 and is approximately equal to 94 dB SPL. The microbar, of course, is a measurement of sound pressure and is one millionth of normal atmospheric pressure. The pascal is also equal to 1 Newton per square meter which is equivalent to 10 microbars.

QUALITY OF A MICROPHONE

The quality of a microphone isn't normally related to either high or low output voltage. If a microphone has a larger magnet it will generate more voltage, but that is only part of the whole story. Is the output linear? Undistorted? Usually, however, a more efficient use of materials does produce a higher output regardless of comparable element size.

High-output microphones are desirable for reasons of "reach" and a good signal-level-to-noise ratio. If the microphone output is too high for close-in use or loud signal pickup with highly sensitive amplifier inputs, the problem can be solved by using voltage reducing pads. High microphone output levels can be accommodated for optimum results. It cannot be restored when it isn't there at the start. In other words if it has high signal output, you can reduce that output if you need to do so.

DISTORTION

There are no commonly accepted standards for the measurement of microphone distortion and so this characteristic generally is not covered in a spec sheet. Probably the most serious aspect of distortion is poor transient response. Ideally, a mic diaphragm should respond only when a sound pressure is applied with movement ceasing instantaneously upon removal of that pressure. In short, the diaphragm should be inertialess.

The transient response of a mic is the time it requires for the output of a mic to reach 90% of its peak with the application of a dc sound pulse. The rise time for a condenser mic is about 15 microseconds but a dynamic mic takes about three times as long. Velocity mics are approximately 25 microseconds.

MICROPHONE OVERLOAD

Commonly, *distortion* in a sound reinforcement system and during recording is blamed on the microphone. Someone talking directly into a mic at practically lip level can produce up to 100 dB sound pressure level. Loudness impacts on the microphone diaphragm located just a few inches from a screaming vocalist or a loud or amplified musical instrument may reach an average sound pressure level of 120 dB, possibly rising to 130 dB on sound peaks. A loudness of 130 dB is the *threshold of pain*. It is the point at which you begin to *feel* sound—in the form of pain—in addition to *hearing* it.

A good-quality microphone will not overload at these pressures. The distortion of a well constructed microphone can be held to 0.5 percent, or even less, with sound pressures of up to 130 dB. Some professional microphones can be used in the mouth of a trumpet, which can produce a loudness of up to 146 dB. Even under such extreme working conditions, the output of the microphone remains clean, that is, *undistorted*. However, such high sound pressure levels result in extraordinarily high signal output levels from the microphone. This high signal output voltage, fed into a following tube or transistor amplifier, can overload the amplifier, driving it into distortion levels.

An amplifier or recorder input sensitivity rating describes the absolute, maximum amount of voltage or power it can handle before it overloads and distorts the signal. A comparison of various input sensitivities reveals a surprising variation. Some amplifiers may have a −60 dB input signal rating meaning it will handle a signal of this level before overloading. Others can tolerate up to −22 dB output from the microphone without distorting. Unfortunately, there is no standard, and there cannot be, since microphones are never used at the same distance. Also, sound-pressure sources are of infinite variety and loudness.

As an example, consider a sound pressure level of 94 dB applied to the diaphragm of a microphone, such as the AKG Model D-190. This will now produce a catalog output rating of −53 dB. But if the sound pressure is raised by 36 dB, the microphone output will be the original output of −53 dB plus 36 dB more. The output is now −17 dB. Smaller negative numbers mean higher output, getting closer to the 1 volt or 1 milliwatt reference point, or 0 dB.

We now have an output signal of −17 dB. Assume, now, that our amplifier has a −60 dB input rating. This means our microphone output is 43 dB higher than this amplifier input rating. Even if we were to use an amplifier that can handle −22 dB without distorting, the output from the microphone is still 5 dB above that figure.

What will happen depends, in part, on whether the amplifier receiving the signal from the microphone is a tube amplifier or a solid-state unit using transistors. Tube amplifiers will give a small warning in slightly increased, sometimes tolerable, distortion when the input overload point is reached. If the amplifier uses transistors it will sound clean to the overload point and will then go into sudden, often violent, distortion.

Many professional recording consoles supply input attenuation with selectable degrees of voltage-reducing pads or electronically variable input-stage gain sensitivity. Some portable, high-fidelity recorders have a provision for modular, immediately replaceable microphone input amplifiers of different sensitivities. In-line pads are also available for use between the microphone cable and amplifier. You can use such pads when you know in advance that the input sound levels are going to be extraordinarily high, as in the case of a rock concert. With rock music, microphone working distances are often very small. When recording classical music, sound pressure level decreases, since the microphone working distance is usually much greater.

You can buy signal attenuation pads, for use between the microphone cable and amplifier, that supply a typical 10 dB or 20 dB voltage reduction from the microphone. The pad can be mounted in or near the amplifier and switched out when not needed. It is much better to use such an attenuation pad rather than selecting a low output microphone to compensate for close-use, high-loudness applications. The disadvantage of choosing a low output microphone is that you lose the advantage of better signal-level-to-noise ratios in normal use, and you could end up with a poorer frequency response. It is always a safe procedure—a *good* procedure—to buy the best microphone possible for all applications. You can always control end-system response and microphone output voltage with various accessories. Reducing the microphone input level to the amplifier is easily done, and solves the problem of "microphone overload."

MICROPHONE SPECIFICATIONS

Microphones are often bought on the basis of frequency response and polar pattern only, but this isn't enough. While a microphone is a separate component, it is a part of a complete electronic amplification system, and this system can never be better than its weakest link. When reading microphone specs, look for information over and beyond frequency-response and directivity characteristics. Microphone specs usually list microphone output level and loading characteristics. You must have this data to make certain that the microphone and its following amplifier will work well together as a team.

There are three basic requirements which should appear in the microphone spec sheet. These are:

☐ The *output* of the microphone, in either a voltage or power ratio. Often the output is in the form of decibels (and that is always a ratio or comparison with some reference level), such as −55 dB.

☐ The *internal impedance* of the microphone, supplied in ohms. The actual value isn't critical, and is sometimes indicated as either low impedance or high impedance.

☐ The amount of *sound pressure* applied. As indicated earlier, this can be supplied in several different ways, such as 1 mW/10 microbars. This corresponds to 94 dB sound pressure level (SPL).

European manufacturers often specify microphone output voltage in terms of millivolts per microbar when the microphone is unloaded—when the microphone is not connected to the following amplifier input. American manufacturers specify output voltage in maximum power output, as expressed in dBM when the microphone is loaded with its characteristic impedance—when the microphone works into an amplifier whose input impedance is similar to the impedance of the microphone. (DBm is simply the decibel referred to 1 milliwatt). The sound pressure applied is often expressed as referenced to 1 mW/10 dynes/cm^2. This is equal to 1 mW/10 microbars and this, in turn, is equal to 94 dB SPL.

When a microphone is to be used with a very short working distance, you should know the sound pressure level which will produce a specified level of distortion. Above this point the microphone diaphragm will produce excessive distortion. Distortion above 1.0 percent is unacceptable and the sound pressure levels reaching the microphone diaphragm must be kept below this point. For condenser microphones, the self-noise level of the mic and its dynamic range are important specifications.

MICROPHONE SPECS AS GUIDELINES

Microphone specifications are useless unless they can be interpreted into meaningful guidelines. Most low-impedance microphones are in the 150-ohm impedance classification, with an actual impedance ranging from 100 to 300 ohms. Such microphones are specified with 94 dB SPL applied. When the microphone output voltage is specified as −50 dBm output, that microphone is 10 dB more sensitive than one rated at −60 dBm. Since conversational speech at a 1 foot distance from the microphone is about 65 dB SPL, the 94 dB SPL used for a microphone rating is typical of a pop vocalist or instrumentalist. Some vocalists, however, practically put the microphone inside their mouth and extremely high sound levels are produced. As an example, a professional condenser microphone with a specified output of −40 dBm produced a measured output of +5 dBm from a vocalist at mouth-touching range. Such a high signal output practically demands the inclusion of a line-level input control to reduce the signal voltage below the overloading point of the following amplifier.

It is important to recognize when excessive sound pressure level produces distortion. In 99 percent of the applications, the input amplifier connected to the microphone will overload long before the microphone diaphragm reaches its limits.

IMPEDANCE

The impedance of a microphone is the combination of resistance, capacitance, and inductance values. Of these, only the resistance value within the relevant frequency range determines the microphone's basic impedance. Impedance, like resistance, is opposition to current flow. Either resistance or impedance can be compared to friction. A certain amount is not only desirable, but essential. Extremes, such as close to zero friction or unusually high friction, are undesirable. If the friction of a road is almost zero, there is no way in which a car can move forward, a condition that sometimes arises on a highly iced road. If the road friction is unusually high, such as a road that has many layers of wet clinging mud, excessive energy is used in moving the car forward. While resistance and impedance are both measured in ohms, impedance includes the electrical opposition not only of resistors, but other components, such as coils and capacitors.

Microphones are alternating-current generators and, as such, have both internal resistance and electrical impedance. Their source impedance is usually measured at 1 kHz (1000 Hz). And to emphasize the fact that the microphone is really an ac generator, you could connect a very sensitive, very-low wattage, electric light across it and this light would flicker *on* and *off* as it received sound from a voice or instrument.

All microphones have a certain amount of impedance. Impedance in electronic components, though, can be a frequency-sensitive function, and to say a component has an output impedance of 200 ohms simply means that this is its impedance at a particular frequency.

Low-impedance, moving-coil dynamic microphones have typical impedance values in the range of 50 to 250 ohms, more usually 150 to 200 ohms. Microphones with a 200 ohm impedance, such as AKG dynamics, require more skill in design and manufacturing but are more efficient than those having lower impedances. An additional benefit is that the impedance is less affected by changes in frequency.

Since all moving-coil dynamic microphones are inherently low-impedance types, a transformer is needed for conversion to high impedances, such as 25,000 ohms and higher. Some dynamic microphones have built-in transformers to supply this required impedance transformation. Such mics often have a provision for bypassing the transformer for alternate connection directly to the moving coil inside the microphone. Such mics, then, are low/high impedance types, depending on the setting of the switch that controls inclusion or exclusion of the transformer.

If the microphone does not have a built-in transformer, in-line accessory transformers are available and microphone cables with a transformer at the amplifier input enables a quick impedance change by substituting a low-impedance cable.

The impedance of a mic is that measured across its signal output terminals. Sometimes this impedance is referred to as the impedance that is "seen" when "looking back" into those terminals. In terms of impedance, mics are usually referred to as either low or high impedance types without an indication as to the specific value of impedance. However, as indicated above, low impedance usually covers a range of 50 to 250 ohms while high impedance indicates some value between 25,000 ohms and one or more megohms. Low impedance mics are preferable when long output cables are to be used without excessive loss of signals in the treble range. High impedance mics are desirable if cable lengths are to be fairly short. This is to take advantage of their open circuit output voltage which is sometimes as much as ten times that of low impedance types.

It is particularly important to connect each piece of equipment to its recommended load impedance. Recommended load impedance, as its name implies, is that value which the device should "see" when looking forward into the circuit it is driving. Usually a minimum recommended load will be specified, meaning that the unit will operate properly as long as it sees a load of any impedance above this limit.

INTERNAL IMPEDANCE

The internal impedance of an active device is not necessarily related to its load. It is that value which we would "see" looking back into the output terminals of the device. Internal impedance tells us, among other things, how much the gain will vary with changes in the external load. As stated earlier, active line level devices do not operate properly below a minimum load impedance, a value often above the device's internal impedance.

Impedance Matching

The output of the microphone is connected to the input of a preamplifier. Like the microphone, the preamp has a certain amount of impedance and, since the signal produced by the microphone is fed to the input of the preamp, what we are talking about is its *input impedance*. It is sometimes thought that the input impedance of the preamp should be identical with that of the microphone. In practice this never happens. You might have identical impedances at one particular sound frequency, but not at others.

There are certain conditions under which impedance matching is desirable, as in the case of power amplifiers and speakers. Here we want to transfer the maximum amount of audio energy, in terms of watts, from the power amp to the speakers, and so impedance matching is significant. Microphones, though, are voltage delivering devices. Since microphones are matched for optimum voltage transfer, the loading of any microphone should be as little as possible. This means that the impedance connected to the microphone should be as high as possible, or at least three times the microphone impedance. Thus, for a low impedance microphone rated at 200 ohms the load impedance should be a minimum of 600 ohms.

Impedances are sometimes simply indicated as low or high, so connect low impedance microphones to low impedance inputs; high impedance microphones to high impedance inputs.

If you divide all microphones, regardless of type, into two categories, *low* or *high* impedance, then you simplify your microphone buying decision. Further, any number of tape recorders or preamplifiers give you your choice. They supply a high-impedance microphone input, a low-impedance microphone input, or both.

Selecting the proper microphone impedance versus the input impedance of a mixer, amplifier or recorder is done to maximize the microphone output signal, to preserve the full frequency response and to minimize pickup of unwanted signals. In general, for optimum performance, the actual equipment input impedance should be five to ten times that of the microphone.

Microphone impedance is specified as a rated number followed by the actual impedance in ohms. Common ratings are 150 ohms (but the actual impedance may be from 75 to 300 ohms); 600 ohms (actual impedance may be from 300 to 1,200 ohms); 2,400 ohms (actual impedance from 1,200 to 4,800 ohms) and high impedance (actual impedance greater than 10,000 ohms).

Constancy of Impedance

Components such as capacitors and coils are frequency sensitive—their opposition to the flow of current changes with frequency. This characteristic is not true of resistors, a component whose opposition to current flow is not dependent on frequency. The output impedance of the most widely used quality microphones for high-fidelity sound, and that includes dynamic and condenser mics (and ribbon mics as well) is almost purely resistive. This means that the output impedance of these microphones remains fairly constant over the entire audio frequency range.

Professional recordists generally select low-impedance dynamics, since they can then have the option of using long cables without being particularly concerned about picking up noise or hum. If your amplifier or tape deck has a high-impedance input only, you can get a matching transformer, known as a *line matching transformer*, that will let you couple a low-impedance microphone to the high-impedance input of the deck. But most tape decks are now solid-state and they do not present microphone connection problems.

Line Matching Transformer

Just because the input to a device such as a recorder or amplifier is high impedance does not rule out the possibility of using a low-impedance mic whose output feeds into a low-impedance cable. This is handled by a line matching transformer, physically located at the component receiving the signal from the cable. By using such a transformer all the benefits of a low-impedance connection are obtained despite the high input impedance requirement of the amplifier or recorder. The line matching transformer is a step-up type with its low impedance primary winding connected to the equally low impedance mic cable; its high impedance secondary winding connected to the equipment.

In effect, the line matching transformer converts any low impedance balanced (3 wire) output to high impedance, unbalanced (2 wire) output for direct plug-in into amplifiers or recorders not having low impedance inputs. The advantage of such an arrangement is that it avoids the losses typical of high impedance mics. Effective shielding of the transformer eliminates hum problems.

Low or High Impedance?

Basically, any microphone can be manufactured as *high* or *low* impedance. Low-impedance connections are more desirable and your choice should be a low-impedance microphone connected to the low-impedance input of your recorder or amplifier. The advantage of low impedance is that it means less susceptibility to hum and electrical noise pickup by the connecting cable run over long distances and with no audible loss of high-frequency response over long cable distances. You can use an XLR-type audio connector, which permits adding cable lengths up to 1000 feet without interference, hum pickup, or loss of high frequencies.

Since a low-impedance microphone seems to have so many advantages, why bother with high-impedance mics? A high-impedance microphone will produce a larger output signal than a low-impedance unit for a given sound source. This higher output enables equipment designers to cut costs in amplifier design by eliminating transformers and preamplifier stages. In other words, if a manufacturer sells a combination microphone and preamp, he is more likely to use a high-impedance microphone, since this permits him to cut down on preamp circuitry and reduce manufacturing costs. In the case of "budget type" amplifiers and recorders, a high-impedance microphone may be required in order to be able to supply enough input signal voltage.

The useful distance between the microphone amplifier and the high-impedance microphone is about 20 feet. A greater length of cable will cause a loss of high-frequency response due to cable capacitance. Further, as cable length is increased there is greater susceptibility to hum and radio-frequency interference pickup. Finally, the phone-plug connectors which are used with high-impedance microphone cable don't always provide a positive connection. This results in a static-like noise and can possibly produce intermittent operation.

However, if the connecting cable to a high-impedance microphone is kept to a reasonable length, 25 feet or less, the problems associated with long cable distances aren't ordinarily bothersome. For many practical installations, cable runs of less than 25 feet are common and acceptable. This length is sufficient for use in home recording, conference recording, group-sound reinforcement systems, television news reporting, and many similar applications.

For the amateur recordist, the choice of which impedance microphone to buy is usually dictated by existing equipment. If the equipment has a low Z (low impedance) input, get a low-impedance microphone. High impedance input? Get a high Z mic. If it has both, get a low-impedance microphone.

When we get into the area of professional recording, long microphone cable runs become necessary. Professional applications require long microphone cable runs in studios, auditoriums, and in any large-area reinforcement systems applications.

Microphone impedance has no relationship to microphone quality or price. A high-impedance microphone could be cheaper or more expensive than a low-impedance mic. The shape, weight, and durability of the microphone are also completely unrelated to its impedance. Complaints of poor results with microphones are almost always due to improper connections or wrong impedance matching despite the fact that the microphone user has a choice only of low or high impedance.

Sometimes the use of a high-impedance mic will result in noise and hum which may be difficult to locate and eliminate. These problems may be due to nothing more than the inadvertent grounding of the mic housing or its stand. These should not be allowed to touch other metallic items which are grounded, such as the surface of conduit or pipes, or metal sheeting, or the metal housing of other components.

The recommended load impedance for low impedance microphones is usually 50 or 200 ohms, to face amplifier input impedances of 500 ohms or higher for minimum degradation over the full frequency spectrum.

For high impedance microphones it is usually 50,000 ohms (50 k ohms) to face input impedances of 150,000 ohms (150 k ohms) or higher for best results and no loss of sensitivity.

Old amplifiers with high impedance inputs sometimes require a higher voltage level from the microphone. If this is the case, a transformer has to be used between a low impedance microphone and such an amplifier to convert the lower voltage level from the microphone to an acceptable level for the amplifier input stage. Such transformers are available from any hi-fi dealer combined with a suitable length of connecting cable.

CABLES

Cable is the word used to describe the wire (or wires) used to lead the electrical signal produced by the microphone to some device that will work on the signal, such as an amplifier. Cables can be high impedance or low. A high-impedance cable, also called *unbalanced line* (Fig. 7-12), consists of a wire (conductor) that passes lengthwise through the center of the cable. This wire is surrounded by some type of insulating material with a flexible metal braid put on the outside of the insulating material. Since the braid is made of metal, it can be used as a conductor, and in a technical sense, unbalanced line really has a double conductor. The braid also works as a shield to keep interfering voltages, such as hum or other electrical noise, from reaching the central conductor, sometimes called the "hot" lead. The *braid*, also referred to as *shield braid*, is the "cold" or ground lead. The shield braid, however, isn't always effective in preventing signal interference.

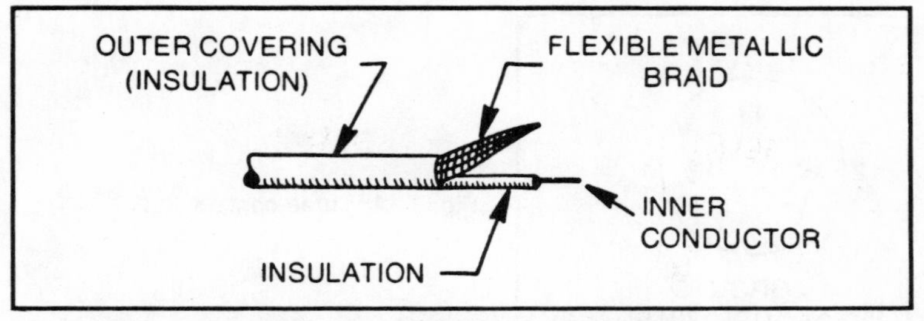

Fig. 7-12. An unbalanced (high impedance) microphone cable.

BALANCED INPUTS

A balanced line (Fig. 7-13) consists of two inner conductors instead of one. Both conductors, running the length of the cable, are fully insulated from each other. A flexible metal shield also surrounds the insulation, but in this case the shield, made of flexible metallic braid, is not part of the signal path and is used as a means of keeping interfering signals out.

In a balanced line, the only purpose of the shield is precisely that—as a shield to keep the magnetic fields of other signals, such as those generated by noise voltages, from reaching the two signal carrying conductors. This shield is connected to pin 1 of the connector. The other two pins, pin 2 and pin 3, are connected to the pair of signal carrying wires. Generally, the wiring arrangement is such that a positive pressure on the diaphragm of the microphone results in a positive voltage on pin 2.

CONNECTORS

Connectors can be 3 or 5 contact types according to IEC 268-14B (International Electric Standard). Three pin XLRs are often used for low-impedance balanced or unbalanced lines. The connections for the three-contact XLR are shown in Fig. 7-14. The drawing in Fig. 7-15 illustrates male and female XLR connectors. In addition to providing good electrical contact, they also supply a good grip on the cable and are intended for heavy

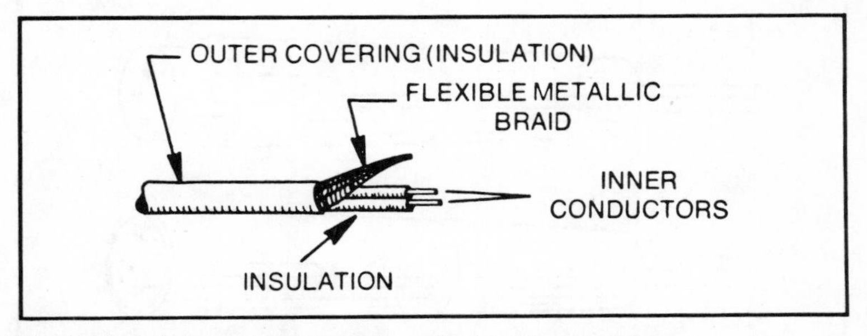

Fig. 7-13. A balanced (low impedance) microphone cable.

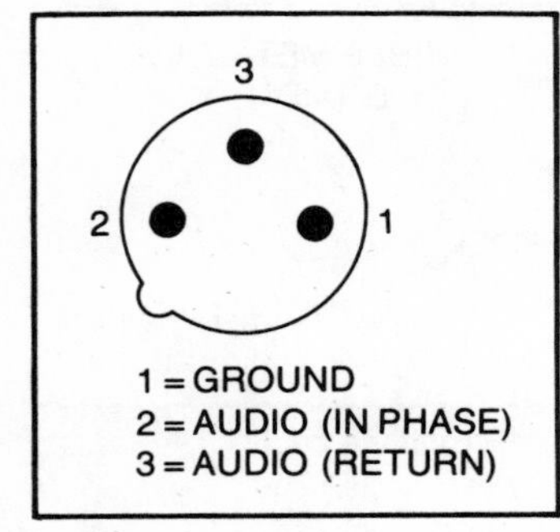

Fig. 7-14. Three-contact XLR.

duty use. Five-contact XLRS, shown in Fig. 7-16 are used mainly for stereo microphones.

The three contact DIN (Deutsche Industrie Norm or German Industry Standards) and the five contact DIN are also used. The connections for these are shown in Fig. 7-17. There is also a twelve contact DIN.

Figure 7-18 shows the phone plugs used for mono or stereo. Phone plugs are also used for microphone connections with 1/4-inch plugs used for high-impedance microphones. There are also two- and three-conductor phone plugs. The two-conductor plug is for unbalanced circuits. The three-conductor plug is used for balanced or stereo unbalanced circuits. Still another type of plug is the RCA for use with low level arrangements (Fig. 7-19).

PLUGS AND JACKS

The words plugs and jacks are sometimes used interchangeably although they are not synonymous. A plug is a male connector and is characterized by having pins. A jack is a female connector and is recessed to accommodate or accept the pins of the plug.

UNBALANCED INPUTS

A low-to-high impedance transformer in the amplifier, is required for

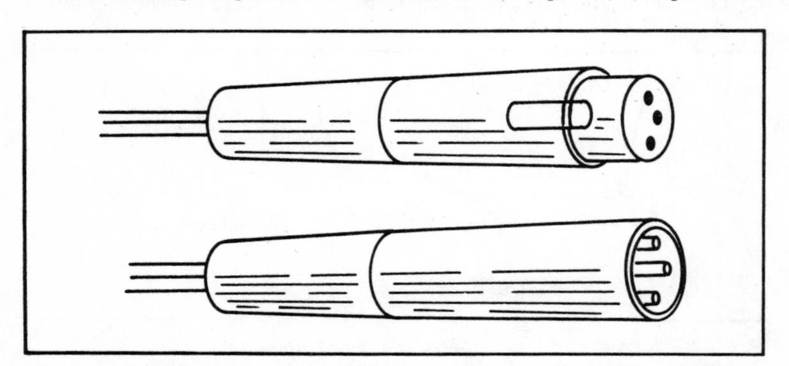

Fig. 7-15. XLR connectors. Upper drawing (female); lower drawing (male).

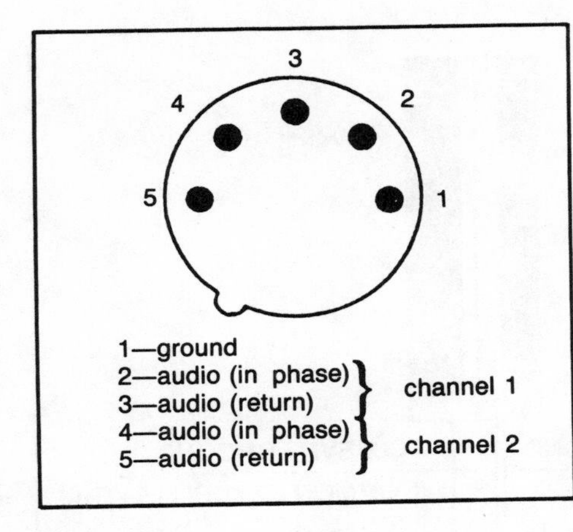

Fig. 7-16. Five-contact XLR connector.

voltage gain and balanced signal isolation of ground circuits to minimize hum and radio-interference pickup. Many tape recorders and some public-address amplifiers provide unbalanced, low-impedance, phone-plug input connections. Such circuitry may be satisfactory for home recording and use in interference-free areas, but balanced, transformer-coupled input circuits are preferable for trouble-free service.

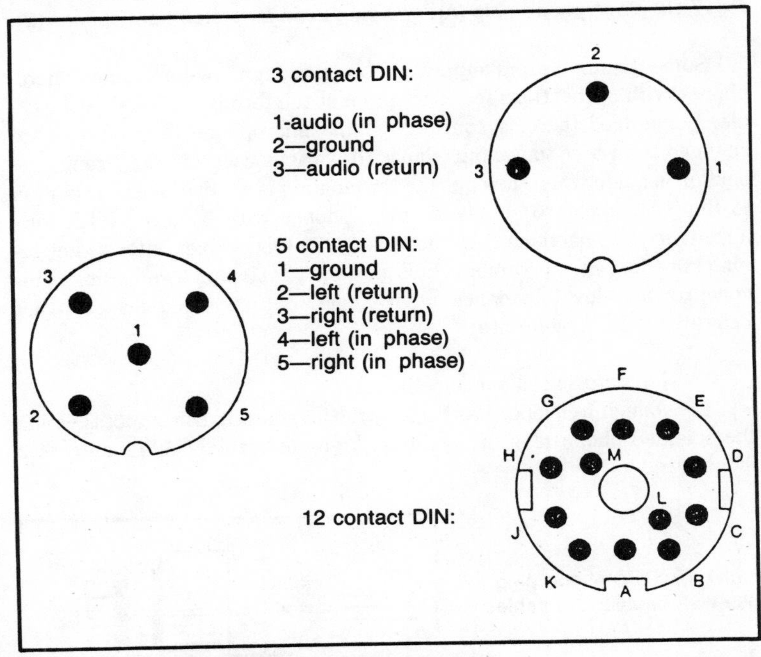

Fig. 7-17. 3, 5, and 12 contact DIN connectors.

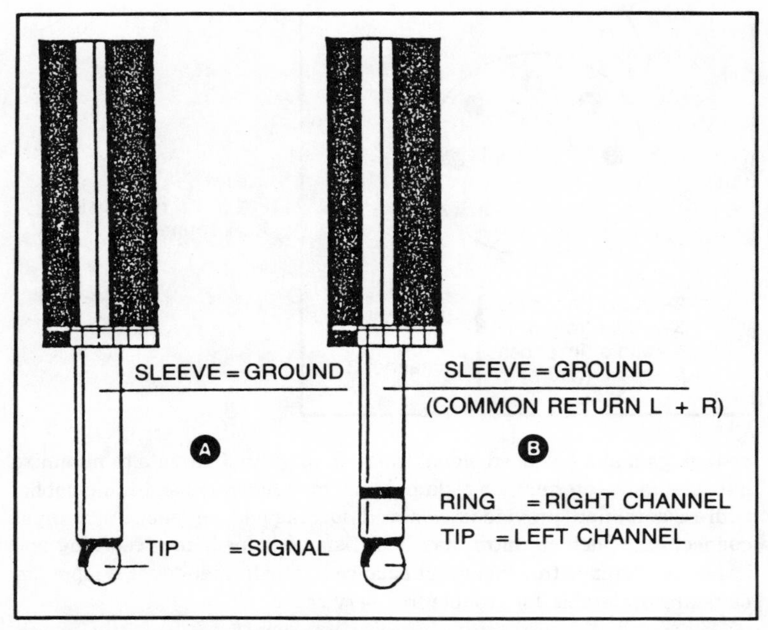

Fig. 7-18. Mono phone plug (A); and plug used for stereo headphones (B).

Some manufacturers supply combination high/low-impedance microphones with connections for user option at the termination end of the cable. Sometimes there is confusion if the microphone terminations are changed from *high* to *low* impedance for various uses and the wrong connection is made when the opposite termination is needed. AKG's solution to this problem is to supply all microphones with balanced XLR low-impedance connections. An accessory 20-foot cable with a low-to-high impedance transformer is built into the phone plug termination. This converts any low-impedance microphone to high-impedance. Further benefits of this system are:

☐ No confusion about impedance

☐ No high-frequency loss because the high-impedance output is within the shielded phone plug, a maximum high-impedance cable length of 2 inches, and

Fig. 7-19. RCA phono plug for use with unbalanced cables.

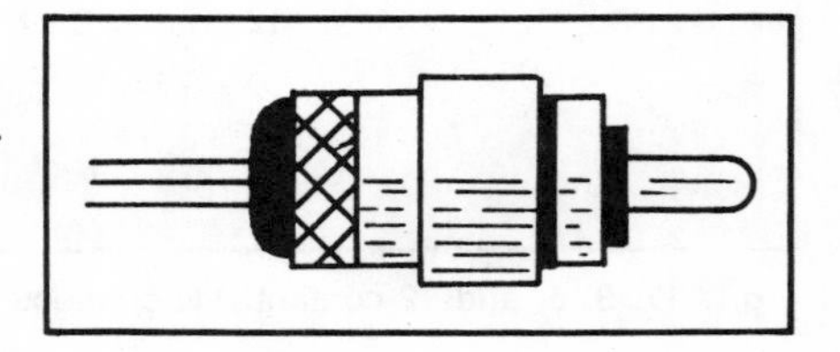

□ Elimination of hum and intermittent connections due to the loss of shielding in a high-impedance cable.

Most consumer tape recorders feature unbalanced low-impedance microphone inputs specified at 150 to 1000 ohms and higher. Any low-impedance mic will work properly with these machines and, as long as the impedance of the recorder input is higher than that of the microphone, exact termination matching is not important. When connecting a three-wire low-impedance microphone to an unbalanced input connector, fasten the wire connected to pin 2 of the three-pin connector at the other end of the cable to the signal or "hot" pin of the unbalanced connector. Connect the remaining two wires to the unbalanced connector shield or ground point. Connections between the second audio wire and the cable shield are necessary for proper operation. After connecting your microphone, should you find the volume weak and the sound quality "tinny," check and make certain that the second audio wire is connected with the cable shield at the ground point on the connector. Cable capacitance will provide the tinny sound if the signal *minus* and *ground* connections aren't fastened together properly.

BALANCED VS UNBALANCED

When interconnecting electronic components, it is essential to consider whether their inputs and outputs are balanced or unbalanced with respect to ground. As indicated earlier, a balanced system is usually preferred particularly in more complex and sophisticated systems due primarily to its superior freedom from interference and avoidance of ground loops. This is achieved by separating the signal and its return from the ground and shielding both paths. Figure 7-20 shows typical input and output variations.

Unbalanced systems utilize the ground as a signal return and require only two terminals. These are variously referred to as "hot" and "ground;" "high" and "low;" or "+" and "−."

Balanced connections require three terminals, with ground being separate from "high" and "low." In preparing balanced cables, it is a sensible precaution to make an ohmmeter check to make sure that neither signal lead has been inadvertently shorted to the cable shield or the shell of either connector. During this test, flex the cable to be certain that an intermittent short doesn't exist.

The positive and negative terminology often used on balanced units refers to the polarity of a signal at any instant, allowing the user to tell if a signal is being reversed in phase in the equipment. This becomes important in multiple channel systems where the same phasing between reproducers must be maintained.

CABLE QUALITY

Like any other component, the word "quality" does not apply equally

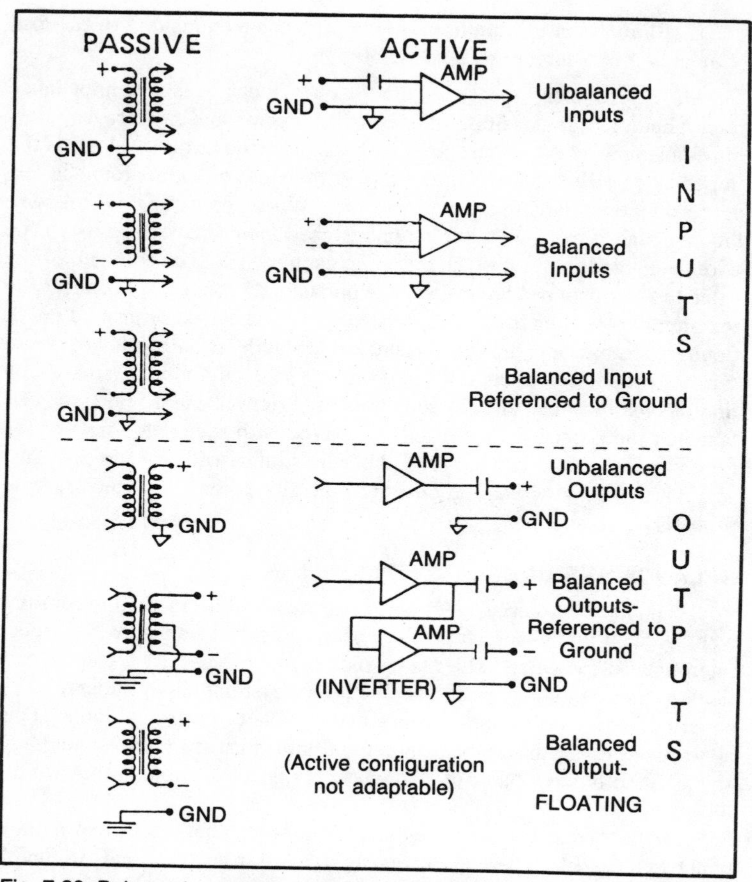

Fig. 7-20. Balanced vs unbalanced connections (drawing and accompanying text courtesy Electro-Voice, the PA Bible).

well to all microphone cables. Consider the shield braid of the cable for example. The best type of braid is a fine mesh and is highly flexible. This is advantageous when cables must be frequently connected and disconnected, or moved. Less desirable are shields made of thin, metallized foil, or a braid in which the mesh is widely spaced. In some cases, the so-called shield consists of nothing more than a spirally wrapped wire. The poorer type shields result in greater susceptibility to hum and electrical noise signal pickup. Some cables come equipped with a "drain" wire making intimate contact with the braid and used for making a good connection to it.

Since cables have an outer covering, it generally isn't possible to examine the cable to check on braid quality. The best alternative is to read the manufacturer's spec sheet for a description of the braid.

Figure 7-21 shows the polar pattern, circuit diagram and frequency response of a two-way, cardioid, dynamic mic. The unit has two coaxially-mounted, dynamic transducers: one designed for optimum performance at high frequencies and placed closest to the front grille; the other designed for best performance at low frequencies and placed behind the first. Each transducer includes a hum-bucking compensating winding to cancel the effects of stray magnetic fields. Both transducers are coupled to a 500-Hz inductive-capacitive crossover network that is electroacoustically phase corrected. This is essentially the same design technique used in a modern two-way loudspeaker system, but applied in reverse.

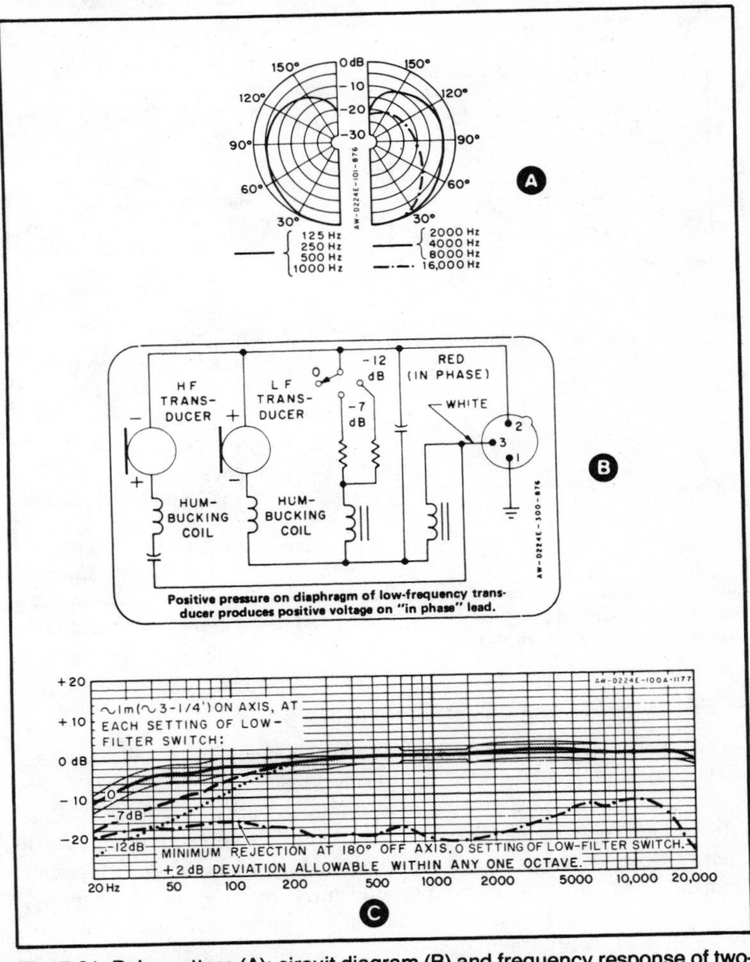

Fig. 7-21. Polar pattern (A); circuit diagram (B) and frequency response of two-way, cardioid, dynamic mic (C).

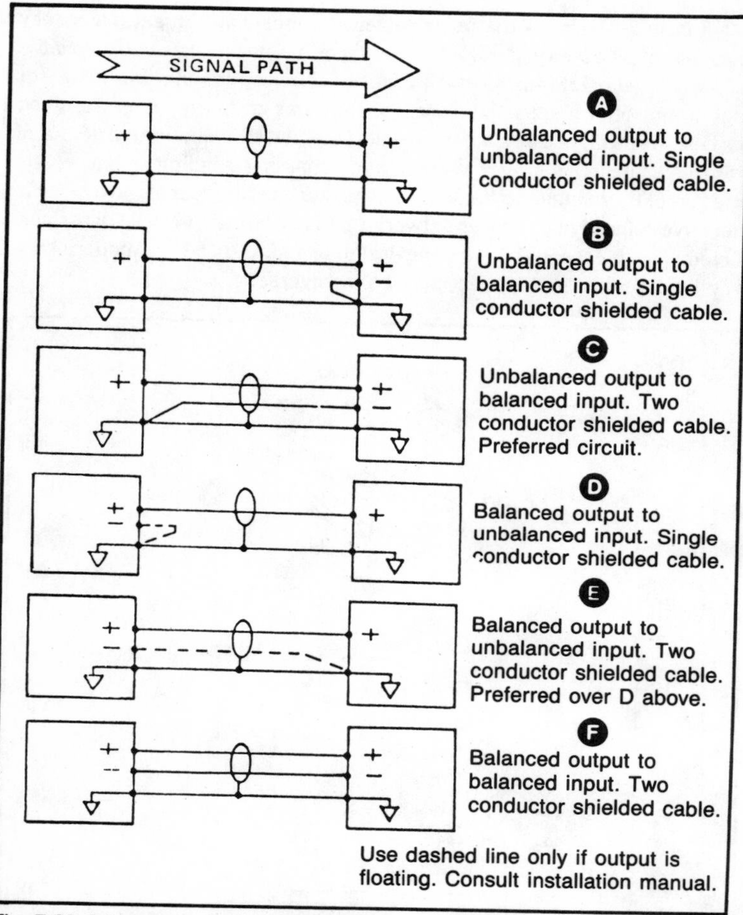

Fig. 7-22. Interconnecting circuits (drawing and accompanying text courtesy Electro-Voice, the PA Bible).

INTERCONNECTING EQUIPMENT

As indicated earlier, some equipment is designed to allow either balanced or unbalanced operation. In other instances, transformers may be added as an outboard module or as a plug-in option. Figure 7-22 illustrates some typical connections. If hum is introduced using circuit B (as when two pieces of equipment are separated by a long distance), circuit C can be used to avoid noise from ground currents in the shield. Similarly, circuit E may be used to advantage but only if the output is not referenced to ground (for example, an ungrounded transformer secondary). Note that in circuits C and E, the "low" signal wire and the shield braid must be tied together only at one end, as shown. Otherwise, ground and signal currents will be combined in the signal path return, defeating the advantage of these circuits.

Circuit F, showing a completely balanced input and output, provides the best interconnection and should always be selected, if compatible with the equipment involved.

LOCATION OF LINE LEVEL EQUIPMENT

It is desirable to place all line level equipment at one physical location, such as in one or more adjacent equipment racks, thus minimizing connecting cable lengths and providing good grounding. This is helpful in minimizing electrical interference.

LINE LEVEL INTERCONNECTIONS

One or two conductor (as required) shielded cable should normally be used for all line level interconnections. However, because the voltage level is typically 40 dB greater than in a low-Z microphone circuit, noise pickup is somewhat less troublesome. In fact, unshielded terminal strips can be used as connectors, particularly when fed from a low internal impedance device. The presence of large grounded areas of metal in the vicinity, such as the chassis, adjacent equipment, equipment racks, etc., also help to minimize any electrostatic interference pickup by exposed signal terminals.

TWO MICROPHONES—ONE INPUT

In some instances, it may be necessary to connect two microphones to one input. In such a case, wire the microphones in series so that the internal impedance of one microphone will not load the other microphone, causing a loss of quality and sound level.

THE SOUND REINFORCEMENT SYSTEM

The simplest sound reinforcement system would consist of a single microphone with its output connected to a preamplifier, followed by a power amplifier and a single speaker. But while it has the advantage of low cost, it is a limited setup.

A more elaborate arrangement would consist of an audio console containing a preamplifier/mixer arrangement which would accept the outputs of a number of microphones. A unit of this kind would have individual gain controls for each microphone, plus fixed line attenuation pads of −10 dB and −20 dB. It would also contain an equalizer to compensate for recording-room acoustics.

Following the audio console would be one or more stereo power amplifiers connected to pairs of speakers. The speakers could have built-in crossover networks designed to separate the audio frequencies into two or more channels for distribution to bass speakers (woofers), midrange speakers (formerly known as squawkers), and treble speakers (tweeters). In some setups, an electronic crossover is connected between the preamplifier and its following power amplifier.

FRONT-TO-BACK DISCRIMINATION

Front-to-back discrimination indicates, in decibels, how much unwanted sound or noise is being rejected to the rear, over the frequency range of the microphone. The frequency response of a microphone may sometimes be supplied in *double graph* form. The frequency response of the microphone is plotted with the sound "head on," that is, at 0° on the polar pattern. Another graph is plotted with the sound directed at the microphone "tail," the 180° point.

Chapter 8 How to Use Microphones: General Applications

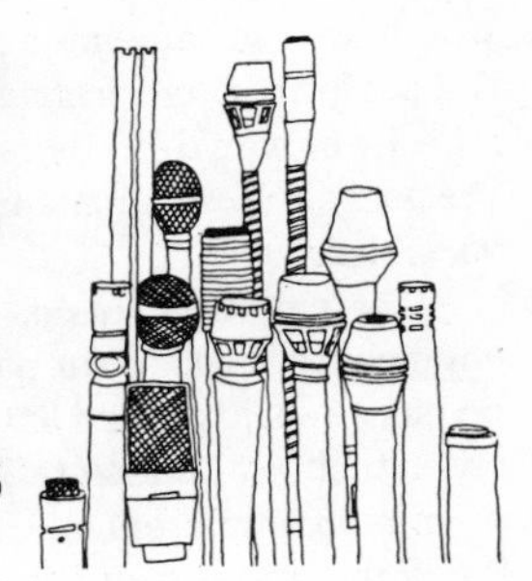

Before you invest in one or more microphones, and to avoid the letdown that comes with impulse buying, there are certain decisions you should make. What type of microphone do you want (dynamic or condenser); what type of sound-collection pattern (omni, cardioid, shotgun, or differential); low or high impedance, and, finally, how much do you wish to spend?

You do not need to be an engineer to be able to read microphone specification sheets, to look at and understand polar patterns, to examine sensitivity numbers and frequency response curves. However, you cannot plug spec sheets into your sound system. Technical measurements and response-curve information in microphone catalogs and spec sheets are prepared by engineers for engineers and technically oriented people. But the information that is supplied is often just for comparison purposes. However, you cannot *hear* a microphone catalog or spec sheet. In the final analysis, it is microphone performance that counts.

Does this mean spec sheets have no value? On the contrary. They can give you an excellent guide as to what you may expect from a microphone, but it is the actual performance that is the final criterion. And even here, final performance does not mean that you can put a microphone anywhere and expect it to give you the results you want. Listening to music is a subjective experience, and, while there are some general guidelines about microphone setups, the best way to make microphones specs meaningful is to work with mics in a real, live recording situation.

Microphone manufacturers test microphones in two ways. Microphone performance must be measured according to established standards for comparative technical data in an anechoic chamber—a specially designed room all of whose surfaces are completely lined with sound absorbing material.

This type of room does not generate any sound of its own, nor does it reflect sound. An anechoic chamber, though, has absolutely no comparison with live recording in your own room, for the acoustic qualities of your recording room are entirely different. Microphone manufacturers also check mics in live recording situations, but the problem here is that no two recording rooms are the same. The recording room can and *does* have a tremendous influence on the sound. And even if the manufacturer could manage to simulate your personal recording room, there is no assurance that he would use the same number of microphones, the same types of mics, that the performer and instruments would be the same, that the sound pressure level would be the same, and that the mics would be positioned in precisely the same manner. This represents quite a large number of variables.

The situation sounds hopeless, but it isn't. You don't need to be an engineer to read a polar pattern, or to understand that one microphone can be more sensitive than another, or to comprehend frequency response and other technical characteristics of microphones. You can have a true appreciation of microphone specs, and such an appreciation and understanding is important. But it is just a *first step*. In the final analysis, after you buy your mics you must learn how to use them and live with them. A quality microphone can give you the sound you want, but that doesn't necessarily mean you will get the sound immediately and automatically. You and your microphones must form a partnership or a marriage. Call it what you like, but *cooperation* is the key word. So cooperate!

ANALOG VS DIGITAL

In nearly all phases of audio technology, professional and consumer alike, traditional analog (continuous signal) recording, processing and signal transmission methods are rapidly being replaced by digital (computer-like pulse signals). This change is motivated by the fact that the digital signal is relatively impervious to the numerous ills which have always degraded analog signals. Such problems as poor dynamic range, noise, distortion, and wow and flutter which are incurred while processing, recording, mixing down, mastering, playing back and amplifying . . . virtually every stage in the process of producing, storing and listening to music . . . are eliminated by the digital process. Ultimately, this technology will be used throughout the recording-playback chain, completely eliminating the problems of analog technology from the process.

There is one component, though, that will be exempt from the conversion of components from analog to digital—the microphone. This does not mean that some day the microphone will not contain an analog to digital converter. The output of the mic will then be in digital form, but the primary conversion step of audio energy to electrical will be analog, unless, of course, some way will be found or some transducer will be devised whose output will be the direct digital representation of input sound energy.

PROBLEMS OF MICROPHONE PLACEMENT

What we hear, musically, depends on our location with respect to a musical instrument or to a group of them. A performer will hear sounds out of his guitar, for example, not noticed by the audience. We are accustomed to hearing music at some distance from the orchestra and instruments are supposed to project sound across the intervening space. However, whether for sound reinforcement or for recording, microphones are often positioned extremely close to instruments, or may be mounted directly on them. The sound pickup by the microphones is not typical of the sound heard by the audience. Close micing can produce sound which may seem unnatural to an audience, that is, it will be sound to which they aren't accustomed.

Correct mic placement involves a number of factors. These include determining the distance of the mic from the source (the working distance). The correct angle of the mic with respect to the source, the distance of the mic from one or more sound reproducing elements such as speakers, the distance and the angle of the mic with respect to other mics (if any), and the distance of the mic from main reverberant sources such as a wall, floor or ceiling are some of the variables making optimum mic positioning quite often a matter of experience.

DESK STAND MICING

Some desk stands permit a microphone to be tilted all the way down to the desk surface to take advantage of non-multi path pickup to provide cleaner sound in many situations. (Fig. 8-1) The drawing at the left indicates the two possible sound paths to the mic. Low- to mid-frequency changes can become very noticeable as the user moves toward and away from the mic. In the drawing at the right, high-frequency sound shifts become unnoticeable.

For the radio interview in a room with good acoustics, one omni mic set on the interview desk equally distant between two persons can work well. While two cardioids back-to-back have the effect of an omni, you may

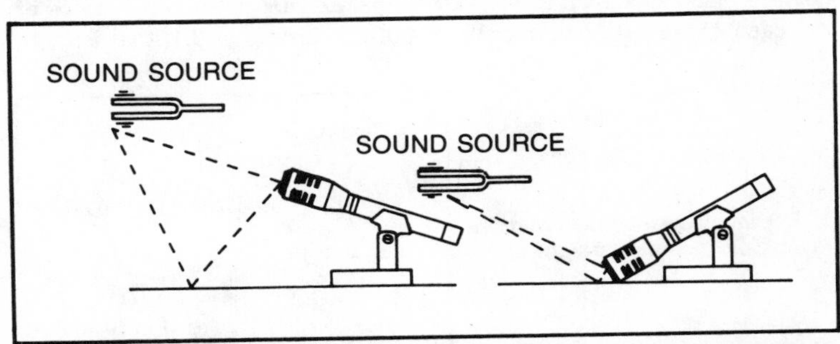

Fig. 8-1. Reception of multipath signals can be minimized if desk top mic can be tilted.

Fig. 8-2. Interview recording using cardioids back-to-back.

find that this configuration (Fig. 8-2) is not always successful. Because of mouth-to-mic distance differential, the voice takes more time to reach the further mic. Sound waves reaching both mics tend to cancel (phase cancellation) and voice transmission may sound artificially hollow or tinny.

If an omni is not available, use one cardioid (Fig. 8-3) between the two persons. Because people tend to look at each other while talking, and with both persons angled approximately the same degree off-axis from the microphone, both voices will be well balanced for true voice perspective. Point the cardioid straight up, toward the ceiling, or down toward the desk with the desk acting as a reflecting surface, as shown in Fig. 8-4 A and B, or set it firmly through a hole drilled through the desk top, as in Fig. 8-4C. In each of these positions a cardioid mic is used "omnidirectionally" but while this technique serves the purpose, an omni dynamic mic would be preferred.

Another possibility, as shown in Fig. 8-5 would be to have two cardioids facing in opposite directions, placed capsule-to-capsule and wired out-of-phase. These will respond with a figure-eight pattern, rejecting sounds that reach them from the sides. For radio interviews in low-frequency noise areas, try taping two cardioids together, capsule-to-capsule on the same plane. (Fig. 8-6). Again, by wiring one mic cable out-of-phase, the resulting differential pickup pattern will reject much of the ambient noise. Only sound presented directly and in close proximity to one mic will be transmitted. If the proximity effect resulting from this technique is too great, two omni mics can be substituted.

In round-table discussions, possibly with four people, try using two separate mics, as shown in Fig. 8-7. Separate the guests into groups of two, each group with one mic. With a distance ratio of 3:1, that is, with

Fig. 8-3. Interview recording with single cardioid pointed straight up.

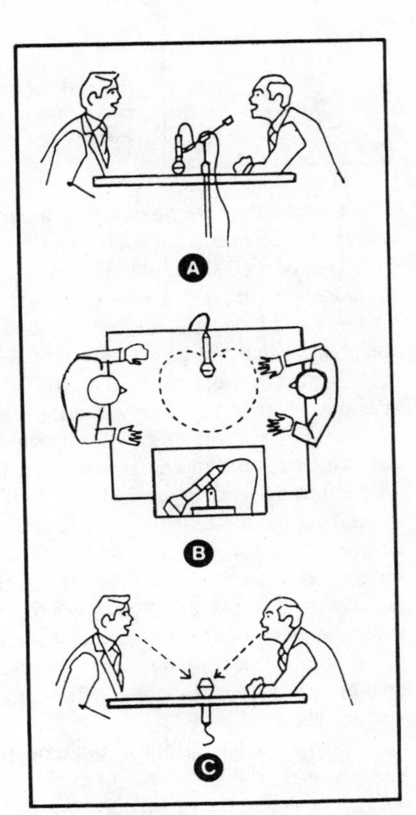

Fig. 8-4. Interview recording with table top acting as a reflecting surface.

each person perhaps two feet from his mic and six feet or more from the second mic there will be 10 dB of separation between mics. In this way there will be negligible negative effect due to voice carry-over (leakage) between microphones.

SOME DO'S AND DON'TS OF RECORDING

It is very easy to become accustomed to noise and, unless you make yourself deliberately conscious of the existence of a noise or noises, you

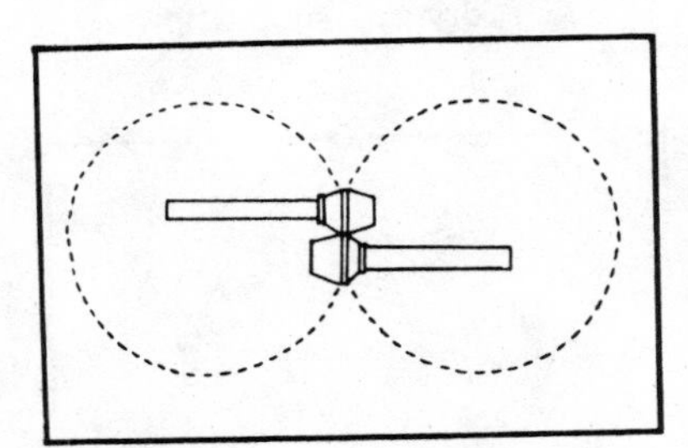

Fig. 8-5. Interview recording using cardioids in opposite directions, capsule to capsule, and wired out-of-phase.

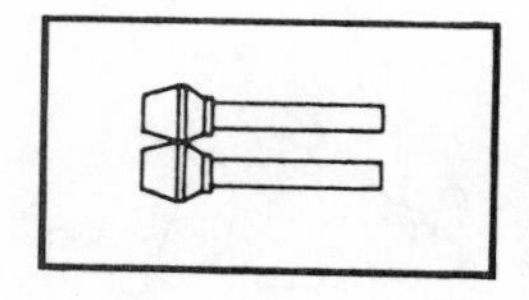

Fig. 8-6. Cardioids in low-frequency noise area, capsule to capsule, wired out-of-phase.

may not be aware they exist. In some instances, it is the absence of noise, rather than its presence, which we may find disturbing. We expect to hear the noise of a closing door. We expect an electric fan to hum slightly or to make some kind of a fan-blade sound. We expect to hear footsteps on a bare, wood floor. And we are so accustomed to the sounds we make when we breathe that we aren't aware of them.

The reason for this is that the human brain has the ability to discriminate, to pick and choose. But the microphone does not. What it "hears" within its pickup range it will reproduce as an electrical signal. That signal will be amplified and recorded on tape right along with your music. The difference is that you will now be listening rather critically to the music and so, for the first time you will be hearing noises you never noticed before; first, because you are now conscious of them; second, because they have been amplified. As a first step in the use of microphones, stand still and listen. Listen hard. You may now hear street noises, or noises elsewhere in your home—a voice, the telephone, the ringing of door chimes, the movement of people in other rooms, or whatever. You may hear your own breathing. And then you will realize that the world is a noisy place—and so it is.

You may also have one or more people in your recording room for various reasons—the performer or performers, someone to help you in making the recording, or just an observer. But the greater the number of people, the higher the noise level. People breathe, they move, and chairs move with them. They sneeze or cough or do some throat-clearing.

The only one that should be holding a microphone would be a vocalist who must move around. For home recording there should be no reason why this microphone cannot be boom mounted. Mount all your microphones in booms having decent suspensions. It doesn't do your recording any good to have people bumping into booms, so a general word of caution for everyone to remain in position once the recording has started is in order. Movement, however, is part of most musicians' repertoire, so make allowance for just what it is that the musicians will do. A dry run will possibly give you some idea of what to expect.

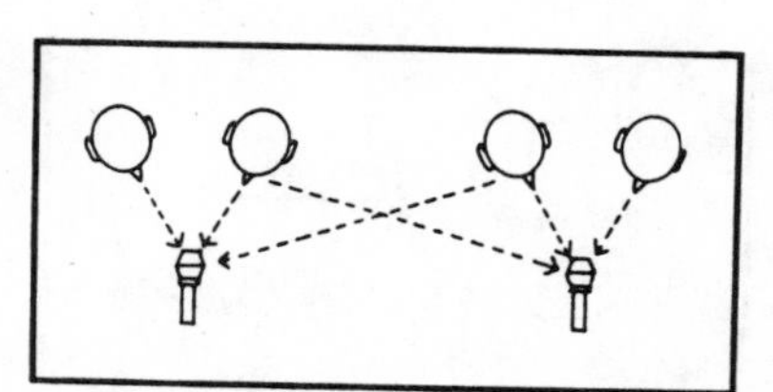

Fig. 8-7. Pair of mics used for interviewing four persons.

Use a windscreen on your microphones for outdoor recording, or indoors when you have a singer doing some really close-in vocals. If you are troubled with "popping" in vocals, try having the singer slant his or her sound output across the microphone head instead of right into it.

You may find that cardioids tend to be "bass happy" when brought close to the sound source. This is something to remember if you want more bass response, or, conversely, if you want less.

By and large, microphones by their very nature are delicate instruments, dictated by the "delicate" jobs they are called on to do—reproducing the broadest possible spectrum of sound waves. Unfortunately, most users aren't aware of this "delicacy" and microphones are often roughly handled, or, in fact, are grossly mistreated. Some microphones are especially made to tolerate such casual treatment, using a capsule suspension that eliminates handling noise.

MECHANICAL VIBRATION

The same sensitivity a microphone has to sound, a desirable characteristic, also penalizes that mic by making it sensitive to mechanical vibration. As a consequence, unwanted sounds produced by vibration can be generated, as, for example, by a mic stand being jostled or thrown over, by vibration as a result of stamping on the floor, whether accidental or as part of a performance, or even by handling of the mic. Not all mics have this characteristic to an equal degree but directional mics seem to have it to a greater extent than others.

Ideally, a mic should have little or no sensitivity to being handled. Whether it does or not depends on its construction, something that is part of the mic's internal structure and cannot be seen by the user. Manufacturers try various damping methods and spring suspensions to minimize vibration, particularly if the mic is to be hand-held.

THE LOW OUTPUT MIC

A recordist will sometimes deliberately select a low-output mic so as to compensate for close-up use in high loudness applications, possibly to avoid the use of an attenuator pad. This is an incorrect approach since it means the loss of the advantage of a better signal-to-noise ratio and creates the possibility of a poorer frequency response. As indicated earlier, reducing the mic level to the following amplifier is easily done.

HEADROOM

For the recordist, dynamic range is important for it establishes the limits, upper and lower, of sound recording. The greater concern is with the upper limit, the largest amount of sound that can be used, the greatest sound level that can be applied without accompanying distortion. The normal operating level of a system is taken as the reference. If a system can supply without distortion, a signal that is 10 dB higher than its normal level, then it is said to have a headroom of this amount. Above this figure, the sound will be distorted.

There are two ways in which a recordist watches for distortion. One is to watch the meters on the recording instrument and the other is to listen. However, it is easily possible for some percussive instruments, such as the piano, to produce tones having large amplitudes exceeding the headroom figure, but which have such a short time duration that the meter pointers cannot follow them.

Oddly, the distortion that goes onto the tape does not remain constant but increases with successive generations being mixed down to the final tape, at which time the distortion may be very obvious.

STEREO RECORDING

Stereo recording requires the use of a minimum of two microphones,

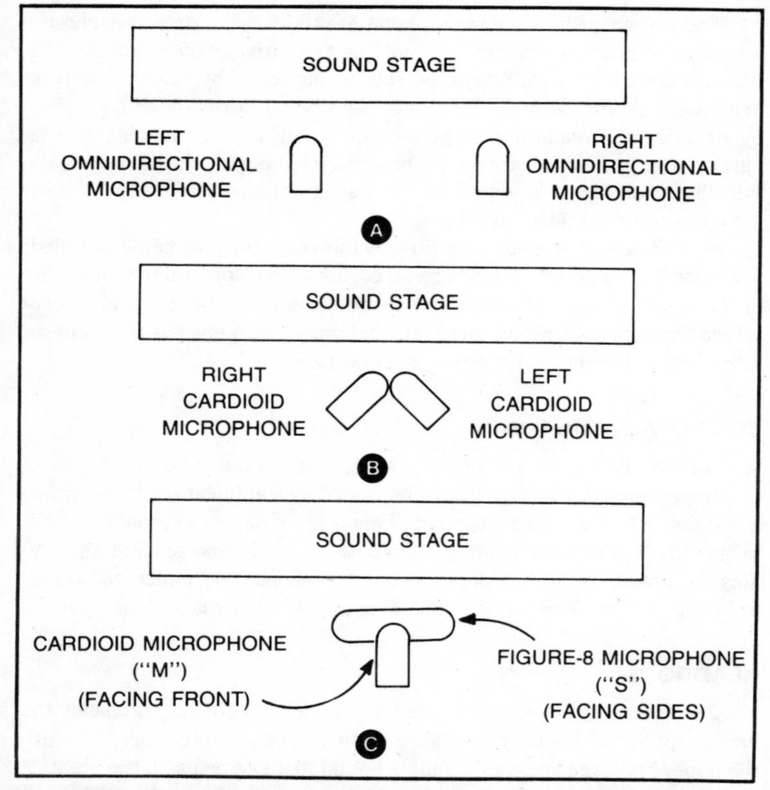

Fig. 8-8. Stereo microphone arrangements. (A) Spaced stereo mics; (B) X-Y stereo recording and (C) M-S stereo recording.

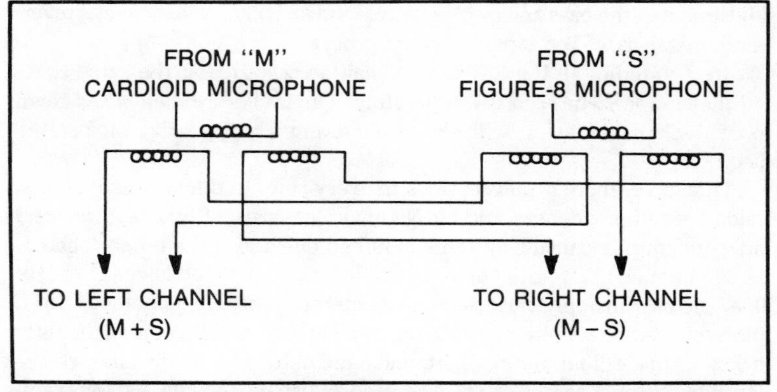

Fig. 8-9. Matrixing network for supplying left and right channel stereo sound from an M-S microphone setup.

but these can be arranged in various ways as shown in Fig. 8-8. In drawing A, there are a pair of omni mics, one at the left, the other at the right, facing the sound stage, a technique sometimes referred to as spaced—microphone stereo recording. Another stereo recording method is shown in drawing B and is known as X-Y stereo recording. In this technique, a pair of cardioids are used with the right mic facing toward the left and the left mic facing toward the right. The final method, illustrated in drawing C, is called M-S stereo recording. One of the mics is a cardioid (M) with its front facing the sound stage. The other mic is a figure-eight (S) with the mic positioned so it faces the sides of the sound stage. To be able to obtain stereo from the M-S microphone setup of Fig. 8-8C, a matrixing network, as illustrated in Fig. 8-9, is used.

HOW MANY MICROPHONES?

The minimum number of microphones, of course, is *one*, but one-mic recording, except for voice interviews, means you are going nowhere. This doesn't mean you can't record music with one microphone—you can. But with just one mic, your output will be mono, not stereo. Furthermore, with just one microphone, there is no chance for experimenting with microphones having different polar patterns. The only thing you can do with a single microphone is move it around to see what effect it will have on the recording.

Professional recording studios use a number of microphones and, at the present time, it seems as though they are using an individual microphone for each particular instrument. The output of each of these individual microphones is brought into a separate track on tape, using a multichannel tape recorder— with as many as 24, or more, tracks. In the studio, the instruments are very close to the mics, and in some instances the microphones are in the instrument. As a consequence, the sound output is initially very dry, with extremely little reverberant sound. This makes adjustment of the bass and treble controls and an inclusion of artificial reverberation a part of the tape-recording process.

In the studio, all the instruments can be recorded at the same time, but they may also be recorded separately. The tracks are then *mixed* down to a two-channel master, with the sound ultimately appearing on a stereo disc.

Of course, there are exceptions to every rule. In the professional mic field, a stereo condenser mic by Neumann consists of two separate and independent mic capsule systems mounted one above the other. The upper element can be rotated up to 270° with respect to the lower. This enables the user to apply various intensity stereo recording techniques. Both mic systems are remote controllable and can be switched independently in 9 steps to cardioid, figure-eight, and omni patterns with six characteristics in between. The mic can also be used as two mono units, as, for example, when two mics with differing directional patterns are needed in the same place. Axis of maximum sensitivity is at right angles to the mic body. The mic is intended for stereo recording applications where the original sound picture, in its natural acoustics, is to be reproduced.

With two or, preferably, three mics, you are set for home recording. You can mix sounds—and with the typical home recorder what you mix is what you get. But you still have an advantage, since you can always rerecord. In a professional recording studio, time is money. Recording, mixing, equalizing, and adding reverb are all part of a professional routine. In your home, with your own equipment and your own mics, you can experiment as much and as freely as you want.

There is a big difference between listening to live sounds and listening to recorded sound. Recording does not necessarily mean that your goal is an absolutely faithful reproduction of the musicians' sound. You can create all sorts of sonic impressions. A competent recordist can be regarded as much of an artist as the performers.

TRACKS VS CHANNELS

These two words—*tracks* and *channels*—are so related to each other they are sometimes regarded as being synonymous. Channels means the number of tracks that are recorded or played back at any one time. Thus, *stereo* is the same as two channel, *quadraphonic* is the same as four channel. Two-channel sound indicates that one channel will be reproduced by the left speaker and the other channel by the right speaker.

The number of tracks means the number of paths on the tape on which sound can be recorded. Open-reel tape, for example, has four tracks. When used for stereo recording and playback, one of the tracks is for left-channel sound, the other for right-channel sound. You listen to these two channels with the tape moving in one direction. However, since the tape has four tracks, and you have used only two of them, you can use the other two, also for stereo recording, except that now the tape will move in the other direction.

You can use open-reel tape for quadraphonic recording because it does have four channels. However, you pay a price, and that is you effectively cut recording and playing time in half. If you want to record quadraphonic sound on open-reel tape you will need all four tracks. Two of the tracks will be for front left- and right-sound; the remaining two tracks will be for back left- and right-sound.

Cartridge tape has eight tracks. You need a pair of tracks for stereo and 8-track has four pairs of two channels for stereo, two pairs of four channels for quadraphonic sound.

Cassette tape has four tracks, supplying two pairs of stereo channels, with one pair in one direction, the other pair in the other direction.

MULTI-TRACK RECORDING

Multi-track recording means recording different parts of the music separately until the output of an entire musical group is down on tape. A variety of recording techniques is possible.

Good multi-track recording can't be done on a hit-or-miss basis. You must have some sort of plan or guide. You'll find it helpful to start with the drums, which supply the beat, giving the correct number of bars that will be needed. You'll also need a "tap in" signal, a signal that will be the clue to the first musical beat.

One of the great advantages of tape, of course, is that you can check your progress as you go along. After recording rhythm and bass, you can then *dub in* the melody.

TAPE DECK INPUT

Not all tape decks are designed for microphone input. Thus, a deck may be equipped with so-called line or aux (auxiliary) inputs but these are specifically intended to be worked from the tape output terminals of an AM/FM/MPX radio receiver or possibly from a mixing console. For the most part, these inputs aren't satisfactory for direct mic connections.

There are some tape decks, though, especially those in the higher price ranges, that do have mic input terminals for connection to low-impedance mics. In general, they are intended for unbalanced inputs with the connection being made via a two-conductor phone plug.

FLANGING

Most people think that the ideal way to reproduce sound would be to record the sound as an exact duplicate of that going into a mic. The problem is that in real life such sound isn't what we normally hear nor are accustomed to, for sound is considerably modified by its environment. In some cases, extraneous background sounds are not only wanted but are essential. Dry sound may seem dull and uninteresting compared to sound which has been reinforced by reverberation.

Efforts are always being made on the part of vocalists and musicians to call attention to their voices or instruments. But the recordist is also involved and can use various techniques to give sound a distinctive character. One of these methods is flanging, so called since it is done by touching the supply reel on a tape deck. This changes the rotational speed of that reel and that, in turn, modifies the frequency of the sound. This is a crude method but we now have electronic flangers. These have the advantage of not only supplying smoother control but a type of control which can be repeated so as to get desired flanging effects.

WORKING DISTANCE

Working distance is the amount of separation, in inches, feet, or any other unit of distance measurement, between the sound source and the diaphragm of the microphone. Sound pressure level decreases 6 dB every time the distance is doubled. Six decibels is a four-to-one ratio. If you place a microphone 1 foot from the instrument of a performer and then decide to change that separation to 2 feet, the sound pressure level in the vicinity of the diaphragm of the microphone will be one-fourth its original value. If you now move the microphone to a distance of four feet (doubling the distance), you will get another 6 dB decrease in sound pressure level.

The decrease of SPL by 6 dB every time the working distance is doubled is applicable to dry sound only and does not include reverberant sounds. It is much more applicable to out-of-doors recording, where there is little or no reverb, than to an indoor micing situation, where the walls, floor, and ceiling can supply substantial reverb. The amount by which sound pressure level will drop as the microphone is moved away from the sound source is greater for a "dead" room, a room with little reverb, than it is for a "live" room, a room whose surfaces reflect sound rather well.

MONITORING

A musician has at least two important considerations. The first is the *quality* of sound output of his own instrument. The other is the *relationship* of his sound to the sound of all the other instruments in the group. In effect, then, what he tries to do is to listen to both.

Depending on room acoustics, the total sound at the listener's ears can be quite different from the dry sound in the immediate vicinity of the performers. If the sound is being picked up by a number of microphones, if it is being mixed, then another variable has been included in addition to room acoustics. Add to this the fact that the loudspeakers being used to bring sound to all parts of the audience can supply their own sound coloration.

Stage monitors are sometimes used to direct sound to the performers, letting them hear the results of their own contribution to overall sound. Conceivably, the performers could hear the audience speakers, but these are generally positioned facing the audience and so speaker sound may not be that audible to the performers. Speakers positioned to direct amplified sounds to the performers are known as *stage monitors.*

While monitoring is sometimes done with headphones, tape playback is often done through speakers. The problem is that headphone sound and speaker sound are quite different. A lot depends on the kind of headphones you use. If the headphones don't provide a good acoustic seal for the ears, you will be listening to extraneous sound as well as any recorded sound. While open-air phones are fine for playback listening, they aren't suitable for a recording session.

MICROPHONE PLACEMENT

It is generally not possible to modify the acoustic environment in which recording is to take place. If you are going to record an orchestra, the number of people in the audience and the noise level they produce are beyond control. Not much can be done for an in-home recording situation unless you want to make a determined effort to change the acoustic characteristics of that room. However, a variable is available that gives you some control of recorded sound—the *placement* of the microphones.

Two extremes for microphone placement are possible. One is to mic close in, the other is to have the performer and microphone widely separated. The first arrangement helps eliminate room acoustics; the other includes room acoustics in the recording.

Like any other extreme situation, adoption of such microphone recording techniques does not produce satisfactory results. Recording very close is equivalent to putting your ears very close to a sound source. This is an unnatural listening position, for you will hear the pure, direct, dry sound of the instrument or voice sounds just as though you had decided you wanted to do such close-in listening. We are much more accustomed to sound at spectator distance.

This does not mean to say that mics aren't handled in this way. Professional recording people do use mics close to screaming voices, inside bass drums, in string bass bridges, near the mouth of brass instruments, and just over the strings inside a piano, for example.

Consider what would happen if you put your ears in such a position. Not only would the sound seem unnatural, but with high sound pressure levels you would suffer physical discomfort, if not acute pain.

What happens when a microphone is worked this way? All the microphone does is generate a very high output voltage. Unless there is a pad in the line preceding the preamplifier input, it is possible that the high voltage produced by the microphone will overload the input-signal handling capabilities of the preamp, leading to *distortion.*

This doesn't mean you cannot work a microphone close in. You may want to produce certain sound effects or you may want to emphasize the dry sound of an instrument. To some listeners the rasp of strings being played or the sound of sliding fingers along steel or gut wires adds realism. But if this is what you want, then use a microphone attenuation pad in the cable to reduce microphone output voltage to a level the following amp can handle. You can always remove the pad for more distant microphone use when the output voltage won't be so high. The attenuation pad is resistive and if correctly designed and constructed will not affect frequency response. Professional recording equipment in studios provides built-in microphone voltage attenuation circuitry so as to be able to control microphone output voltage satisfactorily. Remember that no good-quality microphone will overload and produce audible distortion when compared with that possible in the rest of the reproduction chain. A noise-cancelling, differential-design microphone is the only type which must be used very close to the source and isn't suitable for distance pickup. Such mics are useful when it is necessary to reject distant sounds, primarily in noisy commercial environments for more intelligible voice communications.

MICROPHONE ANGLE

The angle of a mic is taken with respect to the source. The mic can be pointed at the source, above it, or below, to the left or to the right. It does not automatically follow that the mic must be pointed directly at the source and this should be regarded only as a reference or starting point. The optimum angle depends on the radiating characteristics of the source and the polar characteristics of the mic being used.

THE ACCENT MICROPHONE

In recording, it may be advisable to pick up the sound of one of the performers using a single mic for this purpose and also a pair of mics positioned at a greater distance for recording accompanying sound. The single microphone is sometimes referred to as an accent mic. However, it does not follow that just a single accent mic may or will be used. There may be several, depending on what the recordist wants to accomplish. Usually, though, an accent mic is used for a solo performer with musical accompaniment.

MICROPHONE COMPARATIVE ANALYSIS

An omni collects sound from all directions with equal sensitivity. You can use an omni between two performers or surrounded by a group, knowing that all sounds will be picked up with equal loudness. However, the same microphone will be equally sensitive to room acoustics in all directions. This has its advantages but also its disadvantages. In a normal listening pattern we are accustomed to hearing total-surround sound. However, when listening to music many people resent distracting extraneous sounds. In the total family of variations in microphone sound-collecting patterns, and this includes omni, cardioid, supercardioids, and shotguns, all will "spotlight" when used close in and pick up over a wider area with increasing distance from the source. And so even with an omni the amount of ambient sound picked up depends on the working distance of the microphone. Keep in mind that an omni, a well-constructed omni, will pick up sound equally well from all sources, but that not all sources have the same loudness level. An omni, worked in at very close range to an extremely loud sound source, will reproduce that sound source in proportion to its level and the ambient sound in proportion to its level. But when two sounds of greatly different level, such as loud direct and weak ambient, are recorded, the much louder sound tends to cover or hide the weaker, a condition known as *masking effect*. With a relatively loud direct-sound source, with the omni having a very short working distance and much weaker background sound, the background sound may not even be heard in playback. This sound hasn't disappeared; it has just been covered or masked. You can use this recording technique if the only microphone available to you is an omni and you have a recording situation in which background noise cannot be eliminated and would be distracting.

Similarly, when using a single omni to record a group of singers and you want equal distribution of sound, position the stronger singers somewhat back from the microphone, the weaker singers in more closely. There is no reason for vocalists to form a perfect circle around the omni, for you can position them to get the recorded sound you want. Recording is more than just a matter of sticking a microphone in front of performers. Since you ordinarily cannot change the acoustic setup of the recording room, you still have the option of positioning the performers and the microphone to overcome or compensate for acoustic recording conditions. It does take some experimentation, but fortunately tape is a medium that lets you do just that, over and over again, as you wish. And just as a practical demonstration, you can record with an omni pointed away from a voice or full orchestra, with the connector-end pointed toward the sound source, and you will still pick up the direct sound as though the microphone were directed normally at the performers. Further, you cannot, during playback, be aware of the backward positioning of the microphone, or even if the omni had been positioned sideways. With an omni, it is *working distance* that counts, not the way the microphone is positioned.

The omnis are a rather neglected type of microphone. They should be more popular than they are, for they add considerable flexibility to any recording situation, whether in a professional studio or in the home. Cardioids are more often used for recording, far outnumbering omnis and shotguns. Omnis are usually a second choice and will continue in this capacity until users learn how to take advantage of their versatility.

The basic directional cardioid microphone is an acoustic computer. It is programmed for maximum sensitivity in the forward direction and gradually reduces its sensitivity a little to sounds arriving at 90° and becomes relatively weak to any sounds arriving from the rear of the microphone. Thus, the cardioid has a preferred direction for maximum loudness pickup. There is a common misconception that the cardioid shuts off rear sound much as you would use a door to get isolation from noise. Actually, the cardioid is just relatively insensitive to sound arriving from the connector end. Useful sound is received over a very wide angle in all forward directions and it collects far less sound from the rear. The sound shutoff isn't perfect, though, and it would be unreasonable to assume that the microphone forms a perfect barrier against rear sound. Because it does collect far less sound from the rear, however, the cardioid can suppress more room sound and noise than an omni in the same position subjected to an identical sound field.

One of the other advantages of the cardioid is its better coverage at longer working distances. Live classical and small group recordists working in stereo often use a pair of cardioids because of this characteristic.

Because the cardioid's front-to-back sound signal pickup is greater than the omni's, you can work the cardioid in close and take advantage of masking effect to eliminate background sound level when such level cannot be controlled otherwise.

Microphones with a bidirectional, figure-eight pattern are available, but aren't often used in recording. You can get two-direction monophonic pickup with the two sound sources on opposite sides of one omni or cardioid pointing vertically up between them. You can also use two cardioids, tail-to-tail at 180° for bidirectional stereo pickup.

Supercardioid, and shotgun pattern microphones are just variations of the basic cardioid. Forward coverage may be tighter, sound collected over a narrower angle, but microphone placement is governed by the same application of sound-collection control described earlier.

TWO MIC RECORDING

There is not reason for having any doubts about using just two microphones or having just a simple two-channel recording ability, nor should you be unduly concerned about any lack of bass and treble equalization except on playback. You can get quite good results with a pair of quality microphones and a quality open-reel tape deck.

The professional studio technique of cramming a microphone close to every sound source and remixing all this later into pseudo-stereo is necessary to save time and cost and to get the performers in and out of the studio in the shortest possible time. In the professional studio special effects can be created, and the final playback becomes an art form of its own. However, this isn't related to a natural performance. Against this is the fact that in-home tape decks, especially top-level units, are of professional quality, and a number of them permit recording sound-on-sound, sound-with-sound, echo, and other effects. Various outboard devices, mixers, and equalizers, are now available, and in-home recording can be done more professionally. Your high-quality in-home tape deck is capable of the same performance as studio equipment. Studio recorders do use wider tape to carry more channels, but the quality of the open-reel tape you buy is the same as that used by the studio.

ADVANTAGES OF IN-HOME RECORDING

In-home recording has a number of advantages. Professional studio time is rather expensive—$75 an hour, or more. Using in-home recording, you can have as many "takes" as you wish. You can experiment as much as you want, and then, with experience, you can get a final recording that is precisely to your taste. In your home you can move mics around, position the performers, and even modify room acoustics somewhat. Finally, your master tape will quite possibly be a one-generation copy, whereas in the studio several generations are necessary from studio master to final recording.

A multiplicity of microphones, when correctly selected and positioned, can supply very gratifying results. But using a large number of mics, that is, more than two, does create problems not inherent in the use of two microphones, and it also has the potential for more distorted and unnatural sound. But for in-home recording, you can have immediate playback, and you can evaluate results at once.

ACOUSTIC PHASE INTERFERENCE

Using a pair of microphones to pick up a lecturer at a podium can result in some difficulties. Assume you have one mic facing the speaker directly and another mic parallel to the first, a short distance away, possibly a few feet, facing some instrument. What we overlook here is that it takes sound a finite time to travel and the voice of the speaker will reach the second mic a short time—a very short time—after the arrival of the same sound at the first mic. This means the sound pressures are out of *step* or out of *phase*. The effects will be additive or subtractive—that is, the sounds will reinforce each other or will oppose. The overall effect will depend on the frequency of the sound at any moment and the amount of separation of the mics (Fig. 8-10).

MICROPHONE PHASING

When signals are in step or in phase they reinforce each other. If they are completely out of phase and both signals have the same amplitude or strength, they can cancel each other. There are all sorts of possibilities that can exist between these two conditions—you can have a pair of waves that reinforce at certain points of the wave and oppose each other at different points.

In Fig. 8-11, sound from the performer at the left can enter the micro-

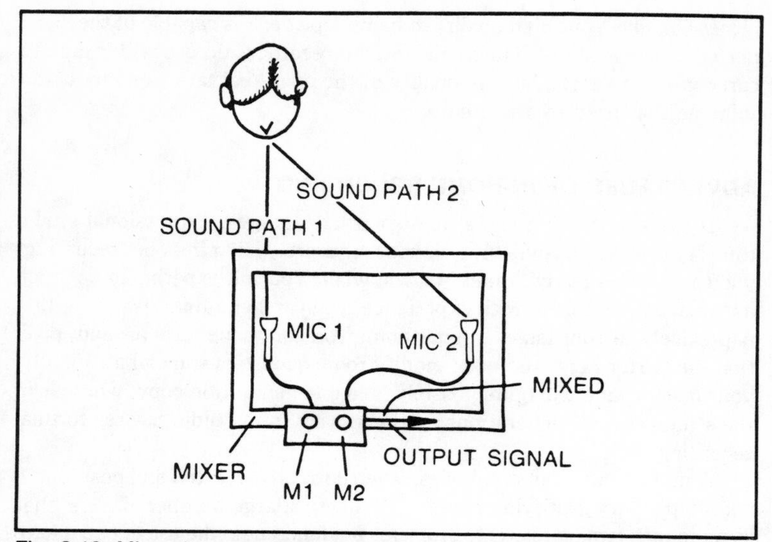

Fig. 8-10. Microphone positioning leading to acoustic phase interference.

phone at the right, and vice versa. Each microphone gets sound from two different sources. However, the sound paths do not have identical lengths and one sound will arrive at a microphone slightly delayed or behind in time of arrival of the more direct sound. Because the two sounds do not reach the microphone diaphragm at exactly the same time, we will get a combination of sound reinforcement and cancellation. The combined output of the two microphones will result in peaks and dips in the frequency response.

To avoid this problem, make sure the distance between microphones is at least three times the distance between each performer and his individual microphone. If the separation between performer and mic is one foot, then set up the mics for a separation of not less than three feet.

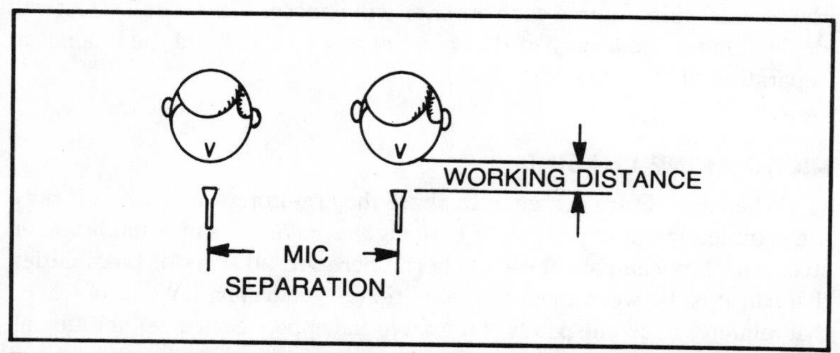

Fig. 8-11. When recording a pair of performers, microphone separation should be at least three times as great as the working distance.

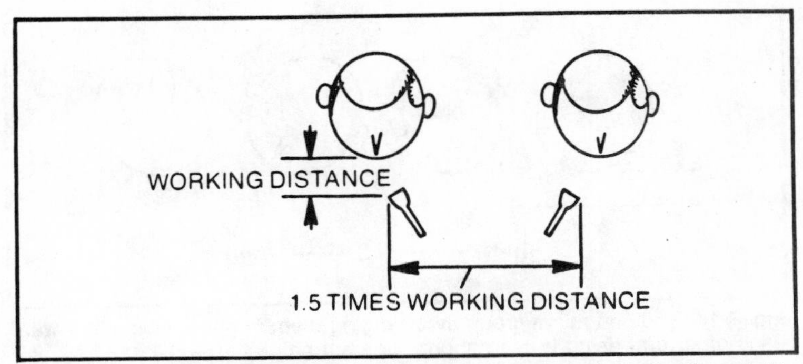

Fig. 8-12. You can minimize sound cancellation by using a pair of cardioids angled away from each other.

However, there may be certain micing situations in which a pair of performers must be close to each other. You might, for example, have a male and a female voice and would prefer micing these individually. If you use cardioids, you can bring them closer together by angling them away from each other (Fig. 8-12). In this case the separation between mics need be only 50 percent greater than that from performer to microphone. If the distance between the performer and microphone is one foot, then the separation can be brought down to 1 1/2 feet.

The same consideration applies when using four mics to record four performers, as shown in Fig. 8-13. Now the chances for phase cancellation have increased. Thus, microphone number 2 can receive input from three performers, starting from the left. In a case of this sort, while the mics can be separated fairly exactly, the distance between performers often changes, adding to phase cancellation effects. A better micing arrangement would be to have the performers work in pairs, facing each other, as in Fig. 8-14, with each pair using a common microphone.

When you have a pair of microphones angled in on a single performer (Fig. 8-15) do not separate the mic heads. Instead, angle the heads in toward each other as shown in Fig. 8-16.

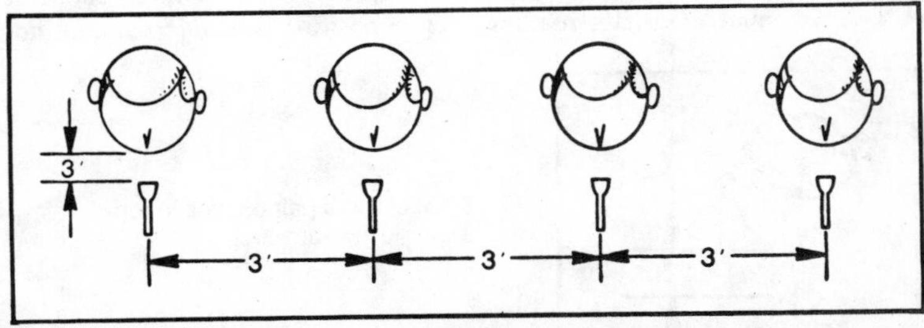

Fig. 8-13. With a larger number of performers, the 3 to 1 separation rule still applies.

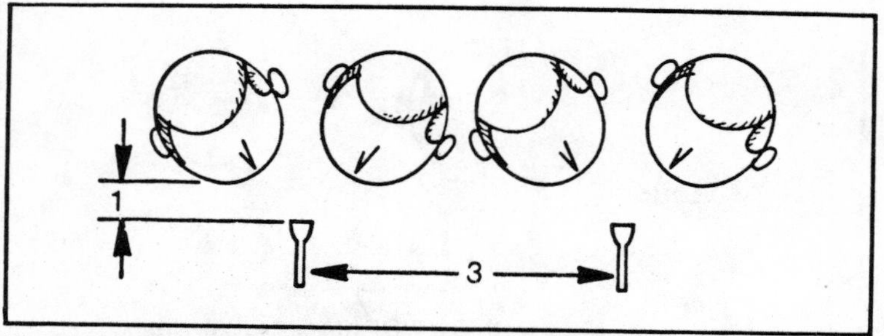

Fig. 8-14. You can mic a group by arranging them in pairs, facing in toward each other and using a microphone for each pair.

Phase Considerations

Two microphones are considered in phase when a common sound source reaching both microphones produces a combined output higher than a single microphone. In other words, if a positive pressure were applied to each microphone, a positive voltage would be present at the same output connections on each microphone.

Assuming equal signal strengths, if one microphone is connected in reverse phase from the other, their output voltages would cancel when mixed together. Theoretically, we would have sound going into the microphones, but no voltage coming out of them. In the laboratory it is possible to simulate this condition and the voltage output is indeed zero. In a live situation, the two microphones might not have identical sensitivities, but in any event there would be a substantial reduction in signal output.

Checking Microphone Phasing

Since there is no standardization of microphone phasing within the industry, microphones from one manufacturer can be out of phase when used with microphones from another manufacturer.

Checking the phasing of two microphones only requires a channel mixer that can handle two or more microphones and a VU (*volume unit*) meter. Using voice as a sound source, set the volume level of one microphone at a time for equal VU meter readings. Then position both microphones ad-

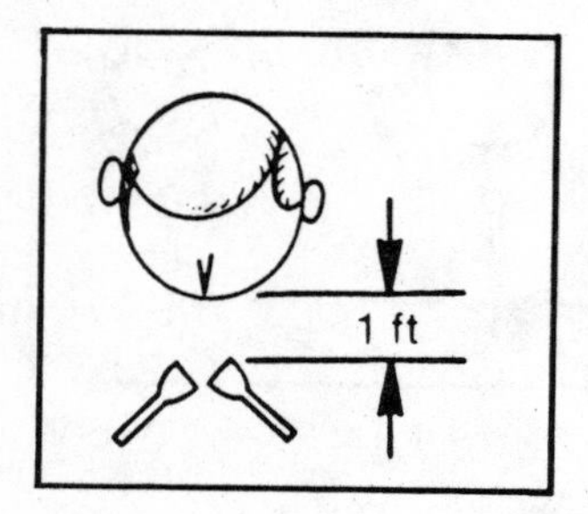

Fig. 8-15. Using a pair of mics to record a single performer.

Fig. 8-16. When using a pair of microphones to record a single performer, angle the mics inward, close to each other. The separation should be much less than the working distance.

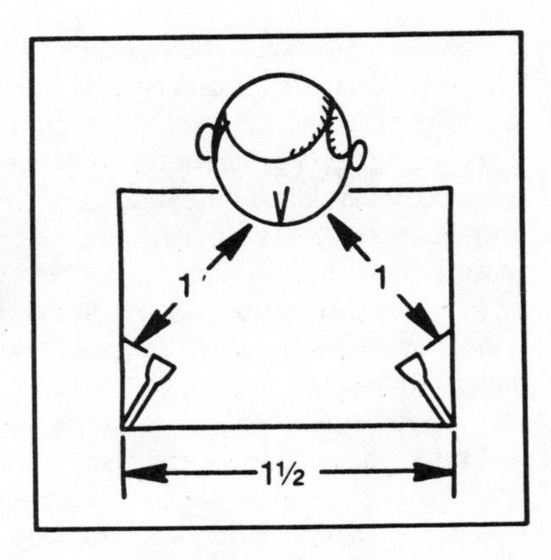

jacent to each other and speak into the area between them. If they are in phase, the resulting signal should be 3 dB higher than that produced by one microphone. To see if this is so, switch one microphone off and talk into the area between the two microphones. Make a note of the VU meter reading. Now switch on the unused microphone so you have both mics working. Talk and note the VU meter reading. If the microphones are in phase there should be a 3 dB increase.

Don't expect to get precisely 3 dB. It's difficult to get an exact reading when a meter pointer is moving. Try to use the same voice level at all times during this test and try to aim your voice at the same "in-between" spot. And of course, your lips should have the same distance from the microphones.

If, instead of getting an increase in dB reading when both mics are on, you get a decrease, one of the microphones is out of phase with respect to the other. You may not get precisely 3 dB less in signal output, but the reduced flick of the meter when both mics are connected, compared to that of a single mic, will be enough of an indication that the mics are out of phase. Since phase is relative to a reference, choose one of the microphones as a reference and phase all the other microphones to it. If you are trying to get three microphones (or possibly more) in phase, just keep repeating the test. Then select any one of the microphones as your reference and connect all the other microphones so they are in phase with the one you have chosen. It makes no difference which one you choose as your phase reference. You can also do the same without a VU meter, by listening to the sound. Mono mix the microphones for dramatic evidence of phasing.

Importance of Correct Phasing

Why bother with phasing microphones, particularly when it is unlikely that complete sound cancellation will occur, and particularly when the sound input levels may supply more than adequate microphone output levels?

Phase relationships between microphones in a multi-microphone recording session can wreak havoc with the recorded sound unless certain guidelines are observed. If microphones are in close proximity to each other, a recording session can become an exercise in frustration. When recording instrumentalists, for example, such as a trumpet player and a sax, the trumpet player's microphone can pick up the sax player's sound, delayed in time depending on the distance between microphones, producing a cancellation of various frequencies in the sax player's sound. If many microphones are used, the combined sound can be badly distorted by these phase cancellations.

A good rule of thumb to use in avoiding phase cancellation is the 3 to 1 rule. The distance to the nearest microphone away from yours should be three or more times the distance from you to your microphone to avoid phase cancellation. For the best sound always use the minimum number of microphones that will give you good sound. What you should follow in good recording practice when using a number of microphones is to make sure the microphones are wired in phase and then space the microphones properly.

Phase cancellation is not only produced by dry or direct sound but can be due to reverberant sound. Sound reflectors, such as walls, tabletops, and microphone stands, can reflect sound back into the side or rear of a microphone, causing phase cancellation. It is easy to run a quick check, taking just a few minutes, to determine if you have this problem. Mount the microphone in the position in which you expect to use it. Talk into the microphone and take a VU reading. If the mic was positioned on a tabletop, lift it about a foot above its surface and repeat your test. If you get a reading of a few decibels more, reflections from the tabletop are causing phase cancellation. The solution in this case, as in other recording situations, is a matter of common sense. If the phase cancellation is due to reflection from a wall, just move the microphone away from it.

Putting Phase to Work

In recording, a fault can sometimes be converted to a benefit. The important thing about out of phase micing is to recognize the fact that it can exist and to know how to take steps to correct it. However, you can use in-phase and out-of-phase microphones to achieve special pickup patterns. If, for example, you are working in an environment with an extremely high noise ambient, you can sometimes improve matters by connecting the microphones out of phase to get noise cancellation. With two cardioid microphones placed back to back and connected in phase you can get an omnidirectional pickup pattern. You can use the same two microphones, connected out-of-phase, as a bidirectional or figure-eight unit.

There are, of course, a few precautions. The addition or cancellation by phase requires that both microphones be of the same model and closely matched in output and response.

Chapter 9 How to Use Microphones: Musical Instruments

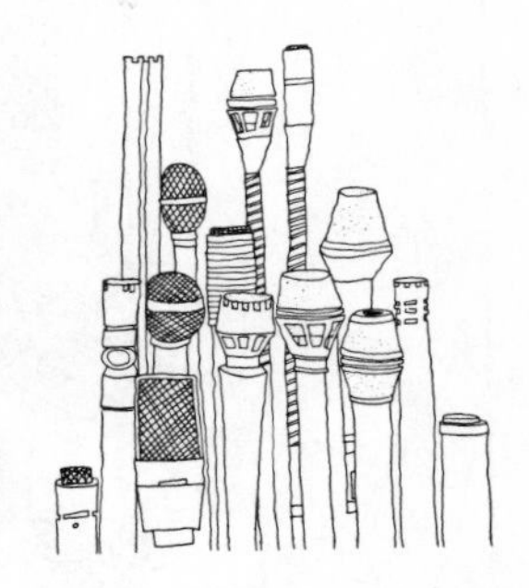

The preceding chapter contains generalized information about microphones, but now we can get down to specifics in the sense that we can discuss the direct application of microphones in selected uses, in this chapter, *musical instruments*.

SOUND COMBINATIONS

There are so many possible vocal and instrument combinations that it is impossible to describe the microphone setup for each and every situation. Enough examples are supplied in the following pages to give you general guidelines for almost any kind of recording problem. It isn't feasible to supply precisely detailed instructions concerning microphone positioning, for there are too many variables. Room acoustics, the number of performers, the instruments used, the various combinations, the number of the types of microphones available will all affect final recording results (Fig. 9-1). There is no way in which you can have an exact "cookbook" approach to recording. However, you must start somewhere and that is the exact purpose of this chapter—to give you your starting point. In any event, even if you should get fair results with your first effort, don't accept it. Tape permits you to record as often as you want, and the first rule in recording is *experimentation*. The playback sound will tell you if you are headed in the right or wrong direction in your recording sessions.

THE PROBLEM OF THE SOLO INSTRUMENT PERFORMER

One common recording technique for the solo instrument performer is to use either a single or a pair of cardioids to zero in on instrument sound

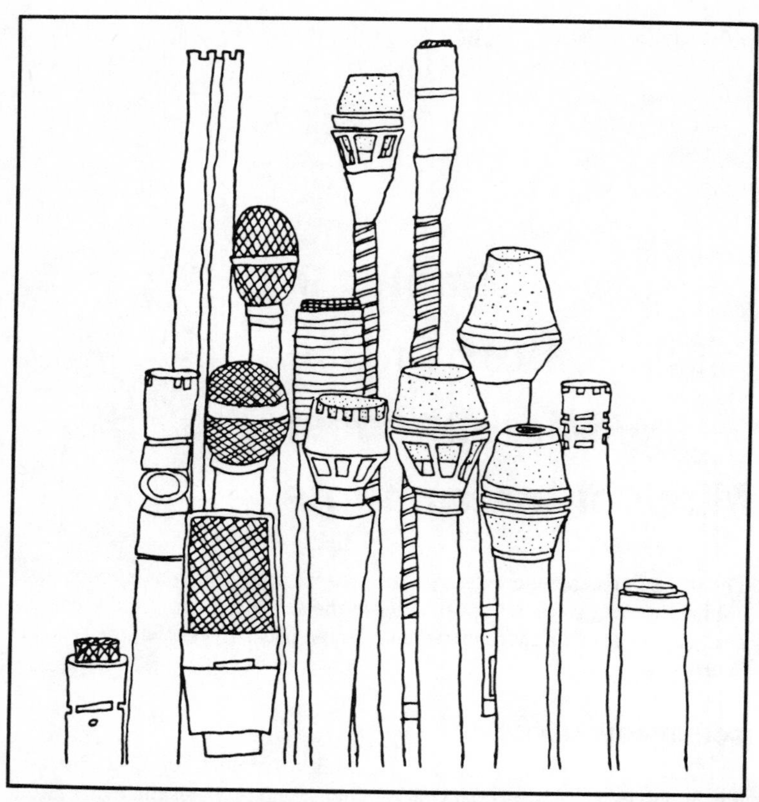

Fig. 9-1. What microphones you select and the way you use them will determine the kind of recorded sound you will get.

and to cut down background noise as much as possible. Microphone placement can become critical if you are concerned with creating an exact illusion. If you mic in very closely, listening to the reproduced music can supply an impression of huge instrument size (Fig. 9-2). Moving the microphones back away from the performer emphasizes stereo effect, supplying a mental image of a *duo* rather than a *solo*. It can also create the idea that the performer is moving from side to side. A way of overcoming this difficulty is to use three microphones: the two cardioids mentioned previously plus a third, either a cardioid or an omni.

RECORDING USING ONE MIC AND TWO SOUND SOURCES

The position of the mic will depend on several factors: do you want equal sound levels from both sources? Does one of the sound sources, whether voice or instrument, have an inherently lower output than the other? Do you want one of the sound sources to act as the musical background for the other or do you prefer the effect of an equal sound duet? What you do want in the way of output depends on the position of the mic relative to each of the sounds. Once you obtain the balance you want, you can adjust the overall sound level.

RECORDING USING TWO MICS AND ONE SOUND OUTPUT

Single sound output doesn't necessarily mean a solo vocalist or instrumentalist. Single sound may be a composite sound of a choir or instrumental group. The advantage of using a pair of mics is that it is basic to the production of stereo sound. With a stereo pair there is the problem of phasing. It is necessary for the sound reaching one of the mics to be in phase with the sound reaching the other mic. Out of phase sounds have a cancelling effect; in phase sounds reinforce each other. Audio waves have peaks and dips, so these should reach each of the mics simultaneously. The problem of phasing may be due to the use of different mics or, if the same mics are used, to their positioning.

THE FLUTE

The flute supplies a rich mixture of middle and upper frequencies. The sharp attack of the flutist's tongue and the corresponding wind sound supply an unmistakable instrumental character. For flute recording you might

Fig. 9-2. Close-in micing can supply an impression of a "larger-than-natural" instrument size.

Fig. 9-3. Recording the flute.

try a two-way microphone. This provides uniform sound quality off axis and when one is placed midway down the flute, facing the player and about a foot away, you can get a good balance of flute and wind sound without being overly breathy. You can move the microphone closer or farther from the flutist's mouth, depending on the degree of attack and breath sound you want (Fig. 9-3).

To reproduce the distinctive sound of the flute, the microphone should have a smooth, wide-range response across the full range of sound. The microphone should be capable of clean, effortless sound transmission and should supply a slightly bright, "open" sound to lend characteristic airiness.

To reproduce the flute's sound without distortion, position the microphone at right angles to and about 1 foot away from the instrument, or position the microphone midway between the mouth and the flutist's finger position to lessen "over-blow" effect. Place the microphone slightly below and looking upward at the flute so as not to be in a direct line with the air stream.

The sound from the flute is delicate and can be easily lost particularly when accompanied by other instruments. When used as a background for the female singing voice it can be incomparably beautiful, but even here there must be a balance between voice and flute, otherwise the flute can be overwhelmed. For this reason the flute requires special microphone handling.

For flute pickup, place the mic as close to the instrument as possible. However, when too near the mouth of the flutist you will hear breath noise for with this instrument the musician blows across the mouthpiece. To overcome this difficulty use a mic having a built-in breath noise control.

It isn't easy to pick up the flute perfectly because the sound energy is projected both by the embouchure and by the first open fingerhole. Therefore, it would be ideal to use two microphones: one for picking up the range to 3 kHz placed along the flute player's line of sight (0°), and the second mic for the range above 3 kHz to the right of the flutist (90°).

It all depends on what you want in the way of a flute recording. Some recordists feel that "breathing" is an integral part of flute sound and makes the sound "more natural". Others regard breathing as an interference with flute music. A similar situation occurs with violin playing, with some objecting to the scrape of the bow across the strings, and others insisting that it adds to a richer performance. If you prefer not having the breathing sound try using a windscreen. Either a cardioid or omni will work well, particularly if the mic is one recommended by the manufacturer for vocals.

THE PICCOLO

The piccolo is a comparable instrument and can be regarded as being flute-like. Compared to the standard orchestral flute, the piccolo produces sounds that are one octave higher in pitch.

THE KICK DRUM

Before you record the drum, ask yourself how you want it to sound. You may want it to be as "natural" as possible, but then you may prefer a fuller, heavier bass, or more bass definition, or you may want a tighter drum sound. Your choice of microphone and microphone placement will work together to give you the sound you want to project.

Remove the drum head and put the microphone inside the drum. Try it for sound. Or put the microphone on top of a blanket inside the drum head for the characteristic sound preferred by many professionals in recording. You may also like the sound that results from using a pad of paper towels where the hammer hits the drum to lessen boominess. Don't try to decide in advance; instead, experiment until you get the kind of sound you want.

To get best results you'll need a sensitive microphone, one that responds well to the instrument's timbre, tone delineation, and definition. If you get rattling or buzzing problems with the drum, put masking tape across the drum head to damp out these nuisances.

Drum sets consisting of a snare drum, tom-toms, splash cymbals, hi-hat, and bass drum are common in combos and large dance orchestras. The drum sound will often spill over into other instrumental mics and may be suitable for recording. If you want a sharp drum sound, pointing an omni condenser microphone down from a boom at the snare drum gives a sharp attack. You get an exciting drive to music when you use this setup in combination with a peak limiter. You can also use a cardioid or omni dynamic microphone in the same position for good results, putting the microphone at the side of the drum head, close to the skin. In this position you can capture the full harmonic texture of the bass drum. You can dampen or muffle the bass drum head as you want for special effects. See Figs. 9-4 and 9-5.

There are other ways of recording the drum set. You can, for example, hang two microphones over the drums, left and right, to get stereo sound. You can also use a separate microphone for each drum. Boom-mount

Fig. 9-4. Recording tom-toms.

Fig. 9-5. Recording the drum set.

one over the tom-tom and make certain it is high enough to pick up the cymbals. The microphone over the snare should be sufficiently high to pick up the hi hat. Or you can try using a separate microphone for the hi hat.

If you are planning to record the bass drum by putting the microphone inside, make sure the mic isn't centered because you'll pick up more overtones if it is positioned over to one side. If you are going to mount the microphone off center in the rear you will be working close to the foot pedal. Drum sounds normally override the speaking of this pedal but your mic may pick up some of the pedal sound. If it does, you must move the microphone away. Figure 9-6 shows how to record congas and bongos.

REED INSTRUMENTS

Figure 9-7 shows a few of the more commonly used reed instruments, including the flute, oboe, clarinet, bassoon and alto saxophone. Starting with middle C, the flute has a fundamental range of three octaves. The fundamental range of the oboe is three octaves from B. Sometimes the note A of the oboe is used for orchestral tuning. The clarinet has a fundamental range of three octaves from D. A comparable instrument, the bass clarinet, has the same fundamental range as the clarinet, but one octave lower. The bassoon is a double reed instrument using a tube that is 93 inches long. Because of this size the tube is folded back on itself. The fundamental range of this instrument is three octaves. The contra bassoon is a similar but longer instrument with the tube folded back three or four times.

There are various types of saxophones with each having a fundamental range of about 2-1/2 octaves.

Fig. 9-6. Recording bongos (left) and congas (right).

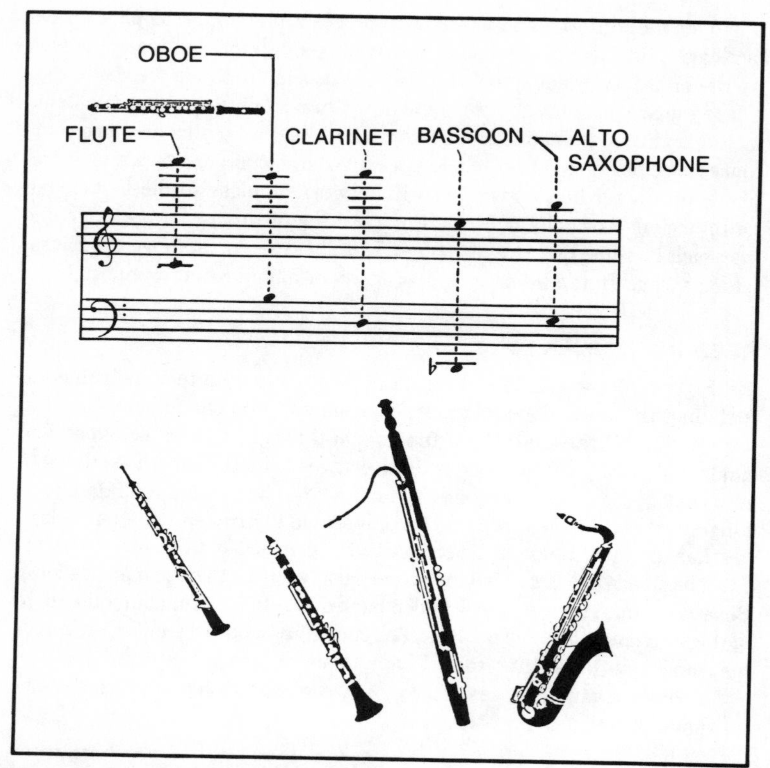

Fig. 9-7. Tonal range of reed instruments.

REEDS AND BRASS

Reed instruments such as the clarinet, sax, and oboe, and brass instruments such as the trumpet, trombone, and tuba, (Fig. 9-8) are usually recorded as instrumental sections in larger orchestras and as individual performers in small combos. Putting the microphone near the holes of reed instruments will result in the rich tonal coloring of the instrument since woodwinds generate more from holes than the bell. Cardioid microphones for solo instruments and two-way cardioids for reed sections give good results.

In large orchestras, where it is desirable to have a separation of sound control between reed sections and the brass sections directly behind them, using a bidirectional microphone or out-of-phase cardioid pair placed back to back will give a full reed sound without brass pickup. The reason for this is that the dead or null side of the bidirectional microphone or figure-eight pattern will be toward the brass.

Brass-instrument soloists often play to the side of a cardioid microphone to reduce the high-frequency bite of the instrument. The reduced high-frequency response at 90° off axis on a single-element cardioid is useful in controlling the brilliance of brass instruments.

You can get a good section blending of instruments when recording brass sections by using cardioid microphones or omnis placed several feet from the section. Of course, if you want greater separation of instruments, use closer microphone placement, as required. Such placement will provide excellent detail to the music, while more distant microphone placement will supply a greater blending of instruments.

It is commonly thought that the sound of reeds all comes out of the bell of the instrument, but this is only partly true. Move up close to a saxophone while it is being played and you will hear some of the sound coming from the finger-hole area. Most of the reed sound comes from the holes or stops. This is more true of the saxophone than any other woodwind. The sound is fairly well distributed between the finger holes and the bell.

Fig. 9-8. Mic positioning for the trombone, trumpet, and tuba.

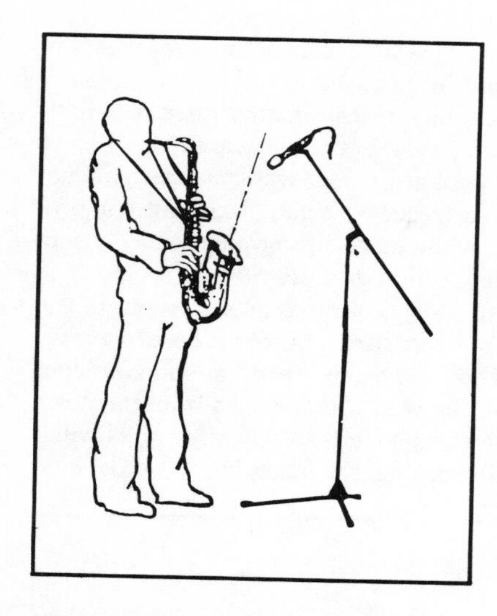

Fig. 9-9. Recording the saxophone.

THE TUBA

To record the tuba mount the mic on a stand as indicated in Fig. 9-8 with the head of the mic pointed down toward the center of the bell of the instrument. While a cardioid could be used, an omni would be preferable here since the tail of the mic is pointed upward, away from the audience, with less chance for noise pickup. The tuba can produce considerable bass tone energy, hence close micing isn't necessary. Tuba instrumentalists generally remain in a fixed position so once the correct location has been found for the mic, little additional experimentation will be needed.

THE SAXOPHONE

To record the sax, put the microphone about halfway between the topmost finger holes and the bell of the instrument, with the mic positioned above the instrument and the head of the microphone facing down toward it. (See Fig 9-9.)

Actually, microphone placement for the saxophone depends on how much key noise you want. If you consider the noise produced by the keys as characteristic of the saxophone sound or the mood of the number, point the mic at the middle of the instrument. However, if you regard key noise as a nuisance, direct the mic toward the front edge of the bell (not into it as this would result in too much wind noise). In both cases the working distance should be 8 to 12 inches.

While there are eight different kinds of saxophones, the four most widely used are the soprano, alto, tenor and baritone saxes. The saxophone is a single-reed instrument and so is sometimes considered in the same category as the clarinet. However, it can also have a conical tube or a flared bell and so in this respect is more like the oboe.

THE CLARINET

When recording a clarinet (Fig 9-10A) you will note that almost all the sound comes from the finger holes. Knowing the location of the source of the sound should help in microphone positioning, for you will get the true character of the instrument by aiming the microphone at the main sound source. Try aiming the microphone along the halfway point of the finger holes.

THE HARMONICA

The harmonica, mouth organ, or western mouth organ, is actually a reed instrument. Its sound output, compared to other reeds, is quite low. Try putting the microphone as closely to the instrument as possible and then listen carefully during playback to determine if you can note any breathing sounds made by the performer. Use a microphone having proximity effect to help bring the bass tones up a bit. (See Fig. 9-10B.)

THE STRING INSTRUMENT FAMILY

Figure 9-11 shows a few members of the string instrument family, including the violin, viola, cello, double bass and harp. These aren't all the string instruments since there are many others such as the acoustic and electric guitar, the banjo, the ukelele, etc.

The violin is characterized by having four strings and two f-holes. The fundamental range is four octaves. The viola has heavier strings than the violin and has a fundamental range of more than three octaves. Overall, this instrument is about 2-1/2″ longer than the violin. The cello is still larger and has a three-octave range. The double bass, also known as the con-

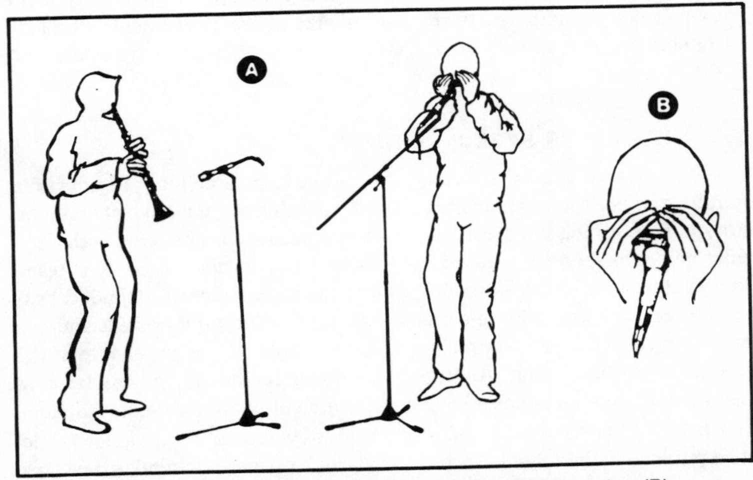

Fig. 9-10. Recording the clarinet. (A) Recording the harmonica (B).

Fig. 9-11. Tonal range of some string instruments.

trabass, has a fundamental range of about three octaves and its strings can be bowed or plucked. Completely unlike these instruments, the harp has 46 strings on a triangularly shaped supporting frame, with a relatively small sounding board compared to the overall size of the instrument. The fundamental range is greater than that of the other string instruments and is 6-1/2 octaves.

THE DOUBLE OR STRING BASS

The plucked or bowed strings of a double bass, a member of the violin family, provide the fundamental tones and rhythm for an instrumental group. In recording low-frequency instruments, such as this one, with close microphone placement, there is a tendency of recordists to confuse proximity effect in a single-element cardioid microphone with extended bass response. Two-way microphones do not exhibit proximity effect and can be effectively placed a few inches from the *f* hole on the upper side of the bridge of the instrument. Another microphone technique you can try is to use a small, personal microphone such as a lavalier. Wrap the microphone in a layer of foam padding and insert it in the opening of the upper *f* hole. The purpose of the foam is to serve as a shock mount to minimize the transmission of mechanical noise into the microphone. You can hold the padding in place with a rubber band. Another method is to mount the mic on a small tripod in front of, and facing, the bridge. (Fig. 9-12.)

To capture the true sound of the bass, think about a small microphone with a neutral, yet full response quality over the entire audio range. The microphone should be small enough to mount in the bridge of the instrument. Properly placed with a foam wraparound, the cardioid pattern should respond well to the mellow sound of the instrument, while reducing "plucking" and "buzzing" sounds from the strings. Some microphones are suited for this application. They may incorporate a specially designed built-in bass rolloff filter, with either normal or rolloff response easily controlled by a recessed switch in the shaft of the microphone. In situations where the bass response may seem too heavy, simply flick the switch for less bass in the rolloff position. Proximity effect is reduced and feedback problems are effectively diminished by proper use of the rolloff switch as well.

THE ELECTRIC BASS

The electric bass, or Fender bass, as it is sometimes called, needs amplification to join other instruments. Try a cardioid or, preferably, a two-way cardioid directly in front of the bass amplifier speaker on axis with the cone of the speaker. Be careful, though. With this kind of positioning and with the electronic amplification supplied by the electric bass, it will be easily possible to overdrive the amplifier being supplied with signal by the microphone. You won't damage the microphone, but this is a situation that lends itself easily—too easily—to overload distortion on the part of the amp. You will probably need to use a microphone attenuator pad to prevent recorder or mixer overload. Another method is to use a transformer-type direct box at a point where the bass guitar pickup feeds its amplifier for clearest sound. The purpose of this transformer is to provide a stepped-

Fig. 9-12. Recording the double bass. Use acoustic foam or sponge as a shock mount. The microphone doesn't rest on the instrument but is suspended. Or else mount the mic on a small stand and have the mic face the bridge, as shown in the drawing.

down voltage suitable for direct connection to a microphone input on the mixer.

THE BANJO

For this instrument, center the microphone on the banjo head. This is the area of maximum sound. Since the banjo has greater sound output than other instruments, such as the acoustic guitar, you may need to pad the microphone down to keep from overloading the amp or mixer.

Another technique for the banjo (or the acoustic guitar) is to mount the microphone on a boom that lets you adjust the microphone into the precise position you want. It's also a good idea to take the musician into your confidence, explaining why you are positioning the microphone the way you are and just what sort of results you are trying to achieve. You can, for example, get a sound-swelling effect by having the musician move his instrument back and forth in the microphone area. Also, the musician can accentuate certain tones for emphasis by bringing his instrument close to the microphone at times. In this way, the electronics becomes part of the musician's technique. With the help of the microphone, the musician can get sounds he could not possibly obtain by playing without microphone pickup.

The frequency response of the microphone will have an effect on banjo or guitar results. If you use a microphone with a frequency response rise somewhere around 1 kHz or 2 kHz, you will find that the music sounds more brilliant.

ACOUSTIC GUITAR

The acoustic guitar is another of those instruments whose sound output is delicate and whose volume output is so low that its tones can easily be lost among more "pushy" instruments. There are a number of different guitar types including the classical, the jumbo, also called the dreadnought, and the f-hole arch top. Figure 9-13 supplies the tuning range and the pitch range of the guitar.

Some recordists like to emphasize the sound made by the fingers along the guitar strings, with strong arguments in favor of realism. Others regard such sounds as intrusive. While you will have maximum natural pickup by "firing" the microphone directly at the guitar's center hole, playback of the tape may indicate this isn't quite what you want.

In general, you can record the acoustic guitar by putting a microphone behind the bridge, placing the microphone over the neck of the guitar at the point where it joins the guitar body, or by aiming the microphone at the hole of the guitar. An omni or two-way microphone supplies natural bass. Micing in close with a cardioid exaggerates the bass. Figure 9-14 shows a method of recording the acoustic guitar using a mic stand.

An alternative mic mounting method for the acoustic guitar is shown

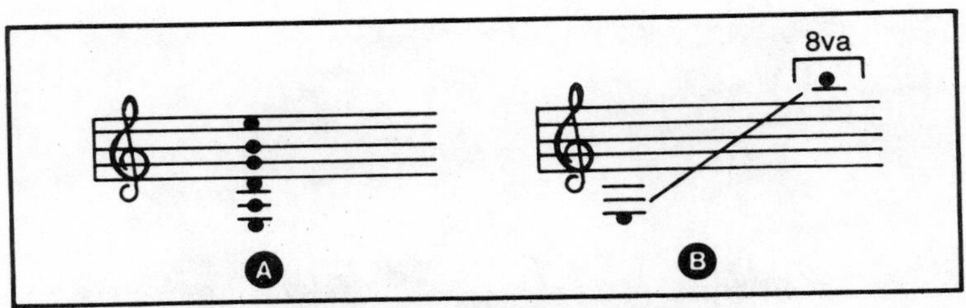

Fig. 9-13. Tuning (A) and pitch range (B) of the guitar.

in Fig. 9-15. Here we have a miniature dynamic mic equipped with a clip which fits on the sound hole. Although the photo shows the mic placed near the frets, it can be placed anywhere around the hole so as not to interfere with the fingers of the guitarist. The acoustic guitar is a "soft spo-

Fig. 9-14. A convenient way to record the acoustic guitar is to mount the mic on a stand and have the mic pointing at the hole of the instrument or slightly below it.

Fig. 9-15. Miniature dynamic mic mounted on sound hole of acoustic guitar. (Shure Brothers, Inc.)

ken" instrument and so mounting the mic this way gives the guitar a chance to compete with other instruments and voices.

There are several factors that determine the recording of any instrument. The first consists of the acoustics of the room; the other is subjective, that is, just what kind of sound are you looking for?

The second factor may be the more difficult of the two for sometimes the recordist may not be able to verbalize or to be precisely specific about what he (or she) wants. It would be helpful to have on hand one each of the three basically different mics: a dynamic, a condenser and a ribbon. With the dynamic, you may notice a greater emphasis in the bass region. The condenser may give greater frequency coverage and so you may note treble tones you did not know existed. The ribbon mic may supply what you consider to be better transient response.

Aside from the microphones, mic placement will affect the resulting sound. The guitar, like other string instruments, has a round radiation pattern. Thus, the hole of the guitar supplies lows, the resonant body furnishes tones in the midrange and the strings supply the highs. Finally, the sound you get will be determined by the ratio of dry to reflected sound. Since these are a large number of variables it is necessary to try and try again, and then to depend on your experience to get closer to the sound you want in less time in future recordings.

How Many Mics Per Instrument?

The usual concept in recording the acoustic guitar is to consider just a single microphone. However, it is possible to use two or three mics, depending entirely on the kind of results you want to achieve.

Thus, in Fig. 9-16 we have a vocalist with a two-guitar accompaniment. The vocal is the important part of the composition and the microphone is positioned to pick up the voice of the singer. The mic will also pick up guitar sound, but to a lesser extent. In this example, the softness of the guitar accompaniment is somewhat overcome by using two instruments.

Figure 9-17 shows a solo vocalist with guitar accompaniment, but using two microphones, with one mic aimed at the hole of the guitar, the other at the singer's mouth. In this arrangement the performer has greater control over the sound. There is no loss of guitar background when the vocalist stops singing. With this setup the performer has a better choice of vocal solo, guitar solo, or a combination of both.

A lot depends on whether you are recording a solo guitar or a guitar plus accompaniment. If it is guitar with accompaniment, you'll need to do

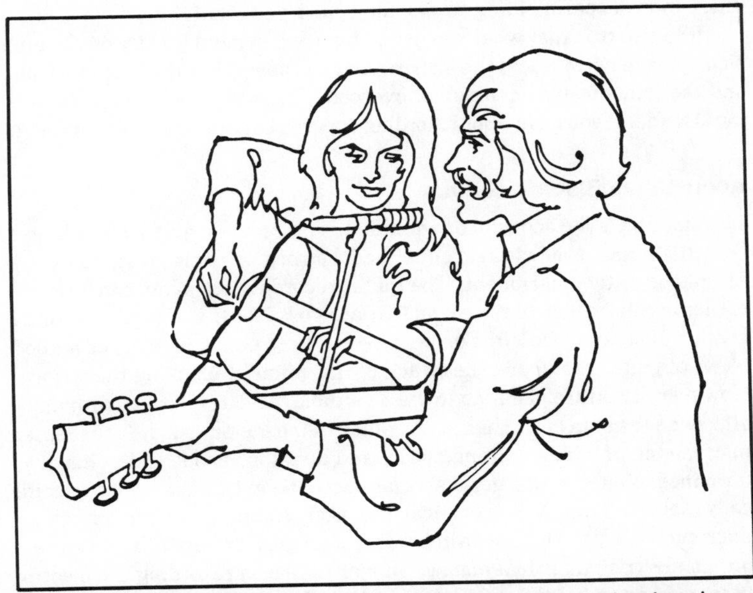

Fig. 9-16. Vocalist with two-guitar accompaniment, using a single microphone.

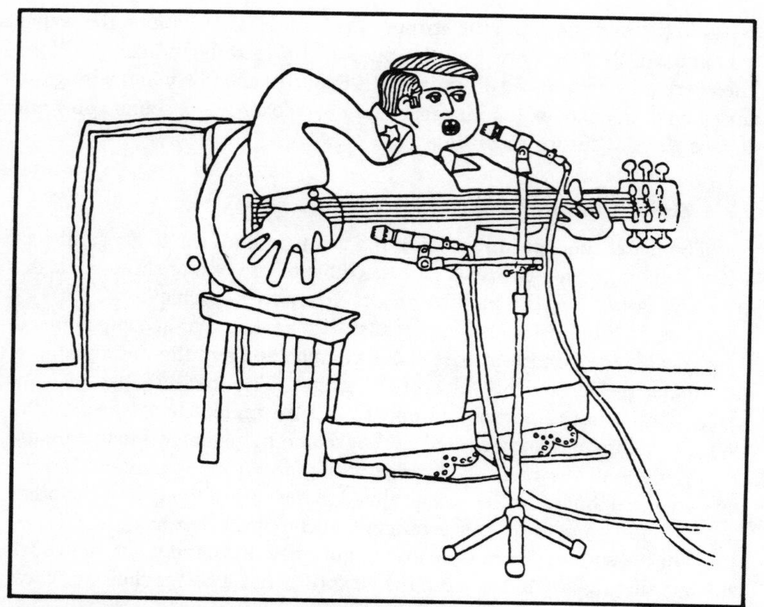

Fig. 9-17. Two-microphone method for recording vocalist with guitar accompaniment.

much more experimenting to get the sound you want.

It is also true that what you get on tape will depend greatly on the musician, his expertise, and the instrument he is using. But the best musician and the finest instrument cannot overcome the way in which the recording is made. If you have prior recording experience, fine. If not, experiment.

Acoustic vs Electric Guitars

Guitars can be acoustic or amplified. The acoustic guitar has a larger body than that of the electric guitar and a strong sound is provided by the resonance of the instrument. The guitar recording techniques mentioned earlier are just a few of the many that are possible. If you mount an omni about a foot from the circular opening of the guitar you will get a good blend of guitar sound and the attack of the guitarist plucking the strings. However, if you are planning to use a cardioid, be careful, since proximity effect can result in an unnatural coloration of the guitar sound. A condenser microphone provides exciting clarity and detail for acoustical guitar performance. You can also get a special microphone contact device specifically designed for the acoustical classical guitar, useful in providing microphone separation when recording a singing guitarist. You can record an electric guitar in a manner similar to that in recording the electric bass, using a microphone in front of the guitar amplifier's cone speaker or via a transformer-type direct box.

Three-Mic Technique for Recording the Acoustic Guitar

You can use three mics to record the guitar, with the main mic, a cardioid, close in to the instrument, and a microphone pair further away (Fig. 9-18). If the main microphone is too close, you may pick up the sounds of the fingers as they attack the strings or slide along the frets. To some listeners such sounds are annoying; to others they supply realism. It is entirely subjective.

Room acoustics will also influence the kind of sound you get, with the positioning of the microphone pair being influenced by the "liveliness" or "deadness" of the recording room.

The spot microphone (Fig. 9-19) must be delicately adjusted. Gain from this microphone must be spare. Always work in the "little" direction since spot overemphasis can wipe out stereo effect.

The positioning of the single microphone will be influenced by the composition to be played. Grace notes on the guitar, while beautiful, can be much quieter than plucked notes, and so you may need to microphone in closer than you would ordinarily.

THE ELECTRIC GUITAR

The music from the electric guitar doesn't come from the instrument

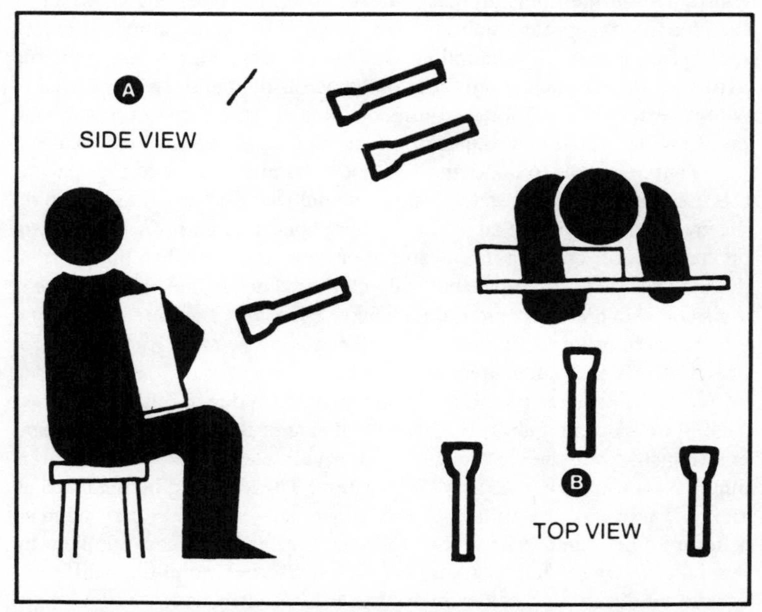

Fig. 9-18. Three-microphone technique for guitar soloist. Side view (A) and top view (B).

Fig. 9-19. Spot mic for emphasizing the drum set.

but, rather, from the speaker connected to the guitar amplifier. What we have, in effect, is double amplification. The electric guitar has a built-in transducer—a microphone—which develops an electrical signal which is then fed into the guitar amplifier. The output of the guitar amplifier is connected to a speaker. The amplifier isn't necessarily a high-fidelity unit and neither is the speaker. Sometimes distortion is deliberately introduced to obtain certain wanted musical effects. This is done through a wah-wah pedal, a fuzz pedal and a special effects box as shown in Fig. 9-20.

What you want to do, then, is to point the microphone at that part of the speaker that produces the loudest sound (See Fig. 9-21). You can do this by disconnecting the input to the microphone amplifier. Turn the microphone amplifier on, turn up the amplifier gain, and run a listening test on the speaker cone. Remove the grille cloth if there is one. You will hear a hissing sound coming from the speaker cone, but one part of the cone may seem to be supplying more hiss than any other part. Aim the microphone at this particular area.

With the acoustic guitar, the sound obtained is due to the kind of wood used in making the guitar, the shape and size of the guitar, and the kind of strings used. In the electric guitar, the quality or kind of sound is determined by the amplifier and speaker system. These should be regarded as an integral part of the guitar. That is why an electric guitar can sound so different when the musician uses the amplifier and speakers supplied by the hall in which he is going to play. Different amplifier + different speakers + different speaker enclosures = different sound.

One of the reasons for micing the electric guitar close in to the speaker box is to avoid picking up stage noises. You may not hear the shuffling

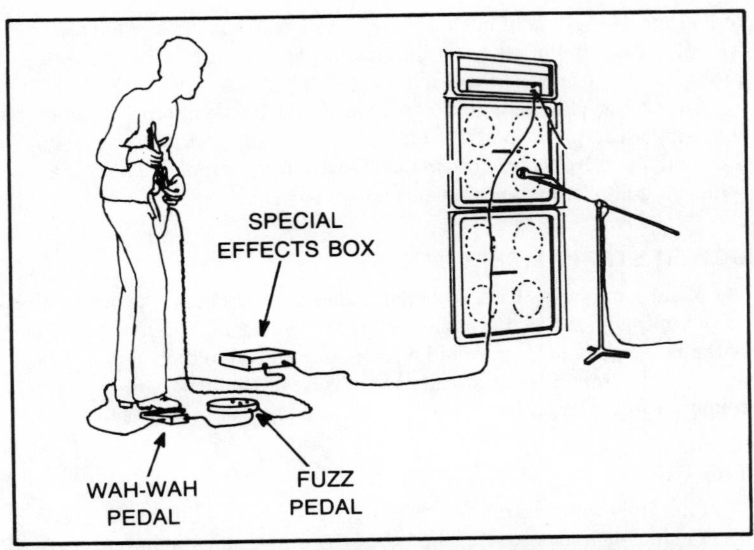

Fig. 9-20. Electric guitarist can supplement the sound of the guitar with a wah-wah pedal, a fuzz pedal, and a special effects box.

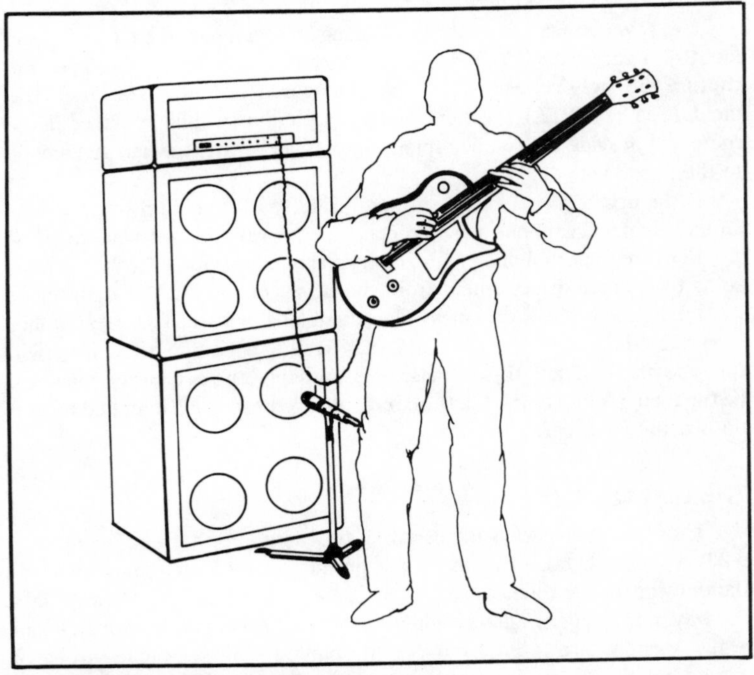

Fig. 9-21. Recording the electric guitar.

of feet, or the pulling of cables along the stage floor, but the microphone will. However, if you are recording in your home you may not have these problems, or at least not to such a degree.

Sound output from a guitar speaker can blast the microphone and its following preamp. It is better to have the guitar amplifier gain turned down a bit and the microphone close in than to have the microphone farther away with the guitar amplifier gain turned up more.

BOWED STRING INSTRUMENTS

Violins and violas have resonant bodies which produce a combination of rich harmonic structure combined with the high-frequency harshness of the bowing action. A cardioid microphone placed to the violinist's side, aimed down from a boom slightly to the side of the bow, will give a rich sound with minimum harshness, depending on side placement.

THE ORGAN

Pipe and electronic organs, as used in churches and theatres, produce a sound in which the room housing the instrument plays an important part in the tonal color of the instrument. Pipe organs produce more mechanical noises nearest the pipes, but some of that noise and the air noise adds to the character and distinction of the instrument.

Use several microphones when making an organ recording. Close placement of a cardioid microphone to the instrument pipes or speakers will capture the high-frequency detail and the percussive effect of the instrument. (See Fig. 9-22.) Another omni or cardioid microphone placed in the room will provide the room ambience and reverberation which gives body to the organ sound.

If the organ is in a church you will find that the church altar makes an excellent microphone positioning point for reverberant sound.

For three-microphone pickup of organ sound, position the stereo mics so as to pick up direct sound from the organ pipes, but put them close enough to obtain the harmonics. The sound level will depend on a number of variable factors: the music being played, the technique of the organist, the type of organ, and the acoustic environment. The two stereo mics will be the main pickup units; the third mic, facing the altar, is intended to pick up broadly reverberant sounds.

THE PIANO

There are many ways of recording the piano. For a grand piano (Fig. 9-23) or upright, raise the lid on the grand and lift it or remove it from the upright. In the case of the spinet, use a cardioid and put it two thirds the way to the upper register facing down into the mechanism from a boom arm suspended 6 to 12 inches above the opening. This positioning will give a good balance of low and high frequencies. Be careful that proximity ef-

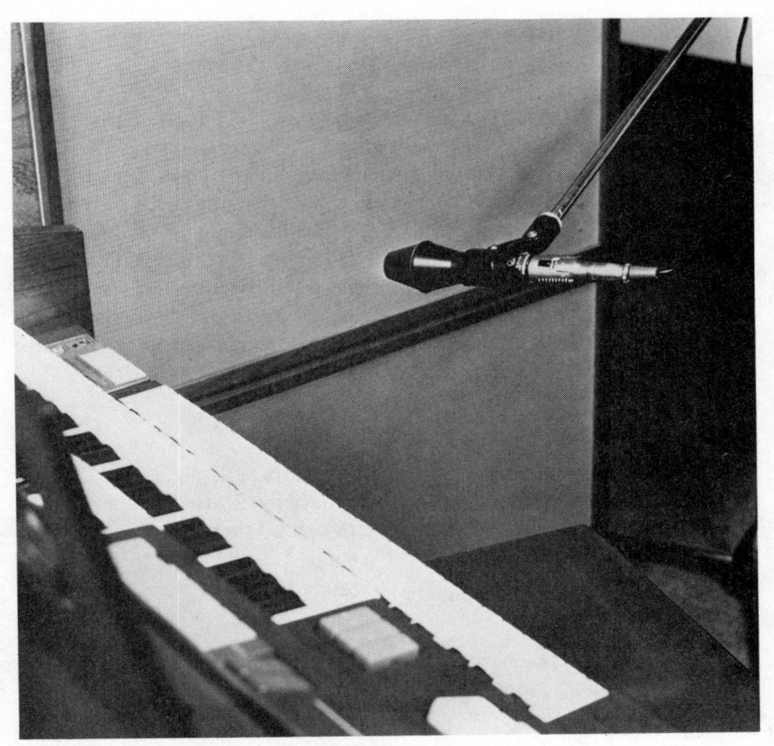

Fig. 9-22. Recording the electronic organ.

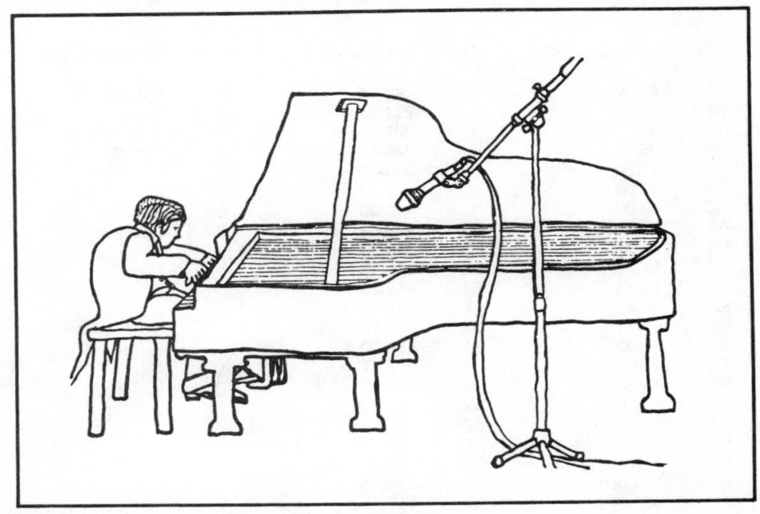

Fig. 9-23. Microphone pickup of grand piano.

fect doesn't add too much coloration to the sound. You'll get the balance of piano attack and tonal resonance you want by raising or lowering the height of the microphone.

There are various techniques recordists use for the grand piano. One is to put the microphone near the hammers, selecting a spot near *middle C*. If you want to emphasize the bass of a particular composition, position the mic toward the rear of the piano in the bass string section. With the piano lid open, and using a cardioid, aim the microphone into the piano. Piano sound is reflected from the bottom of the open piano lid and you also get direct sound from the points where the hammers hit the strings. Work between the strings and the bottom of the lid until you record the kind of piano sound you prefer. A two-way cardioid can be particularly responsive to string harmonics.

FREQUENCY RANGE OF THE PIANO

The tonal frequency range of the piano starts at 27.50 Hz and extends to 4186 Hz as indicated in Table 9-1. From key A4 to A3 is an octave; from A3 to A2 is an octave. Starting with key C3 there are seven octaves to key C4.

Upright Piano

For the upright piano, there are three general techniques you can try. You can put a microphone over the open top of the piano, inside it, or behind the sound board.

Table 9-1. Frequency Range of the Piano.

Key	Frequency (Hz)	Key	Frequency (Hz)	Key	Frequency (Hz)
A4	27.50	B1	246.94	C3	2023.00
B4	30.87	C	261.63	D3	2349.30
C3	32.70	D	293.66	E3	2637.00
D3	36.71	E	329.63	F3	2793.80
E3	41.20	F	349.23	G3	3136.00
F3	43.65	G	392.00	A3	3520.00
G3	49.00	A	440.00	B3	3951.00
A3	55.00	B	493.88	C4	4186.00
F3	61.74	C1	523.25		
C2	65.41	D1	587.33		
D2	73.42	E1	659.26		
E2	82.41	F1	698.46		
F2	87.31	G1	783.99		
G2	98.00	A1	880.00		
A2	110.00	B1	987.77		
B2	123.47	C2	1046.50		
C1	130.81	D2	1174.70		
D1	146.83	E2	1318.50		
E1	164.81	F2	1396.90		
F1	174.61	G2	1568.00		
G1	196.00	A2	1760.00		
A1	220.00	B2	1975.50		

Fig. 9-24. (A) Recording the upright piano. (B) Microphone placement for vocalist with piano accompaniment.

The use of a pair of microphones doesn't always indicate that a stereo recording is being made. Thus, in Fig. 9-24A a pair of left/right stand mounted mics are being used to record an upright piano. The top of the piano is opened and the mics are aimed downward at the strings.

Placement and blending can provide a wide panorama of piano sound, but remember that people don't only hear music on playback, they also get a mental image of the instrument. You may like the sound when reproduced; but if you generate a vision of an 8 foot wide piano, the microphone positioning technique is unrealistic.

Depending on room acoustics and microphone separation of instruments near the piano, you can place a cardioid microphone close for distinctively sharp piano sound or farther away for better tonal blending. Generally, a single microphone placed in the middle of the piano's arc, aimed down at the strings, will give a good tonal blend. For pop recording where close placement is customary, a single microphone aimed down at the second hole from the keyboard will give the necessary balance. When recording a piano-playing vocalist (Fig. 9-24B), correct placement will ensure separation of the vocal signal from that of the piano for best mixing and control.

Three-Mic Technique for Recording the Piano

Put one cardioid mic close to the piano. The piano is a percussive instrument and the sharp rise and fall times of the notes are the beauty of this string instrument. The distinctiveness of the percussive action decreases with distance from the instrument. Do not, however, have the microphone so close that it favors certain strings. The idea is to give equal pickup value to all the strings.

There is no absolute formula you can use for working distance, since not all musical compositions use the keyboard equally. Some compositions, for example, emphasize bass tones. Since the strings for bass tones produce more acoustic energy, you will need to position the single microphone a bit farther away from the strings. If the composition emphasizes treble, move closer in.

Boom-mount the other two omni mics (Fig. 9-25) for sound pickup six feet or so above ground and at least three feet apart. Precise positioning depends on the acoustics of the room. Generally, the larger the room the greater the distance of the mics from the piano.

THE TYMPANI

This is a percussive instrument characterized by sudden starts having a high sound pressure level. Position the microphone 8 to 12 feet away and sufficiently high enough to clear the other instruments to get clear tympani sound.

THE CELLO AND BASS FIDDLE

If you want a sound characterized by brightness, position the mic over the bridge. For a full sound of either of these instruments, aim the microphone into the *f* hole. Experiment with positioning to minimize the sound of bow rasping. At the same time, try to get bass reproduction that is clear and distinctive, not muddy. (Fig. 9-26.)

The cello, more formally known as the violoncello, is the bass tone instrument of a group that includes the violin and viola. Figure 9-27 shows the tuning and pitch range of the cello. The cello is characterized by having a spike at its bottom end which can rest on the floor. The spike can

Fig. 9-25. Three-mic technique for piano pickup.

be retracted when the instrument is not in use. As indicated earlier in Fig. 9-26, the mic can be mounted on the bridge.

ORCHESTRAL BELLS, XYLOPHONE, VIBES

Mount the mic about four to six feet above the keyboard. Then adjust the mic for the best balance of sound with other instruments. Figure 9-28 shows a micing method for the xylophone. Like the piano, the xylophone is a percussive musical instrument. Its construction consists of two rows of wood, in the form of bars with these positioned above hollow resonators. Orchestral xylophones have a range of 4 octaves, and sometimes only 3-1/2 octaves for some of these instruments.

PERCUSSION INSTRUMENTS IN GENERAL

Experiment with positioning of the microphone between two to six feet away from the instrument.

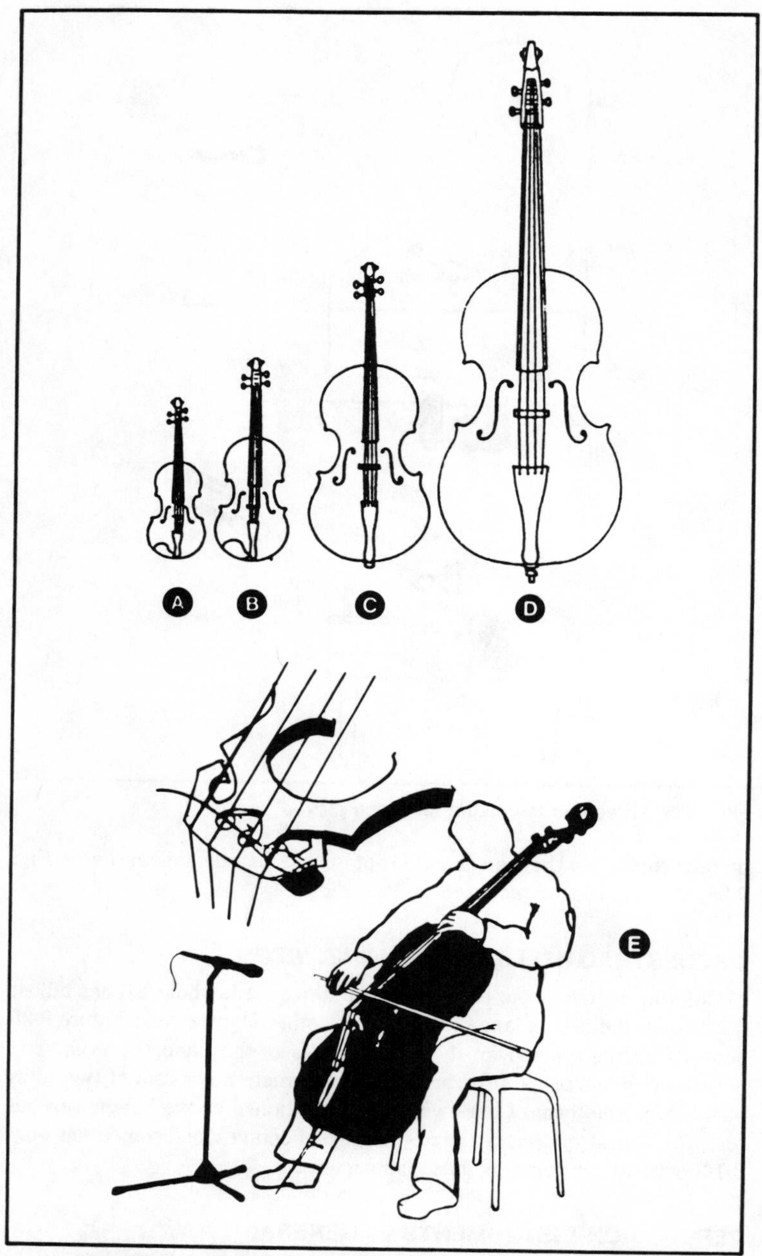

Fig. 9-26. Relative sizes of the violin (A), viola (B), cello (C) and double bass (D). Recording the cello (E).

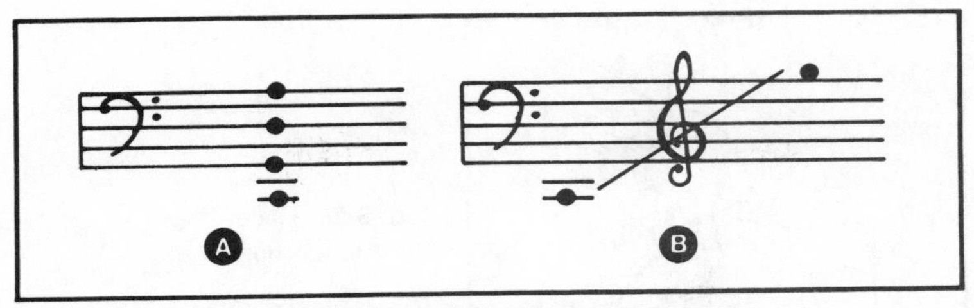

Fig. 9-27. Tuning (A) and pitch range (B) of the cello.

HORNS

Try positioning the microphone two to three feet from the bell to get the full sound of the instrument. You can mic closer in to the instrument, but if you get too close you will hear wind noise. (See Fig. 9-29.)

THE VIOLIN

The cardioid response pattern should be smooth for linear off-axis response without coloration. Try positioning the microphone about two feet

Fig. 9-28. Recording the xylophone.

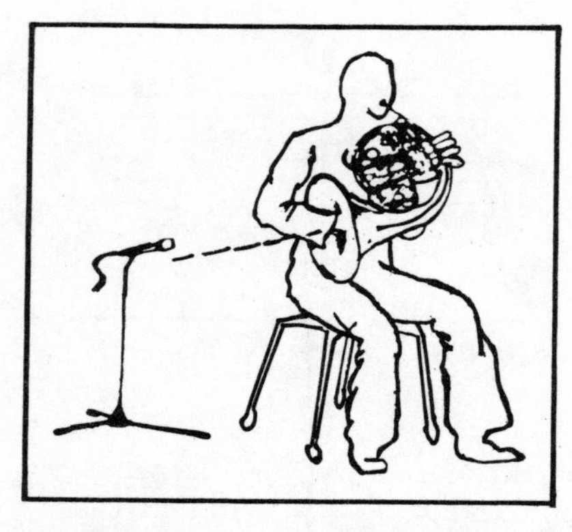

Fig. 9-29. Recording the French horn.

above the violin and direct it down into the hole. You'll find more freedom of movement without change of violin timbre because of any distance differential (Fig. 9-30).

The cardioid pattern should be so smooth that the microphone will respond without tonal coloration when the violin is moved off axis.

For micing the violin you can also use a miniature dynamic unit that mounts on the instrument. The advantage is that it supplies excellent isolation from other instruments and freedom from feedback. As shown in Fig. 9-31 the mic has an expansion mount for string hole mounting, suitable not only for violins, but for violas and cellos as well.

When recording the violin, the recordist should be aware of the special effects that can be produced by the performer. Thus, in a pizzicato action a strong percussive effect is produced since the musician plucks the strings between the thumb and the forefinger of the right hand. The sound is not as loud as that generated by bowing action. Similarly, a delicate sound occurs when the violinist plays grace notes on the E string near the *f* hole, and often these require careful listening. And if the violinist uses a violin

Fig. 9-30. Recording the violin.

Fig. 9-31. Miniature dynamic mic for mounting on the violin. (Shure Brothers, Inc.)

mute tones will also be softer than usual. Figure 9-32 shows the pitch range and the tuning range of the violin.

THE INSTRUMENTAL GROUP—SOME GENERAL SUGGESTIONS

Because some musical instruments have a louder sound than others, some will have a higher acoustic output. Others have their own amplifiers and speakers. Invariably, these instruments are played behind the vocalist's

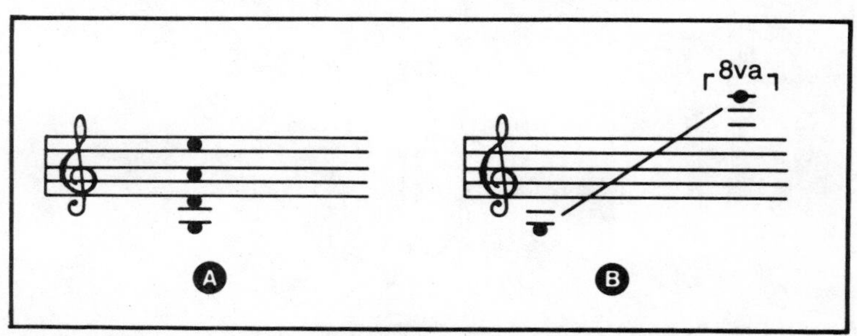

Fig. 9-32. Tuning range (A) and pitch range (B) of the violin.

mic and are amplified twice. The result is inevitable. The vocalist tries to sing louder, or more gain control is used, increasing until feedback occurs.

There will be times when the performers will complain of insufficient volume while the sound will actually be too loud for audience comfort. Consequently, it is desirable to set up the mics and the sound system when a theatre or hall is empty. When the instruments are properly balanced, and it is possible to judge this by taking a center seat in the auditorium, turn up the amplifier gain until just approaching feedback level, characterized by a slight ringing sound. When the audience arrives, considerable "damping down" of the acoustics will occur and the sound will be stable.

STRING QUARTETS

Unlike the vocal quartet, the string quartet (Fig. 9-33) is usually arranged in the form of a semicircle. Visualize the performers located on the outer circumference of this circle with the sound directed toward the center. The center would then be the logical position of the main pickup microphone. If you prefer just one microphone and want to minimize audience noise pickup, use a cardioid. Some recordists, though, feel that audience noise adds realism. While the omni would seem to be a logical choice, either an omni or cardioid can be used. It is the spacing of the microphone from the group that makes the difference.

Not all instruments in a quartet produce the same sound level. If you have a problem of instrumental music balance, move the mic closer to the

Fig. 9-33. Two-microphone pickup of a string quartet.

instrument whose sound must be brought up. Mount the microphone on a boom so it is above the heads of the players. Point the microphone downward so the head of the mic faces the instruments, as shown earlier in Fig. 9-33.

If you want to add a stereo effect use a pair of mics located behind the main pickup unit, somewhat higher, and at least three feet from the nearest musician. Since it is a single, the main pickup microphone is mono; the separated pair are stereo. If the output level of the main pickup is too high, the effect on playback will be monophonic phase distortion and bad stereo. A good starting point is to have the output level of the mono and stereo mics about the same and then to adjust the level down until you get the balance you prefer.

While professional string-quartet musicians often insist that string reproduction must be pure, some prefer a slightly attenuated bass response. They've found that diminished bass plus a very slight rise in both upper midrange and higher frequencies adds an instrumental presence that helps project a more forward string sound. Both microphones in a two microphone situation can be stand-mounted via a stereo bar or may be hung from the yoke of a stereo bar between 10 to 15 feet above the floor level. Angle the mics down at the musicians at about an angle of 45°.

VIOLIN SOLO WITH PIANO ACCOMPANIMENT

Since violinists do not remain in one spot when playing, use an omni, positioning it at a height of about one to three feet above the violinist and about two to three feet in front. Do not set the microphone so that it is either horizontal or perpendicular to the violin strings, but select a position that is somewhere in between.

You can use a pair of stereo microphones to pick up reverberant sounds. Position these two mics (Fig. 9-34) equidistant from the performer and at a greater distance than the single-pickup microphone, but the exact distance depends on the size of the room and its acoustics. You'll need to do some experimenting to get the best position that offers ideal balance between direct and reverberant sound.

VOCAL QUARTET WITH PIANO ACCOMPANIMENT

Unlike the barbershop quartet usually clustered around the mic, the vocal quartet with a piano accompaniment is usually arranged as a single horizontal group in front of the piano (Fig. 9-35). Further, the barbershop quartet usually huddles in closely, forming a rather tight group while the instrument-accompanied quartet tends to spread out. This requires greater microphone coverage and you should use at least two mics. Mount the microphones just above the heads of the vocalists, but not too closely. The mics should be wind-screen-equipped to minimize popping and blasting sounds.

Fig. 9-34. Recording the violin with piano accompaniment.

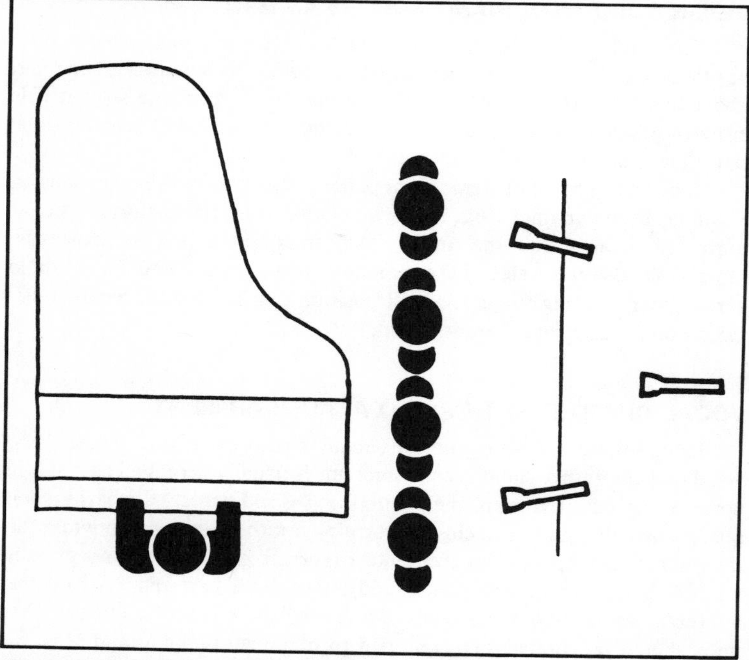

Fig. 9-35. Recording a quartet with piano accompaniment.

JAZZ TRIO AND QUARTET

A jazz trio will often consist of a piano, bass strings, and a set of drums. As in the case of the string quartet, the trio is arranged around a centrally positioned mic (Fig. 9-36). The microphone should be high enough so it is above the cymbals of the drum set. Start with the microphone positioned as centrally as possible, but you will probably find it quite off center when you are finished experimenting. Moving nearer to the drum set will emphasize the rhythm, but if you bring it too close the beat will override the musical theme supplied by the piano. If you locate the microphone too close to the bass strings you may get a thumping effect. Some recordists start with the microphone near the drum set and then gradually move the microphone back for the best sound balance.

If you want to include the acoustics of the room in the recording or you want to supply stereo, mount a pair of mics higher than the main pickup and farther back. All mics should point downward toward the performers. You might also try beginning with the main pickup mic, working back and forth with it until you get the kind of sound you want. This sound will be mono. Then add a stereo pair and adjust their position with respect to the main pickup mic until you get the kind of balance between the main and stereo pair you want. The idea here is to fix the position of the main pickup first and then find the best position of the stereo mics. If you try to adjust all three mics simultaneously you may find yourself frustrated by the enormous number of possibilities. Figure 9-37 shows a mic technique for recording a jazz quartet.

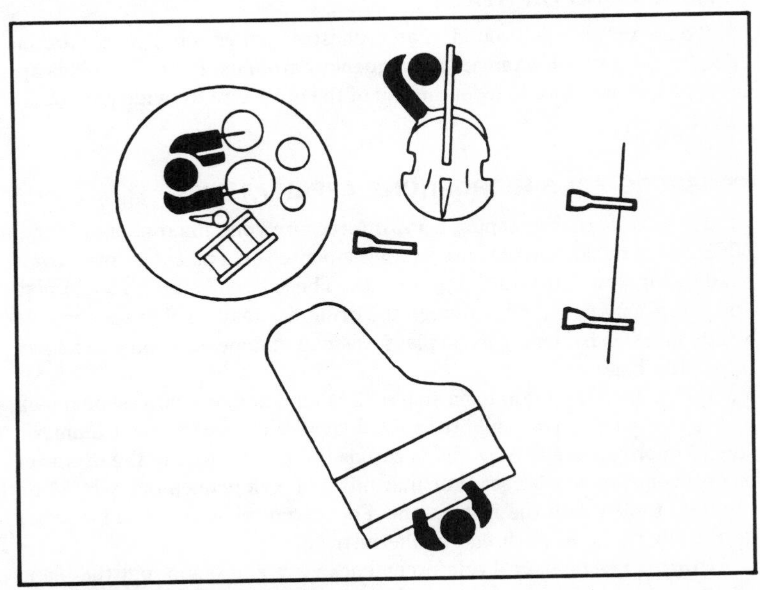

Fig. 9-36. Recording a jazz trio.

Fig. 9-37. Recording the jazz quartet.

SCHOOL BAND ON STAGE

To capture the full sound of an orchestra, professional people are inclined to use two wide-range full-frequency cardioids. Each cardioid is approximately one-third in from the end of the orchestra for wide panoramic sound.

PRESSURE ZONE MICROPHONE APPLICATIONS

Percussion. Try taping a PZM (pressure zone microphone) Model 30GP inside the kick drum with another mounted on a boundary plate overhead facing down toward the performer. These two channels should give you everything you need, although the drummer may need to concentrate a little more on balancing as he plays, instead of depending on a mixdown to get the balance.

For monaural reinforcing a single PZM mounted on a gobo should pick up the entire kit or you could try a PZM lavalier clipped to the drummer's shirt. For stereo use two PZMs on stands three feet apart of the drum set or on booms above the set. For marimbas or xylophones put a PZM on the floor underneath the instrument. For stereo recording, use two mics on the floor, one at each end of the instrument.

Piano. For monaural reinforcement of a grand piano, put the piano lid on the short stick and tape a PZM Model 30GP to the underside of the lid near its center, facing the strings (Fig. 9-38). You can get balance between bass and treble by moving the mic as you wish. You could also place a microphone on the floor directly under the keyboard. Use carpeting under the mic to get a softer sound.

For stereo recording put the lid on the short stick and mount two PZMs to the underside of the lid, positioned on either end. Be careful to keep the bass mic well clear of the hammers. If the grand piano doesn't have a lid you can tape two PZM Model 6LP microphones to the inside of the sides of the piano, one on the treble end, the other on the bass. This may require some adjusting to find the sweet spots, but will give you some fine sound with no mics readily visible.

If you need to eliminate extraneous sound when recording the piano tape a Model 6LP at each end, for stereo pickup to the underside of the lid, whether grand or upright, and close the lid. If you still get some leakage from other instruments, throw a blanket or two over the piano.

The opposite situation exists if you want to reinforce ambient noise. For a grand, put the lid of the piano on the high stick and mount a PZM on a four by four panel of masonite or plywood or clear plastic (Fig. 9-39). Put the PZM in the middle of the panel and mount the panel six to eight feet away from the piano, facing the lid, with the panel parallel to the lid. For stereo pickup, use two PZMs on a four by eight sheet still parallel to the lid.

To record a performer who both sings and plays the piano mount two PZMs on opposite sides of a panel and place the panel about where the music rack would be on the piano. In fact, the mics could be mounted on the rack itself. For stereo pickup, use a longer panel and four microphones.

String Instruments. To mic a solo string instrument place a PZM in the center of a larger plate made of masonite, plywood or clear plastic and position in front of the performer with the mic facing the performer. This isn't the only approach and there are various alternatives. Thus, you can mount a PZM on a boundary plate suspended from a boom above the performer to avoid interference with audience viewing. If pickup of audience noise is a problem in this placement move the mic closer to the audience and tilt the plate back toward the performer. You can also put the PZM on the floor in front of string performers, such as guitarists and cel-

Fig. 9-38. Single mic used for monaural reinforcement.

Fig. 9-39. Mic setup for reinforcement of ambient sound plus piano sound pickup.

lists, who are seated. Another technique which has worked well is to tape a Model 6LP to the music stand. The sound of paper could be a problem using this technique, so be careful. In some cases the PZM can be mounted on the body of the instrument with the body forming part of the boundary. This can be particularly effective with guitars.

Larger string instruments such as bass fiddles can be miced by putting a lavalier PZM inside the *f* hole where the mic's ability to put out undistorted sound even when receiving 150 dB can be used.

String sections can be picked up with the PZM mounted on a plate above and in front of the section, with directional shaping with carpet, if needed, to attenuate nearby instruments. Generally, any PZM technique which works well with a string soloist can be used as is or multiplied to pick up a section. You can also try using PZM lavalier mics with one mic per instrument clipped to the strings below the bridge for reinforcement of a string section.

Chapter 10 How to Use Microphones: Voices

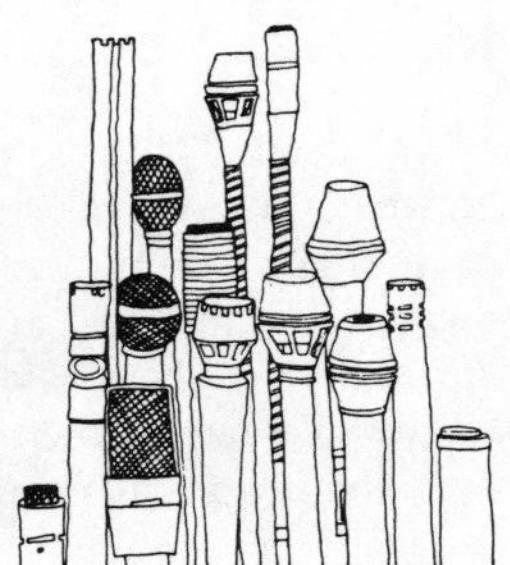

Continuing our discussion concerning microphone applications and techniques, this chapter will discuss the use of microphones when recording vocals, both solo and group, and voices with musical accompaniment.

The recording of the voice encompasses a wide range of techniques and includes studio work, live concerts, amateur, semi-pro and professional artists. But no matter what the level of the performance everyone, without exception, is anxious to have a mic that will augment the voice, possibly making it better than it really is and certainly no worse.

Originally, the music industry was interested in microphones that were rugged and that could take the abuse of a number of different vocalists, most of whom had no concept of how a microphone was constructed and some of whom had no idea of how to use the mic as an aid, rather than a hindrance. Today, the microphone must meet the demand for both ruggedness and high quality. Mics must cover the full range of male and female voices, and must work equally well for all kinds of vocal music: country, rock, folk and jazz. Mics must be capable of being worn, hand-held or stand mounted. In some instances, the mic practically becomes a part of the singer. Thus, a rock singer may use the mic as an extension of the body and so the mic gyrates, vibrates, works out. The mic must tolerate the roughest handling but must still deliver brilliant sound, without hindering in the slightest the performer's freedom of movement. Oddly, of all the components in the recording process, possibly with the sole exception of the musical instrument, no other component must meet such contradictory demands.

FREQUENCY RANGE OF MUSICAL VOICES

The frequency range of bass, baritone, alto, tenor and soprano voices

Table 10-1. Frequency Range of Musical Voices.

	Low Frequency	High Frequency
Bass	87.31	349.23
Baritone	98.00	392.00
Tenor	130.81	493.88
Alto	130.81	698.46
Soprano	246.94	1174.7

is supplied in Table 10-1. Of all these, the soprano has the widest range. However in terms of octaves the range for all voices is fairly uniform and is approximately two octaves, including the soprano. The frequencies indicated in Table 10-1 are in Hz (cycles per second). Note that the tenor and alto voices both have the same frequency at the low end but the high frequency range for the alto is somewhat higher than for the tenor. Interestingly, for low frequencies the bass and baritone voices aren't far removed from each other. Figure 10-1 illustrates the tonal range of the human voice.

RECORDING THE SOLO VOCALIST

There is no such thing as a single, universal microphone suitable for all solo vocalists. The type of microphone to use depends on the performer and the type of singing. For example, a rock performer usually works his microphone as hard as he works himself, and such singers have special problems. The fact is that no matter how hard a microphone is worked, it must have a clean, linear response with balanced bass reproduction, even when used close to the mouth. Because of the high sound pressure levels of rock program material, proximity effect, usually associated with cardioids, must be limited. While the microphone may be designed for very high

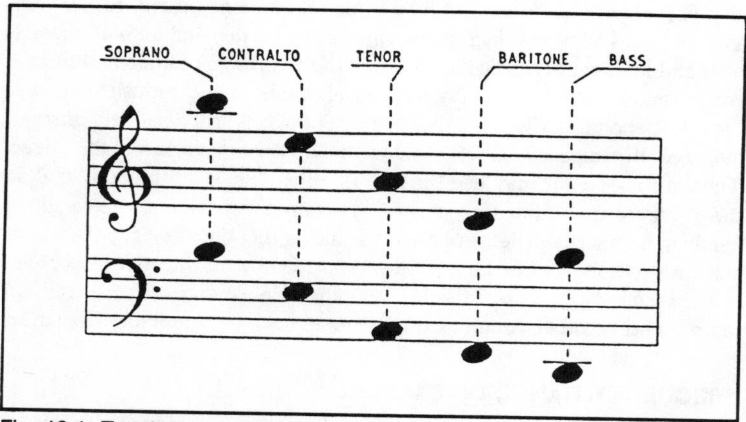

Fig. 10-1. Tonal range of human voices.

sound pressure levels, it should respond well under overload conditions and effectively reject low-frequency sounds that result in feedback.

Also, because of the nature of rock music, such performers may want to consider a microphone that delivers a more brilliant sound. Such a microphone tends to increase intelligibility and allow a more forward vocal projection.

Rock singers are active singers and not every microphone is made to take the punishment. Microphones should have a transducer element that is elastically shock-mounted to resist any response to external microphone or cable vibrations. The performer, no matter how active, needs complete freedom of movement.

Of course, not all vocalists are rock singers. There are all sorts. Some people just like to sing. And while they're singing, they like to move around a little. The trouble is that while they're moving, the microphone is standing still, and that can be a problem.

When the performer must have room to move, most likely he will be about two to three feet from the microphone. The customary thing to do to reach a natural sound level is to boost amplification. But higher amplification means possible feedback problems.

In such situations, you should consider a microphone that is comparatively free from bass coloration. Otherwise, as with many cardioid microphones, bass overemphasis will occur when the microphone is worked too closely, and vocal timbre will change as the performer moves toward and away from the microphone.

For this soloist, consider a two-way double-element cardioid microphone with an extremely wide frequency response. The microphone should have an extremely flat off-axis rejection—it should reject different sound frequencies equally without discrimination to allow for considerably high volume levels before feedback.

Finally, the closer the microphone is to the mouth of the vocalist, the greater the chance of recording *aspiration*, the sound of the vocalist's breathing. This usually happens when the microphone is about one inch from the mouth of the performer. All it takes is a few more inches of working distance to eliminate breathing sounds.

For solo vocalists the hand-held cardioid is best. The relative closeness of the mic to the mouth supplies a higher signal-to-noise ratio—that is, there is less opportunity for ambient noise to get at the mic. The hand-held mic also provides better separation and control between the vocals and accompanying instruments. A mic with a three-position *mode* switch is useful in reducing bass response. The idea here is to supply more vocal brilliance against an instrumental background.

Single element cardioid dynamics are more directional at high frequencies than omnis. If you are micing a vocalist or instrumentalist you can get more sharply defined high-frequency response by pointing the microphone right at the performer's instrument or at the mouth of the vocalist. Conversely, if you want to reduce high-frequency crispness, point the microphone upward or to one side.

With cardioids, vary the distance between mouth and microphone. When singing loudly, move the head back slightly, or move the microphone away. When a soft, intimate mood sound is needed, bring the microphone closer to the mouth.

Possibly one of the most unfortunate situations for a vocalist is using an unknown mic for the first time in a recording session. The best rule for the vocalist is to experiment with a mic and then to make a comparison of the results by listening to a playback (Fig. 10-2A).

Angle of Incidence

The vocalist should not sing directly into any microphone. Doing so will not only pick up excessive breath noise but will also cause sibilance. It is better to sing to one side of the microphone or over its head. The result will be a well-balanced, natural sound (Fig. 10-2B). Further, when using cardioid microphones you'll find you won't pop your p's this way. T's, sibilants and breath sounds will also be better.

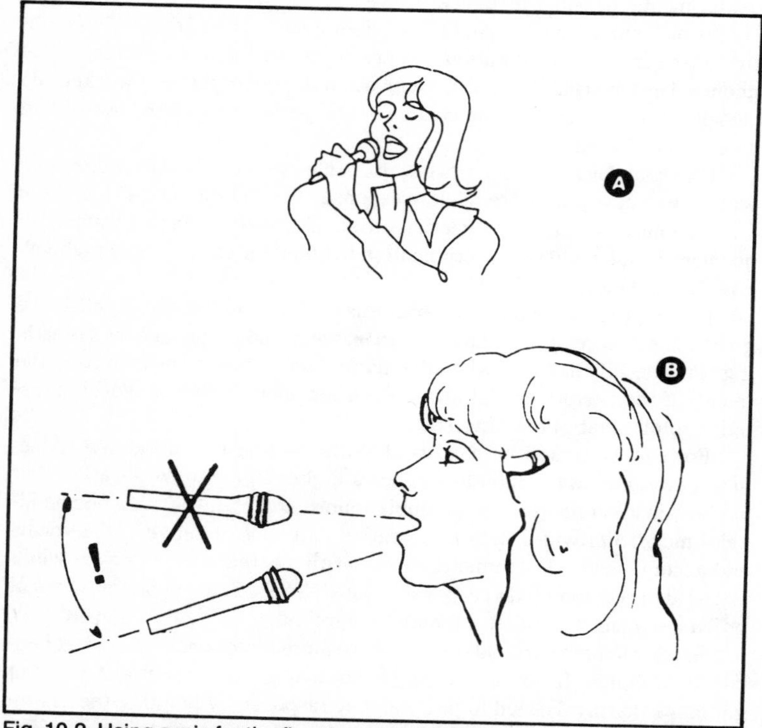

Fig. 10-2. Using a mic for the first time may require some experimentation (A). Angle of incidence (B).

Close-Up Vocals

There are a number of problems associated with close-up vocals. With a loud orchestra, or even with a loud succession of musical passages, the vocalist may try to overcome the background and the result may be preamplifier input overload distortion. *Overload distortion* is a condition that, in this instance, is produced when the output voltage of the microphone is larger than the permitted input signal voltage level of the preamplifier which follows the microphone.

With the microphone placed so close to the lips of the vocalist, it may produce breath pop. Some of the mouthed sounds, such as words using the letters p or t, are percussive. Also, with a hand-held microphone there is always the possibility of producing mechanical noise.

A windscreen will prevent breath pop and, of course, is necessary for outdoor use. The amount of *microphone handling noise*—mechanical noise—is a function of the construction of the mic. This is one of the features you don't see when you buy a microphone, but is something you should ask about when you do.

Distant Vocals

Not all vocalists enjoy holding a microphone, for a number of reasons (Fig. 10-3A). Some feel that the presence of the microphone hides part of the face, and for a vocalist trying to achieve personality projection, this can be a serious consideration. If a vocalist is really capable of "belting it out" the presence of the microphone implies that his or her voice requires an electrical assist, when this may not be the case at all. Also, the vocalist may be occupied with playing an instrument while singing, and a hand-held mic cannot be considered.

You can compensate for the fact that the microphone is now more than six inches away from the artist by turning up the gain of the preamp. This isn't an unmixed blessing, for when you do you also raise the level of the background noise. You may also introduce a condition of feedback (consisting of sound returned by the speakers to the mic) making the amplifier less stable than it should be. At its worst, feedback can produce howling, an extreme case. More subtly, feedback can raise the gain of the amplifier over a limited frequency range, giving tones that come within that range a sharpness they shouldn't have. The microphone is capable of picking up sound reflected from the stage floor or hard walls behind the vocalist. This can change the character of the sound going into the microphone.

With the microphone removed from the performer, one problem, that of breath pop, is automatically eliminated. The microphone that is needed in this situation is one that has a pickup pattern so that the microphone is most responsive to sounds coming at it from the front and insensitive to sounds reaching the microphone from all other directions. This means, then, that the microphone should have a cardioid pickup characteristic. Feedback will be much less critical; but if it persists, try moving the

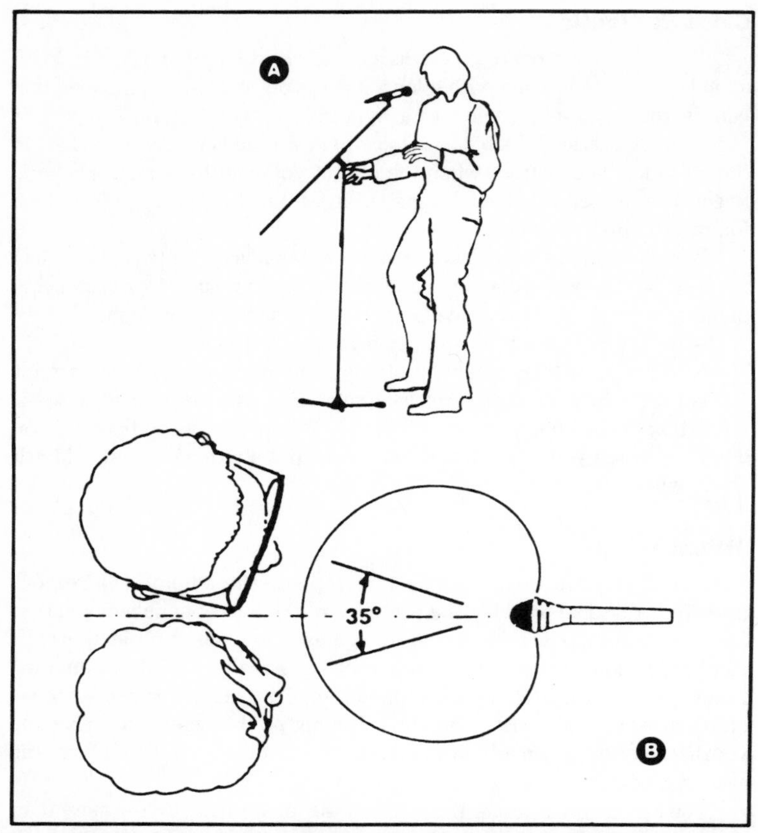

Fig. 10-3. (A) Some vocalists prefer using a mic stand. (B) Maximum angle of incidence for two vocalists.

speakers to a different location, or, if that isn't possible, move the vocalist to a new spot.

Background Vocals

Never let more than two people sing into one microphone. The angle of incidence should not be greater than 35° (Fig. 10-3B) because the cardioid microphone is relatively insensitive to sound arriving from the sides. In order to get the same level as for the solo voice, mixer gain would then have to be very high which would increase the feedback risk.

Vocalist With Piano Accompaniment

Whenever you have a vocalist with instrument accompaniment there is always the possibility that the instrument will override the vocalist (Fig.

10-4). This may force the vocalist to shorten the working distance in an effort to increase the vocal sound level. Recording, however, should never be a contest. Listen carefully to the balance between the instrument and vocal and adjust the instrument level accordingly. As a general rule, instrument level should never rise to more than 50 percent of vocal level. Making the instrument level much too high compared to vocal and background levels will tend to make the recording sound monophonic. The listener on playback should be able to visualize the location of the vocalist with respect to the accompaniment. If the vocalist is too loud compared to the instrument it will appear, on playback, that there is a considerable distance between the two performers. If the instrument is louder than the vocalist, it will seem as though the piano is right up against the vocalist.

When recording a vocalist accompanied by a piano there is a natural tendency to concentrate on the two performers and to ignore the acoustics of the auditorium. Reverberant sound and audience noise do contribute substantially to realism. You can use a pair of omnis, located away from the main pickup microphone and more toward the audience. The main microphone, a cardioid, will be more sensitive to the percussive effects of both voice and piano, while the omnis will add a sense of spaciousness to the sound.

THE BARBERSHOP QUARTET

Because harmony is a blend of separate voices, usually in different keys, each voice must retain its own particular vocal characteristics if harmony is to be maintained at audience level. If the balance of the voices is upset, the result can be quite the opposite of harmony. It's important, then, for such a group to consider a microphone with a linear response across the full frequency range, a microphone that responds clearly and is free of coloration. Further, because such small groups tend to cluster around the microphone, it should respond well to voices that reach it off axis so that individual vocal timbre is maintained.

THE POP GROUP

Use a cardioid for the vocal, an omni or cardioid for the drum set, and

Fig. 10-4. If vocalist uses excessive working distance to mic, accompanist can override the singer.

Fig. 10-5. Recording a small choral group.

a cardioid for the kick drum. For additional instruments use a cardioid or omni.

Keep all microphones very close. You'll need booms to give the performers play room. Use a microphone mixer to control sound balance and add microphone attenuation pads if the mixer input produces distortion. That's quite likely, particularly if the group lets itself go. Use a close microphone setup for feedback control in sound reinforcement.

SMALL GROUPS, ORCHESTRAS, AND CHOIRS

The primary recording technique uses a left/right spaced microphone pair. This is described professionally as A—B time-intensity stereo. The stereo spread is created by the difference in arrival time and loudness from any source to both microphones (Fig. 10-5).

In this arrangement the multiple sound sources, such as small groups, orchestras, and choirs, are progressively left and right, closer to one or the other microphone. The closer microphone collects the sound quicker and louder than the more remote microphone. The same precedence is perceived during playback and recreates the feeling of spacious, horizontal stereo spread.

ROCK GROUPS

In this recording situation, each instrument and the vocalist, if any, all try to outdo each other. Maximum loudness is the order of the day. The only practical solution seems to be to use a microphone for each instrument, with a separate boom or stand for the vocalist recording. Use cardioids. You'll need to mic in closely to keep any one instrument from being overwhelmed.

If a guitar is used in this group it will be an electric, sometimes working into a fuzz box. A *fuzz box* is an amplifier deliberately designed to distort sound or produce unusual sound effects. While some rock music seems to be an exercise in noise production, some is highly musical and beautiful. Listening to rock music is a subjective experience, as it is with all other music.

If you have never recorded before, starting with rock is an extremely ambitious undertaking, not only for the multiplicity of microphones involved, but simply because rock recording requires at least a measure of professionalism. However, if you can record rock and can produce acceptable playback, you can consider yourself to have graduated in the school of microphone use.

For recording rock, if you have a pair of electronic amplifiers, try to separate them widely and arrange for microphone pickup by having one cardioid for each, facing in to the cones of the speakers. For the drum set, mount one or two mics on stands, with the mic heads facing down but still out of the reach of any upward moving drum stick. Of course, all the mics have to be working in phase.

RECORDING A LARGE GROUP, ORCHESTRA, OR CHOIR

There is a simple approach to spacing a left/right microphone pair in front of a large orchestra or choir (Fig. 10-6). Stand on a center line facing the musical group. With the performers in action, move backward and forward, closer and farther away until you like the live sounds as a listener. From this preferred listening position, extend imaginary sight lines angling to the left and right edges of the performers.

You are now at the apex of a triangle. The musical group in front of you is the base. Put the left and right microphones on each angled sight line an equal distance from the base. The left to right spacing of the microphones, at the start, can be equal to one half the distance between yourself and the base line.

This is your basic starting setup. Now move the two microphones closer to or farther from the base line so you can control room reverberation. You can change the left/right microphone spacing to adjust stereo perspective. Keep in mind that the left/right microphones should be closer rather than farther apart to avoid having a hole in center sound reproduction.

Each microphone should be equally as high as possible on fully extended floor stands. Use microphone boom arms on the floor stands for even more height. Highly placed microphones angling downward help establish the loudness balance between the front and last rows of the performing group. Final positioning is achieved when balance and stereo perspective are acceptable on subjective playback evaluation tests.

There is also a less known, but remarkably effective stereo recording technique known as X—Y intensity stereo. The microphone pair (cardioids) is mounted on a stereo bar with the heads almost touching, as shown in the small insert in Fig. 10-6, and crossfiring left and right from one coin-

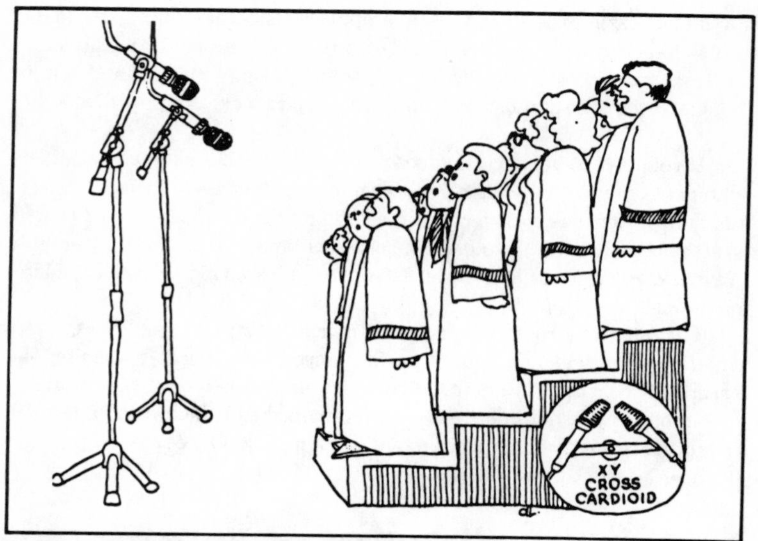

Fig. 10-6. Microphone placement for large vocal group. Inset (lower right) shows arrangement of X-Y cross cardioids.

cident position. Separate microphone tails from 90° to 120°, depending on the kind of group you are micing—that is, whether they are a narrow or wide group. Use on a boom, high in the center, several feet from the group.

When using the X—Y intensity stereo technique remember to reverse connections to your tape recorder input, since the right-position microphone covers the left side of the performing group, and vice versa.

You can also use an omni pair spaced for A—B stereo mounted on booms and shock absorbing stands. Try the mics one-third the group width apart and several feet down in front. Mount the mics as high as possible and angle them down to the last rows of performers. If reverberation is excessive, move the omnis closer to the group or else switch to cardioids.

Microphone pairs for spaced and coincident stereo sound collection should be identical models and have the same pickup patterns. If you use microphones that are different and whose frequency responses and pickup patterns aren't the same, you can get an unbalanced, disoriented stereo image in playback. The sound sources will not relocate correctly in playback and can swim back and forth in the stereo image.

You can use an omni pair or a cardioid pair. However, use only a cardioid pair for coincident stereo bar mounting, since only directional microphones will be most effective. When all sound sources are an equal distance from the two microphones, virtually in the same place, there will be no difference in arrival time, only changes in loudness.

The setup suggested here for music group recording requires a simple mixer for individual microphone level control and final balance in the recording. You can use the same setup for sound reinforcement.

The exact technique in recording large orchestras or choirs will largely depend on the size of the group, the size of the room, and its acoustics. If the room isn't excessively reverberant, a combination of direct sound and room reverberance will add warmth and body to the recording.

The problem of recording a church choir with organ is not so much in getting good sound but in getting too much low frequency organ sound. Here you should consider the advantage of a microphone designed to roll off slightly in the bass region. Low-end response would sound more natural, an advantage in reverberant situations. At the same time, a gradual, slight rise in the upper midrange and high frequencies would add vocal presence.

RECORDING A CHOIR OF SMALL CHILDREN

Use a cardioid pair mounted on a stereo bar. With one boom and a stand you can position the microphones centered and high, several feet from the choir. Separate the microphone tails from 90° to 120°, depending on the width of the group. Crossfire with the microphone heads almost touching.

If the choir has a solo singer, use a separate boom-mounted mic for this voice pickup. Start by recording the choir with the cardioid pair until you get the sound you want. Then bring in the soloist and adjust the soloist's microphone until you get the proper balance between soloist and choir.

GOSPEL, FOLK, AND COUNTRY VOCAL GROUPS

Use omnis or cardioids for best off-axis pickup fidelity of vocalists to the sides of the microphone. Make A—B separation at least three feet. Microphone X—Y on the stereo bar with cardioids only.

You'll probably find that the two-way cardioid is best for effective recording of a group vocal. This microphone supplies a more uniform response over a wider acceptance angle than you can get with single-element microphones (Fig. 10-7).

COUNTRY AND WESTERN GROUPS

Use a cardioid pair with left and right A—B stereo spacing. Keep the mics high and angled down. You will probably need to do some experimenting with microphone positioning. Start with the microphone stands several feet away from the group. Moving the mics closer to the band or farther away will control room sound and balance. Also, try both mics in an X—Y configuration on the stereo bar with one stand in the center.

RECORDING VOCAL SOLOISTS WITH PRESSURE ZONE MICROPHONES

From the standpoint of the recording engineer, the best setup would

Fig. 10-7. Recording Country group. Note contact mic on guitar. Bass uses instrument mounted mic.

be to put a PZM on a large, flat wall, and stand the performer in front of the mic, from two to four feet away. If you can convince the performer to sing to the wall, you'll get some great sound. But since some performers would be really uncomfortable with that you might be able to get them to sing to the control room window with the mic taped to the window.

Failing that, put the PZM on a 4′ × 4′ panel and mount the panel in front of the performer at about knee height, or hang the plate overhead. If you use a clear acrylic panel, this overhead position will also work well for stage reinforcement and not interfere with visibility. In many situations, both recording and reinforcement, the PZM will give you good pickup if you put it on the floor in front of the performer at a distance of about four feet.

In any situation where the soloist is being accompanied you will need to be careful that the PZM and the performer are positioned so that leakage does not become a problem. The PZM has a hemispherical pickup pattern with no axis and will pick up any sounds nearby. Isolation of the performer and mic with gobos or using separate rooms for performer and accompaniment may be necessary. A gobo is used when instruments must be acoustically isolated, as in multitrack recording. A gobo is simply a large, heavy frame working as an enclosure for the musician and is made as sound retardent as possible.

Interestingly, the PZM design does not provide a proximity effect, as frequency response does not change with distance. If the performer depends on proximity for some of the voice personality, it is suggested that you do *not* use a PZM.

Chapter 11

How to Use Microphones: Special Applications

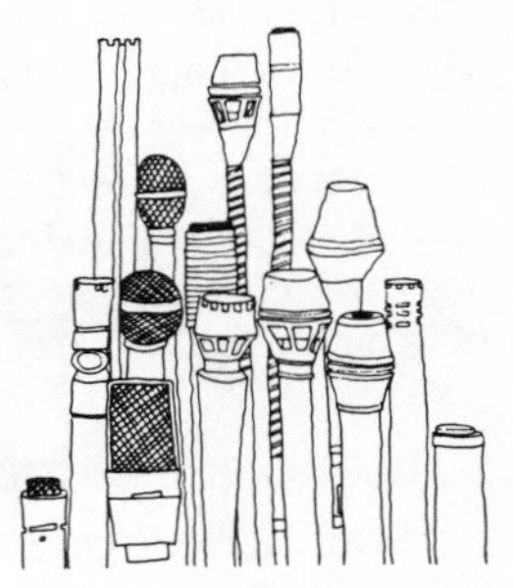

Once again continuing our discussion on how to use the microphone, this chapter covers application procedures and special techniques for using the microphone to capture the spoken word, whether it be a news conference, a speech, or a press conference—as well as recording *natural* sounds. There is also special information on ringing and howling and some final tips on using the microphone.

INTERVIEW AND CONFERENCE RECORDING

If you are recording an interview with one or more persons, an omni microphone or cardioid straight up in the middle of the group will supply the most uniform recording with noncritical microphone placement. This setup can be used for recordings made at a conference table.

If the recording is being made in a high ambient noise environment, cardioid mics can help because of their discrimination against unwanted sounds. Sometimes a stereo recording with two cardioids will provide better speech intelligibility. The directionality of the cardioid helps in speaker identification. If the environment is highly reverberant, a number of mics may be needed to lessen speaker to microphone distance for best clarity.

MICROPHONES AT THE PODIUM

Whether for lectures in an auditorium or church pulpit, microphone placement at the podium can provide the key to successful communications between a lecturer and his audience. Because single-element cardioids characteristically have good on-axis response with reduced high-frequency response off axis, two such microphones are often used on a podium to provide a wider coverage area, particularly if the lecturer has a tendency to move around. Two microphones placed at opposite corners of the podium and aimed at the lecturer provide variations in response when the lecturer moves from one side of the podium to the other. A better placement of the two microphones is in the middle of the podium, with the microphones aimed so that their heads touch at a 90° angle. Wider pickup coverage of the lecturer is provided without phase cancellation.

If one microphone is used, a two-way cardioid placed in the center of the podium provides much better coverage than a single-element unit (Fig. 11-1). If the microphone is placed in front of the system's loudspeakers, it may be necessary to put the microphone on a gooseneck to bring the microphone-to-subject distance closer and reduce acoustic feedback. In selecting a microphone for this use, choose one that has wide pickup angle and a switchable bass attenuator. This feature will provide further control of acoustic feedback.

LOW PROFILE MICROPHONES

Some persons get stage fright when using a microphone, especially for the first time. And, in some instances, when photographs are to be taken, one or more microphones can be intrusive elements. Lavalier microphones can be used, but as an alternative, low-profile mics, such as those in Fig. 11-2 are helpful. These are virtually invisible when placed on altars, lecterns, tables, and similar settings. Thus, the mics can be encased in a white foam "envelope" for use on such surfaces as white marble or linen-covered altars or encased in brown foam to blend with wood finishes.

LECTURE RECORDING

A lecturer seated at a desk can be troubled by reflection of sound from

Fig. 11-1. The microphone at the podium.

Fig. 11-2. Low profile mics are housed in acoustically transparent envelopes whose color matches that of the surrounding surface. (Shure Brothers, Inc.)

the desk top. If the microphone is 12 or more inches from the lecturer and is placed on a desk stand, tabletop reflections can cause phase cancellations and affect sound quality.

In reinforcing or recording conferences you may need to use a number of microphones to get adequate pickup. Excessive room reverberation is a common problem as more microphones are used. You can try out-of-phase pairs of microphones placed back to back.

RECORDING WITH HAND-HELD MICROPHONES

For hand-held use, reduced bass response below 150 Hz is desirable to minimize handling noise and proximity effect. Microphones designed for recording, as well as hand-held use, often have a recessed bass attenuator switch in the microphone shaft. Extended bass is provided in the flat position of the switch and bass attenuation at the other switch position.

Two-way cardioid microphones, with their extended bass response, are not generally recommended for hand-held use. Performers often like the proximity effect of the single-element cardioid. When close-talked the increased output lessens feedback and provides better separation of vocals from instruments. Omnis require less critical placement by performers and do not have proximity effect, but you will get increased stage band pickup and there is always the possibility of quicker acoustic feedback.

ORCHESTRA MICROPHONES

When using microphones to reinforce an orchestra located directly below the performer, place microphones so there is separation between orchestral and vocal sounds. If the orchestra is picked up by the vocalist's microphone, and vice versa, the sound-system operator will not be able to provide adequate separation and control of the sound sources.

In typical symphony orchestra situations where the sound of 100 or more musicians must be captured, "flying" the microphones is most effective and easiest. Physically, the microphones are out of the way; visually they're almost unnoticeable.

Typically, the microphones are placed 12 feet to 15 feet (Fig. 11-3) over the head of the conductor and angled toward the orchestra about 45°. If too low, the microphones will be over-responsive to the front of the orchestra; if too high, orchestral sound will be over-reverberant.

Try positioning each of the two microphones one-third in from opposite ends of the orchestra for a wide panoramic sound that, if recorded, will be noticeably stereophonic. Or try a one-foot mic-to-mic spacing, or cross-firing the microphones (in both cases aimed at the same 1/2-from-end of orchestra points) for a reduced, somewhat more natural perspective. Microphone placement in disc recording situations is a very critical business.

You may decide to highlight the sound of a particular instrument via a third microphone, as a solo French horn or tympani, for instance. But whatever microphone is selected, use it judiciously to pick up only the sound of the highlighted instrument.

GENERAL RECOMMENDATIONS FOR SOUND-SYSTEM OPERATORS

The amount of amplification before feedback in a sound reinforcement system is generally dependent on the microphone, loudspeaker, and room acoustics. In a difficult feedback-prone environment, the minimum number of microphones live at any time will assure minimum acoustic feedback. The acoustic gain available before feedback is reduced by 3 dB each time the number of live microphones in use is doubled. Four live micro-

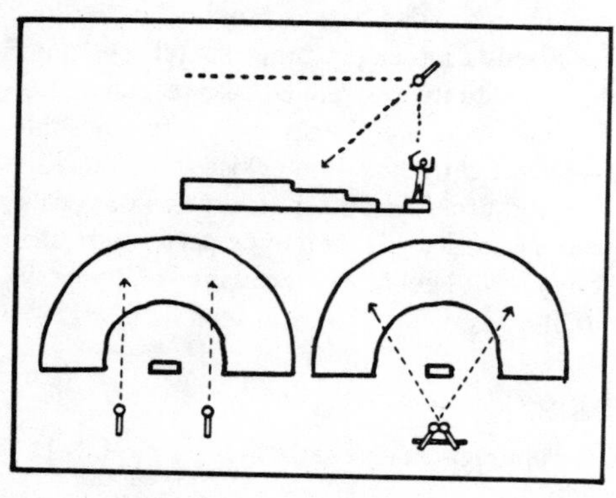

Fig. 11-3. Orchestral recording.

phones will reduce acoustic gain by 6 dB as compared to one microphone.

When the sound system must be operated close to the feedback threshold, try using low-frequency filters, either in the microphone or tone controls in the amplifier. This will minimize feedback. You should always have plug-in attenuator pads for those situations in which the performer practically swallows the microphone.

FILM AND STAGE MICROPHONE APPLICATIONS

Often mounted on a boom, the shotgun microphone allows microphone pickup control over a narrow angle, as required. A common assumption, but a mistaken one, is that a shotgun microphone will produce exceptional results in a typically reverberant room. In such a room, sound emanating from the sides or rear of the microphone will reverberate from the front walls of the room and come in front of the shotgun.

When used on a boom to pick up an actor's voice, a shotgun microphone can also pick up sound reflections from the wall behind the actor, minimizing intelligibility of the actor's words. When used under proper techniques, indoors and outdoors, the shotgun microphone is a workhorse of the film, stage, and broadcast industry. In outdoor use, a foam windscreen will provide a wind noise reduction of about 20 dB. Shock-mount shotgun microphones to booms or other mounts to minimize handling and boom noise.

MICROPHONE PICKUP OF STAGE PERFORMANCES

Tabletops provide sound reflection that cause phase cancellation in microphone pickup of an announcer seated at a table. Stage floors provide similar reflections and, when footlight microphones are mounted on short pipes above the stage floor, phase cancellation can result in supplying the sound with a distance quality. Try putting microphones on thin foam pads on the stage floor instead. This will supply full sound quality without sound cancellation and with minimum mechanical transmission of foot noise. Single microphones placed at intervals across the stage apron or pairs of microphones crossfired at 90°, provide uniform sound pickup with minimum visibility.

MICROPHONE APPLICATIONS FOR SOUND REINFORCEMENT

Microphone selection and placement for sound reinforcement applications adds another problem area—*acoustic feedback*. Generally, in a sound reinforcement arrangement, the microphone, loudspeaker system, and the room play equal parts in the successful or unsuccessful performance of the reinforcement system.

Acoustic feedback is caused when sound from the speaker system reaches the microphone at such intensity and phase relationship as to put the sound system into a condition of positive feedback. This results in a condition of oscillation or ringing. The sound fed back from the speakers to the microphone is in phase with the sound going into the microphone. The two, the *original* sound and the *reinforced* or *amplified* sound, strengthen each other and do so in a continuing process. If kept within limits, the result will be ringing, making the sound extremely sharp. If the feedback continues its growth, the result will be howling.

Sound feedback can occur in either a professional or home recording situation. Further, the ringing or howling will take place at a particular frequency or a small band of frequencies. You may find the condition occurring when a certain note is struck.

No two systems will feedback at the same frequency. A room has its fundamental and harmonic resonances, in addition to an amount of reverberation, that produce their effect on good sound. A microphone with a smooth frequency response on axis and uniform cancellation off axis can produce more amplification before feedback than can a microphone with peaky response and uneven cardioid rejection.

In use, the two-way cardioid microphone with its smooth on-axis response, smooth 90° response, and uniform back rejection at both high and low frequencies (as compared to many single-element cardioid microphones, which have good rejection at mid frequencies only) can provide up to 6 dB more amplification before feedback as compared to a similarly priced single-element cardioid microphone. Proximity effect, peaky high frequency response, and uneven back rejection cause acoustic feedback prevalent in so many sound reinforcement systems.

RINGING AND HOWLING

Where you are working with a three-mic setup, using a pair of cardioids for stereo pickup and a third, more remote omni to supply background sounds for presence, the single omni located too close to speakers can result in feedback. Try turning the speakers into a different position and, if possible, facing them away from the omni. If you have obtained the correct sound balance between the stereo cardioids and the omni, try moving the speakers further away from the microphones. You may also need to turn the amplifier gain down a bit. Positive feedback not only requires a pair of in-phase audio signals into the microphone, but the sound from the speakers must be of sufficient amplitude to trigger the beginning of ringing or howling.

The fact that a sound reinforcement setup has ringing but no howling is no cause for satisfaction. The condition is a highly unstable one and it is possible that the slightest increase in sound strength of a particular tone will throw the sound system from ringing into howling. You can consider a sound reinforcement system as stable only if there is no evidence of ringing at any sound frequency, that is, no matter what tone is being played, no matter what combination of tones are being played, and no matter how loudly they are being played.

RECORDING OUTDOORS

If you plan to do any outdoor recording, a windscreen is absolutely essential. The usual windscreen supplied with a mic will serve in almost all applications,but there may be times when the windscreen is just not enough. In such cases, use a nylon stocking and wrap the microphone head with it. You may need to use one or more stocking layers.

RECORDING MOTION

When recording the sound made by a moving object, you can convey a sense of motion in the recording by using more than one microphone. Put a number of microphones along the path the object will take. Microphone separation will depend on the speed of the moving object. The faster it moves, the more widely spaced the mics should be.

If you are recording a marching band, for example, position one microphone well ahead of the band to create the illusion of arrival. Other mics, spaced at distances of about 50 feet, will help supply the image of a band in motion. You will heighten the effect by using cardioids. The crowd noise level will be rather high, and it is debatable whether you should also include an omni. If you do, keep close control of the background sound level so it does not override your primary objective—the sound of the marching band.

RECORDING NATURAL SOUNDS

You can record natural sounds in various ways. One is to use a parabola or "dish" that works as a *sound collector*. This is simply a large, dish-shaped metal reflector and is quite inexpensive. If you want to pick up bird sounds, arrange the microphones around the area where the birds normally gather. Use two or more cardioids, pointing inward to the feeding area. Stay at a distance to keep from frightening the birds and watch the action with binoculars.

You can use one or more microphones to collect all sorts of outdoors sounds. You can tape such sounds and then, if you have a tape deck that permits recording sound-with-sound, you can add such sounds to the sound of a vocalist, an instrumentalist, or an orchestra, to produce some unusual results. Many outdoor sounds have musical quality. Wind sounds, ordinary street noises, the movement of trains or cars, marching bands, and even crowd noises can sound quite unusual and interesting when recorded and played back in the quiet of your listening room.

RECORDING OFF THE AIR

You can record AM or FM broadcasts by positioning a mic or mics somewhere in front of your speakers, but this isn't recommended. Even using cardioids, the mics will pick up room noise or outdoor sounds. This is one instance in which you can eliminate the microphones and get better results. All you need do is to use a patch cord to connect your receiver's output to the input of your tape recorder. If your receiver isn't equipped with sound-signal output terminals on its rear apron, buy an earphone plug for connection to the earphone jack of the receiver. Use a pair of flexible wires to connect to the plug and attach the other end of these wires to the AUX or LINE input of your tape deck.

SOME FINAL TIPS ON MICROPHONES

Here are a few last suggestions for using microphones.

☐ When using two mics as a stereo pair, use the same model microphone made by the *same* manufacturer. Intermixing two different models in a stereo pair will make it difficult to obtain a correct stereo balance.

☐ Remember that omnis pick up sound from all around and that cardioids favor front sound to rear sound.

☐ A microphone stand with a boom will supply more flexibility in recording.

☐ Very close-in use of the microphone means you may overload the recorder. When this happens, the result is distortion. Back off the mics, but if you must work in close, use microphone attenuation pads.

☐ Try the microphone positioning suggestions given earlier before using your own. *Then experiment.*

☐ The original sound source, whether vocal or instrumental, is direct or "dry" sound. Sound reflected from the walls, floor, ceiling, or other objects is reverberant sound. Adjust microphone spacing for best balance between the two.

☐ Keep left/right mics close to avoid a stereo hole.

☐ On multitrack overdubs there are clear sonic advantages in using the less expensive patternless omnidirectional mic when there is only one performer in the studio and there is no possibility of leakage or feedback.

☐ Mics perform better without obstruction. If you are recording indoors, and are not recording a vocalist, there is no need for a windscreen. A built-in windscreen that cannot be removed can become a liability instead of an asset.

☐ As a general rule, or as a starting point, use the following microphone spacing for single instruments:

strings and woodwinds	19 to 36 inches
harmonica	9 to 12 inches
accordion	19 to 36 inches
brass	6 1/2 feet
vocals	19 to 20 inches

☐ To reduce popping, see the illustrations in Fig. 11-4.

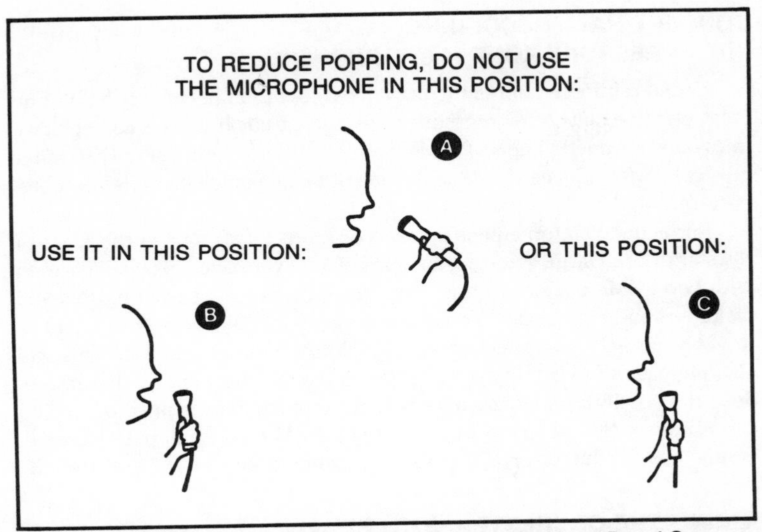

Fig. 11-4. How to avoid popping in vocal recording. Positions B and C can reduce sibilance also.

□ The greatest feedback resistance is achieved by placing the main speakers out of the microphones' pickup area, that is, at the front edge and to the sides of the stage. When singing behind the speakers, though, the vocalist will inevitably hear the voice softer than it is actually being reproduced. Using monitor speakers help, although at the expense of increased risk of feedback. Therefore, never point the microphones directly toward the monitor speakers (Fig. 11-5).

Feedback can also be triggered by resonances (determined by the acoustics of the recording room) particularly in the low-frequency region, i.e. (indirectly) by proximity effect. In this case, it is easy to stop the howling by increasing the working distance, or by turning down the bass on the mixer.

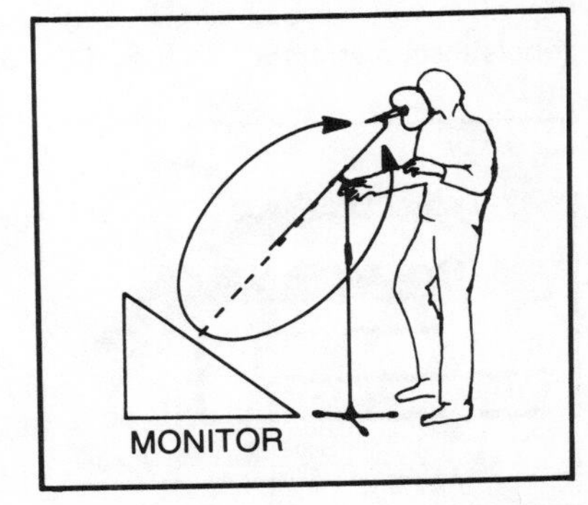

Fig. 11-5. Put monitor speakers out of microphone's pickup area.

CONCERT HALL RECORDING WITH PRESSURE ZONE MICROPHONES

Sound reinforcement is the prevailing condition in concert halls. The most effective setup for pressure zone micing could be the use of these mics on suspended panels. This is more objectionable in opera than other musical performances since the illusion of being somewhere else is so important.

For sound reinforcement put two or more PZMs across the front of the stage. Even large-stage productions of Wagner have been well covered with two PZMs positioned this way. You can also use carpeting to avoid pickup of audience noise. You can use carpeting, as shown in Fig. 11-6, to get a cardioid-like effect to avoid pickup of audience noise. This will also increase the visibility of the mics to the performers and thereby reduce the possibility that the mics will be stepped on or tripped over. You can reinforce the orchestra by mounting a PZM on a small panel directly behind the conductor or PZMs can be mounted on the walls of the pit.

MICING THE ORCHESTRA WITH PZMs

In general the PZM works best for orchestras if you think of it as the ears in your head. And just as each of us hear well with only two ears, you will find that the best orchestra PZM is micing is usually accomplished with a minimum number of mics. The PZM is *not* recommended for micing sections of the orchestra or for micing individual instruments in an orchestra simply because its pickup pattern and reach create a leakage problem which require complicated isolation problems.

For a disc compatible recording mount two pressure zone mics (Models 30GP or 31S are preferable) on 4′ × 4′ panels and suspend them above and in front of the orchestra so the panels form a V. This duplicates the coincident crossed pair configuration which is popular for conventional mics and will provide strong channel separation for stereo recording. If the panels are made of clear acrylic, this configuration can also be used for reinforcement.

If disc compatibility is not essential, mount two PZMs on panels as described, but suspend them in front of and above the orchestra so the mic panels form a straight line facing the orchestra and about 20 feet apart.

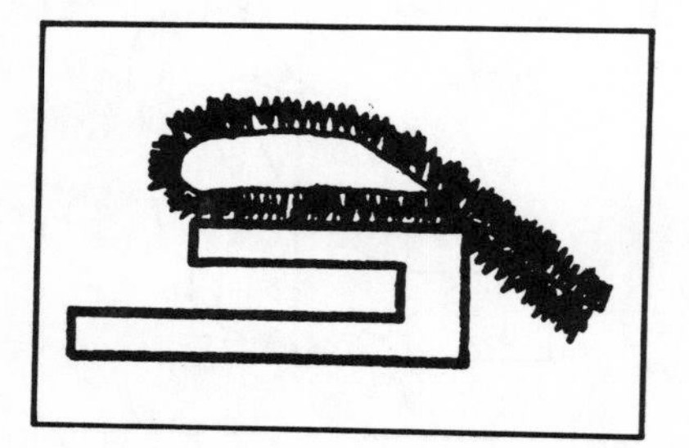

Fig. 11-6. Carpeting used to give PZM a cardioid-like effect.

This seems to be a good setup for PZM micing and has been selected over other conventional mic setups.

You can also place two PZMs on the floor, 10′ to 15′ in front of the nearest players with each mic centered on its half of the orchestra. In some situations a center PZM may be needed, also on the floor, right in front of the conductor. Some pattern shaping may be needed here to reduce audience noise pickup.

Reinforcement and live music recording may require an extra PZM for certain sections or solo instruments which need to be emphasized. Placement will depend on which section. Woodwinds, for instance, can be picked up well with a PZM on the floor. Brass, French horns, and tubas could be picked up off a back wall, particularly if the orchestra is playing in a shell. String sections can best be reinforced by a PZM mounted on a panel suspended above the section.

"For-the-record" tapes, where top sound quality isn't absolutely necessary, sound can be picked up well with very simple PZM setups, such as one on the floor in front of the orchestra, or on the back wall of a shell, or even taped to the proscenium arch.

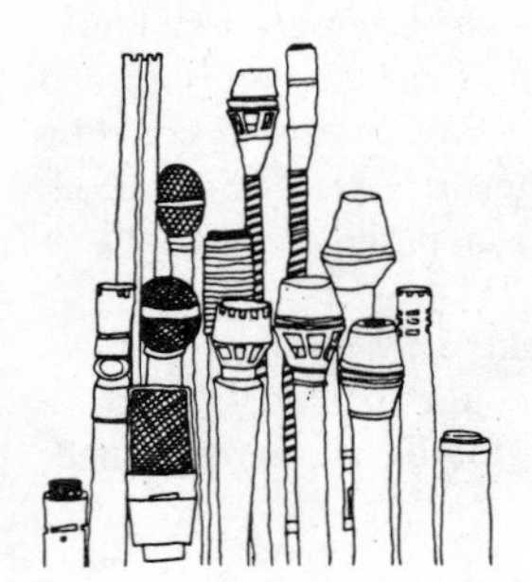

Chapter 12 Selecting the Right Microphone

There are no fixed rules in choosing a microphone or microphones for recording applications. The microphone selection and placement recommendations given in the preceding pages are just a start. Results will depend on your own creativity, patience, and your willingness to experiment. It will also be helpful if you have had some musical training. Also remember that the microphones you use in one recording environment may not necessarily be the best choice for another. The microphone positioning you have found to be suitable in one place may not be the way to record elsewhere. However, the more you experiment with microphones and the more you learn about them, the more you will know about their behavior and their characteristics and the greater will be your ability and confidence in using them.

A UNIVERSAL MICROPHONE?

The manufacturers of quality microphones go to considerable lengths to emphasize the fact that there is no single microphone that can be used for every recording condition. No microphone, no matter how well made, can lay claim to being a universal type. Mics, however, can be categorized broadly as falling into two basic groups: vocal and instrumental. But even here we are dealing in generalities.

In physical terms, microphones designed to be used with musical instruments are often smaller and lighter. In some instances they are mounted directly on or in the musical instrument, or are positioned on a stand or boom. They are made to be as non-intrusive as possible, with the windscreen made rather small.

Vocal mics have their own problems which must be met by the manufacturer. Since such mics are handled they must have minimum handling noise and cable noise. Often they will have a larger windscreen so as to minimize the popping sounds produced by consonants such as p and t, or to handle breathing sounds when the mics are used close in.

A mic designed for instrument use can easily have a different frequency response characteristic than a vocal type. An instrument mic may have an extended bass response and a smoother high frequency output.

A-B TESTING

An A-B comparison is seldom possible when evaluating the major components of a recording system. Mixers, recorders, or even monitor loudspeakers are rarely auditioned side-by-side, but mics are always subject to this acid test on a daily basis.

THE BEST MIC

The best mic isn't necessarily the most expensive. The best microphones are those that supply "final" sound, after all the processing, you find most satisfactory. The best mic isn't always the latest model or one that has just been released by a manufacturer. The values of many mics engineered 40 years ago are still valid today.

HOW TO SELECT A MICROPHONE FOR DIFFERENT INSTRUMENTS

Table 12-1 is designed to supply you with a guide in your choice of AKG microphones for various applications including vocals, guitars, strings, woodwinds/reeds, brass, keyboards, and percussions. Not listed in the chart because of lack of space is the CK-X series, a group of professional audio quality mics in a small package size. This allows the mic to be concealed in stage sets, to hang inconspicuously above a stage or choir, to be mounted in a drum set without being obstructive, to be used as a conference system microphone using a minimum of table space or any other application requiring a small, unobtrusive microphone of high quality.

The D-330BT

The D-330BT is the premier model in the D-300 series and embodies all of the sophisticated design principles that went into the creation of the series. The microphone has a variable bass-vs-distance contour marked bass emphasis or "proximity effect" when used close up; a progressively diminishing bass response when used farther away, and a presence-rise contour for added crispness. It has nine different combinations of bass roll-off presence-rise equalization. This microphone preserves critical signal-to-noise ratios in all of its equalization modes. It has a hypercardioid directional pattern that is more discriminating than a standard cardioid and is

Table 12-1. Selection Guide for AKG Microphones. CMS Is a C460B or C451 Preamp. CK9 Works Only with C451E (Not Listed).

Note: First (1) and second (2) choices are listed, but other microphone types may be used as well depending on individual preferences and priorities as to acoustic characteristics, sound coloration, application, and budget.

Type Nr.	C414EB/ C414EB-P48	C34	C422	CMS+CK1	CMS+CK1S	CMS+CK22	CMS+CK5	CMS+CK8	CMS+CK9	C535EB	C567E	D224E	D202E1	D310	D320B	D330BT	D12E	D130	D109	D125	D190E	D900E	D58E	D590B	D510B	D558B
Polar Response	variable	variable	variable																							
Application																										
VOCALS																										
Soloist	1			2		1	1			1				2	1	1										
Background voc.	1			2		1	1							2	1	1										
Chorus		1	1	2		1	1							2	1	1										
Speech	1				2	2		1		1	1	2	1	2	1						2		PA	PA	PA	PA
Conference			1					2				1	2								2					
Reporter						2		1																		
GUITARS																										
Acoustic	1	1	1	1						2	1	2				1				2	2					
Electric										2		1		2						1	2					
Mandolin	1			1							1									1	2					
Banjo				1								1								1	2					
E-Bass												2	2				1			1						
Zither				1	2			2			1	2									2					
STRINGS																										
Violin	1			2							1		2													
Cello	1			2							1		2													
Bass	1			2							2		1													

							FOR SPECIAL APPLICATIONS (Drama, etc.)			FOR SPECIAL APPLICATIONS (Outdoor & Film)									FOR SPECIAL APPLICATIONS (Interviews & Background Rec.)	FOR SPECIAL APPLICATIONS (Vocal Recording)			FOR SPECIAL APPLICATIONS (Outdoor & Film)			
WOODW./REEDS																										
Flute	1			2								1	1				1									
Clarinet	1			2				2						1												
Saxophone	1							2	2			1		1												
Mouth Organ	1				2				1		1			1		2	2				2					
BRASS																										
Trumpet	1			2				2	1			2	1	2							2					
Trombone	1			2					1		1	2	1	2							2					
Tuba	1			2					1		1	2	1	2							2					
Horn	1			2					1		1	2	1	2							2					
KEYBOARDS																										
Acoustic Piano, etc.	1			1								1	2	2								2				
Electric Piano												1	1	1							2	2				
Leslie Speaker (Top)													1	2				1			2	1				
Leslie Speaker (Bottom)													1	2				1			1					
PERCUSSIONS																										
Snares				1							1		1	2							2					
Toms				1										1		2					2	1				
Bass														2				1								
Cymbals	2			1				1	1		2			2												
Hi-Hat				1					1		2		1	2												
Triangle				1					1		2		1								2					
Vibraphone	1								1				2	1								2				
Congas				1									1	1							1	2				
Bongos				1					1				1	1							1	2				
Timbales	1										1		1	2							2					
Tympani				1									1	2				1								

also uniform with respect to frequency. See Table 12-2.

The D-330BT is a low-impedance balanced output unit fitted with a standard 3-pin male XLR type connector. The microphone is supplied complete with an SA-41 stand adapter and a foam lined vinyl protective case.

Frequency Response. This microphone has a normal, unequalized frequency range (i.e., frequency range unaltered by user adjustments of its low-frequency and/or mid-high-frequency response) of 50-20,000 Hz (Fig. 12-1), accompanied by the following on-axis characteristics: (1) a 16 dB variable bass-versus-distance contour at 100 Hz that ranges from −3dB response rolloff at a working distance of 1 m (≈ 3 1/4 ft) to +13 dB typical proximity-effect boost at a working distance of 1 cm (≈3/8 in.); (2) a fixed presence-rise contour from 1500 Hz to 12,000 Hz that boosts response +5 dB at 5000 Hz.

Directional Pattern. Under the normal, unequalized conditions previously specified, the microphone has a highly uniform hypercardioid directional pattern (Fig. 12-1) with respect to frequency as follows: (1) at a sound-incidence angle of 90 degrees, typical off-axis frequency response

Table 12-2. Technical Data for the D-330BT.

Transducer Type: Dynamic
Directional Characteristic: Hypercardioid
Frequency Range (bass, presence set norm): 50-20,000 Hz
Nominal Impedance at 1 kHz (bass, pres set norm): 370 ohms
Rated-Impedance Category: 250 ohms
Recommended Load Impedance: > 1000 ohms
Sensitivity at 1 kHz:

Open circuit:	1.2 mV/Pa; −58.4 dBV*
Maximum power level:	−58.5 dBm (re: 1mW/10 dynes/cm^2)
EIA Gm:	−156 dBm
High-Z output w/MC-25T, −25TS (optical).	−48.5 dBV at 1 Pa*
Tolerance:	+0, −1.5 dB

Sound Pressure Level for 1% THD:
  40 Hz, 1000 Hz, 5000 Hz: 128 dB
Hum Sensitivity (1 mG field; bass set norm): −143 dBm
Case Material: Nickel-plated die-cast zinc alloy
Net Weight: 340 g (≈ 12 oz)
Included Accessories:
  SA-41 flex. snap-in stand adapter with 5/8-in. −27 thread
  Foam-lined vinyl case
Optional Accessories:
  SA-26 clothespin stand adapter with 5/8-in. −27 thread
  PF-10 foam pop filter (red, blue, yellow, off-white, gray)
  GN-7E, GN-20E modular flexible-gooseneck kits
  KM-series floor and boom stands, stand accessories
  ST-series table stands
  MC-series microphone cable assemblies
*1 Pa (Pascal) = 10 μb = 10 dynes/cm^2 ≈ 94 dB SPL

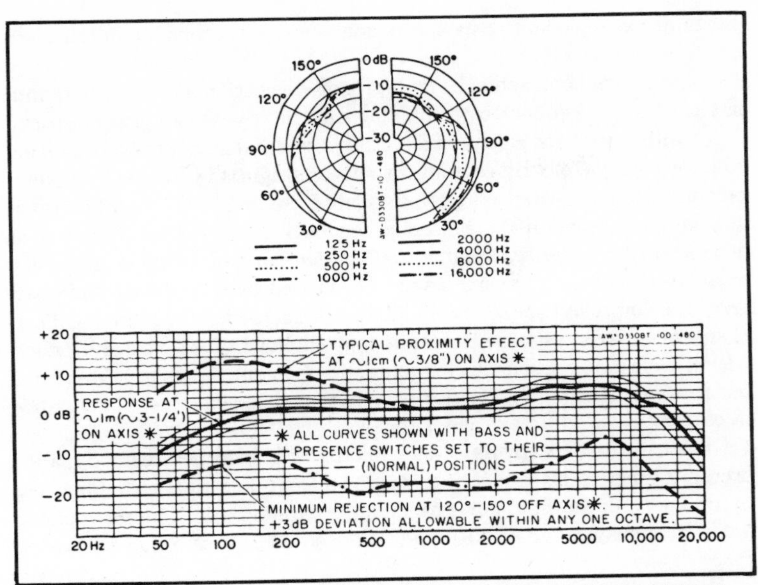

Fig. 12-1. Normal frequency response and polar pattern of the D-330BT.

does not vary more than 5-6 dB from 125-16,000 Hz; (2) at a sound-incidence angle of 120-150 degrees, minimum front-to-rear discrimination exceeds 10 dB at 100 Hz, 18 dB at 1000 Hz, and 15 dB at 5000 Hz. An effective hypercardioid pattern is maintained over the entire frequency range.

Impedance and Sensitivity. Under the normal, unequalized conditions previously specified, and at 1000 Hz, the microphone has an open-circuit sensitivity of 1.2 mV/Pa (−58.4 dBV) and a nominal (true) impedance of 370 ohms. However, in accordance with various industry standards for calculating other sensitivity figures from this open-circuit value, the microphone's rated impedance is categorized as follows: (1) for a maximum-power-level output of −58.5 dBm (re: 1mW/10 dynes/cm^2), the rated impedance is categorized as 250 ohms; (2) for an EIA sensitivity rating (G_m) of −156 dBm, the rated impedance is categorized as 600 ohms. The microphone is capable of handling a maximum sound-pressure level of 50 Pa (128 dB SPL) at 1000 Hz with distortion not exceeding 1%.

The D-58E

D-58E was designed for close-talk use wherever high ambient noise levels, feedback-prone environmental acoustics, wide temperature extremes and moderately high humidity levels are problems. It can be used either gooseneck-mounted or hand-held. In either configuration it is recommended for general radio communications, paging, public address, talkback/intercom and specialized broadcast applications. It can be used as a newsdesk microphone in a newsroom or as a sports announcer's microphone in crowded, noisy stadiums and arenas.

The microphone achieves its high noise rejection and relative immunity to acoustic feedback by combining a dynamic pressure gradient transducer with a front and side-ported housing in a differential design technique. The microphone clearly reproduces speech originating within its recommended working distance (approximately 5 cm or 2″) but greatly attenuates low-frequency on-axis noise and feedback components originating at a distance of 1 meter (approximately 3 1/4 feet) or more. Higher frequency noise and feedback components are attenuated by its hypercardioid pattern. For improved speech intelligibility in narrowband communications channels, on-axis response is intentionally emphasized between 1 kHz and 5 kHz. This rising response characteristic also contributes to better noise penetration in paging applications where the loudspeakers are likely to be in extremely noisy areas. See Table 12-3.

Frequency Response. The microphone incorporates a dynamic pressure-gradient transducer enclosed in a front- and side-ported housing to produce a differential noise-cancelling characteristic. The microphone has a frequency range of 70-12,000 Hz (Fig. 12-2) with an on-axis rising

Table 12-3. Technical Data for the D-58E.

Transducer Type: Dynamic
Directional Characteristic: Hypercardioid
Frequency Range: 70-12,000 Hz
Nominal Impedance: 200 ohms
Recommended Load Impedance: $\geq$ 500 ohms
Sensitivity at 1 kHz:

Open circuit:	0.072 mV/μb; −82.9 dBV
Maximum power level:	−62 dBm (re: 1 mW/10 dynes/cm^2)
EIA Gm:	−154.5 dBm
Tolerance:	+, −2.5 dB

Sound Pressure Level for 0.5% THD:
1000 Hz: 128 dB
Hum Sensitivity: −127 dBm (1 mG field)
Temperature Range: −20° C (≈ −4°F) to +60°C (≈ +140°F)
Maximum Relative Humidity: 90%
Case Material: Nickel-plated brass; steel-wire mesh
Net Weight: 45 g (≈ 1-1/2 oz.)
Included Accessories: Foam-lined vinyl case
Optional Accessories:
GN-7E, GN-14E, GN-20E
Modular Flexible Gooseneck kits
KM-221C flange adapter
KM-237 clamp adapter — for use with
KM-238 clamp adapter — either gooseneck
ST-5 table stand
ST-305 anti-shock table stand
W-20 foam windscreen

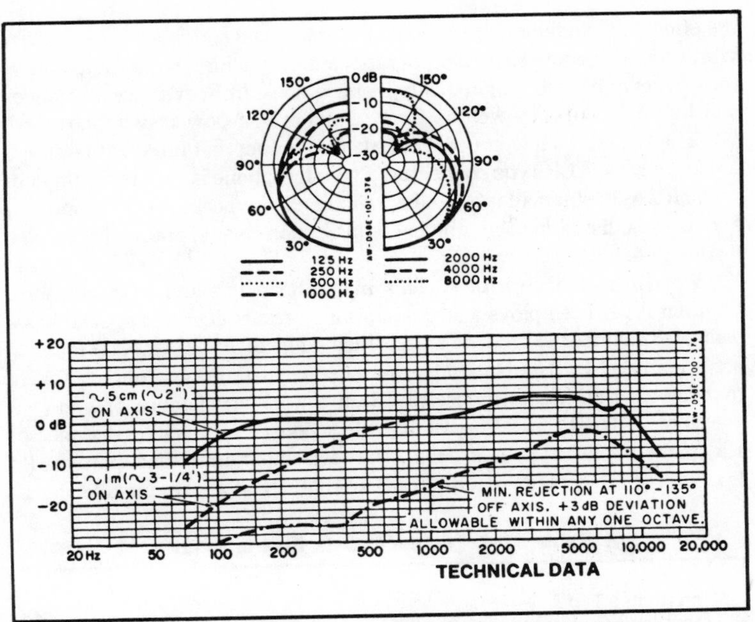

Fig. 12-2. Frequency response and polar pattern of the D-58E.

response between 1000 Hz and 5000 Hz for improved speech intelligibility in narrowband communications channels. When used within a working distance of 5 cm (≈2 in.), the microphone's rejection of on-axis noise originating at a distance of 1 m (≈3 1/4 ft) or more exceeds 16 dB at 100 Hz. The microphone has a hypercardioid directional pattern. The off-axis discrimination exceeds 16 dB at 1000 Hz at a sound-incidence angle of 110-135 degrees, and an effective hypercardioid pattern is maintained over the entire frequency range.

Impedance and Sensitivity. The microphone has a nominal impedance of 200 ohms. The output level is −62 dBm (re: 1 mW/10 dynes/cm^2), and the microphone is capable of handling a maximum sound-pressure level of 500 μbar (128 dB SPL) at 1000 Hz with distortion not exceeding 0.5%. The EIA sensitivity rating (G_m) is −154.5 dBm.

The D-12E

This microphone is designed for sound pickup from the bass drum, the E-flat electric guitar, low-frequency driver of the Leslie, string bass, deep brass and woodwinds, without susceptibility to various forms of low-frequency noise.

The microphone is a cardioid dynamic using a moving-coil transducer having an unusually large-diameter diaphragm coupled to a bass resonator chamber. The transducer element is spring suspended for isolation from the effects of handling noise, mechanical shock and spurious low-frequency vibration. The transducer incorporates a hum-bucking winding to cancel the effects of electro-magnetically induced noise from ever-present power and lighting cables as well as from dimmers and power switchboards.

The D-12E is a low-impedance balanced output unit fitted with a standard 3-pin male XLR-type connector. The microphone is supplied complete with an SA-30 stand adapter and a foam lined vinyl case. Several optional accessories, listed in the technical data section below, are available. See Table 12-4.

Frequency Response. This microphone is a dynamic pressure-gradient type. It employs a large-diaphragm transducer and special bass-resonator chamber to produce a frequency range of 30-15,000 Hz (Fig. 12-3) accompanied by a 10 dB variable bass-versus-distance contour at 100 Hz that ranges from flat response (0 dB) at a working distance of 1 m (≈3 1/4 ft) to +10 dB typical proximity-effect boost at a working distance of 1.5 cm (≈ 5/8 in.). The microphone has a cardioid directional pattern. Its front-to-rear rejection exceeds 12 dB at 1000 Hz at a sound-incidence an-

Table 12-4. Technical Data for the D-12E.

Transducer Type: Dynamic
Directional Characteristic: Cardioid
Directional Reference:
Nickel grille = front of mic; black grille = rear of mic
Frequency Range: 30-15,000 Hz
Nominal Impedance at 1 kHz: 290 ohms
Rated-Impedance Category: 250 ohms
Recommended Load Impedance: ≥ 500 ohms
Sensitivity at 1 kHz:

Open circuit:	2.2 mV/Pa; −53.2 dBV*
Maximum power level:	−53 dBm (re: 1mW/10 dynes/cm^2)
EIA Gm:	−145 dBm
High-Z output w/MC-25T, −25TS (optional):	−43 dBV at 1 Pa*
Tolerance:	+2, −1.5 dB

Sound Pressure Level for 0.5% THD:
40 Hz: 120 dB; 1000 Hz, 5000 Hz: 128 dB
Hum Sensitivity (1 mG field): −132 dBm
Case Material: Steel-wire mesh; nickel-plated brass; black trim
Net Weight: 480 g (≈ 17 oz)
Included Accessories:
SA-40 flex. snap-in center adapter with 5/8-in. −27 thread
Foam-lined vinyl case
Optional Accessories:
SA-26 clothespin stand adapter with 5/8-in. −27 thread
KM-series floor and boom stands, stand accessories
ST-series table stands
MC-series heavy-duty microphone cable assemblies

*1 Pa (pascal) = 10 μb = 10 dynes/cm^2 ≈ 94 dB SPL

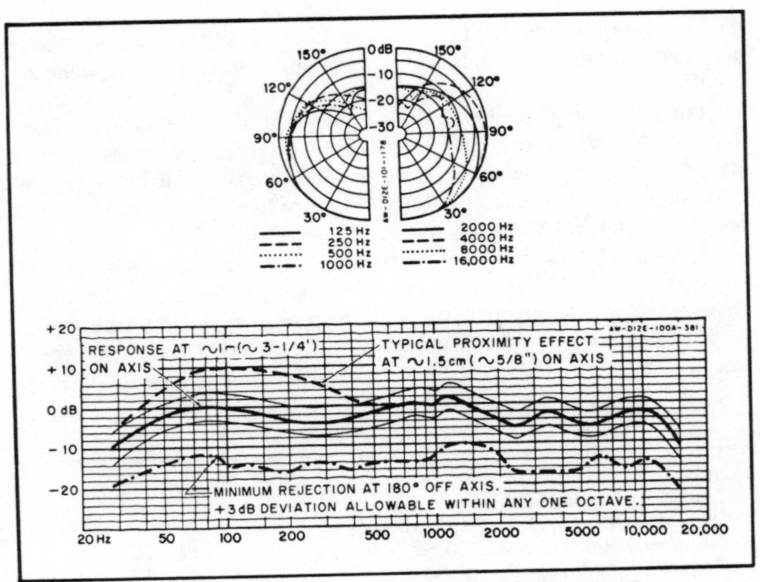

Fig. 12-3. Frequency response and polar patterns of the D-12E.

gle of 180 degrees, and an effective cardioid pattern (Fig. 12-3) is maintained over the entire frequency range.

Impedance and Sensitivity. The microphone has a nominal (true) impedance of 290 ohms. The output level is −53 dBm (re:1 mW/10 dynes/cm^2), and the microphone is capable of handling a maximum sound-pressure level of 50 Pa (128 dB SPL) at 1000 Hz with distortion not exceeding 0.5%. The EIA sensitivity rating (G_m) is −145 dBm.

The CK 67/3

This unit is an electret-condenser lavalier, a phantom-powered unit designed for clip-on applications in TV, film and theatrical or general sound reinforcement. It has an exceptionally wide frequency range and low distortion at high sound-pressure levels. Its extremely small size and nonreflective satin-black chrome-plated finish make it unobtrusive. See Table 12-5.

The CK 67/3 is an integrated, non-modular unit consisting of a microphone "head" (the elastically suspended electret-condenser transducer with its FET impedance-converter preamp, encapsulated in an all-metal housing having a sintered-bronze grille); a 1.3 m (≈4 ft) flexible cable; and an all-metal output module containing an electronic dc-regulator circuit plus an audio-output transformer and standard XLR-type connector. (Should field service become necessary, the integrated transducer-preamp-cable assembly can be removed for replacement simply by opening the output module, unsoldering two leads, unscrewing the sintered-bronze grille from the head housing, and pulling the entire assembly out of the housing.) A head-and-cable assembly only is also available for wireless-microphone applications.

For improved reliability and for prevention of damage from battery leakage, this microphone does *not* use an internal battery. Rather, it uses *external* 9-52V phantom power, which can be obtained from an associated mixer or recorder, or from any of AKG's own in-line ac or battery-operated power supplies.

The CK6713 is a low-impedance balanced-output unit fitted with a standard 3-pin male XLR-type connector. For maximum versatility, the microphone is supplied complete with *four* accessories: an H-20 tie tack and an H- tie bar, each of which holds one CK 6713 head; a W-37 black wire-mesh windscreen; and an H-16 belt clip for use with the output module. An op-

Table 12-5. Technical Data for the CK 0713.

Transducer Type: Electret condenser
Directional Characteristic: Omnidirectional
Frequency Range: 20-20,000 Hz
Nominal Impedance: 200 ohms
Recommended Load Impedance: $\geq$ 500 ohms
Sensitivity at 1 kHz:

Open circuit:	6 mV/Pa; −44.4 dBV*
Maximum power level:	−43.5 dBm (re: 1 mW/10 dynes/cm^2)
EIA Gm:	−136 dBm
Tolerance:	±3 dB

Sound Pressure Level for 1% THD:
1000 Hz: 132 dB
Typical Self-Noise:
CCITT C-wtd: 2.2 μV (equivalent SPL: 25 dB)
IEC 179 A-wtd: 1.4 μV (equivalent SPL: 21 dB)
Hum Sensitivity (1 mG field; 60 Hz): −151 dBm
Case Material: Zinc alloy (head); brass (output module)
Net Weight: 100 g (≈ 3-1/2 oz)

Included Accessories:
H-20 tie tack for one microphone head
H-21 tie bar for one microphone head
W-37 black wire-mesh windscreen
H-16 belt clip for use with output module

Optional Accessories:
H-22 tie bar for one or two microphone heads
18E battery power supply for one mic to balanced input
N-62E ac pwr supply for two mics to balanced inputs
N-66E ac pwr supply for six mic to balanced inputs
MC-series heavy-duty microphone-cable assemblies

*1 Pa (Pascal) = 10 μb = 10 dynes/cm^2 ≈ 94 dB SPL

Table 12-6. Technical Data for the C-535 EB.

Transducer Type: condenser (self-polarized) cardioid (pressure gradient receiver).

Frequency response: 20 to 20,000 Hz ±3 dB from standard curve.
Sensitivity: 9 mV/Pa ≙ 0.9 mV/μb ≙ 61 dBV (with pre-attenuation: 1.8 mV/ Pa ≙ 0.18 mV μb ≙ −75 dBV.

Pre-attenuation: −14 dB.
Bass roll-off: from 500 Hz downward 6 dB/octave.
Bass cut-off: below 100 Hz (12 dB/octave).

Total Harmonic distortion with 500 ohms load and 1 kHz: ≧ 0.5% (50 Pa ≧ 128 dB SPL), ≦ 1% (80 Pa ≙ 132 dB SPL).
Equiv. Noise Level: 21 dB SPL (measured with filter CCITT-C according to DIN 45405).

Powering: Univ. Phantom Powering (9 to 52 volts according to DIN 45596).
Dimension: 45/25 ϕ × 183 mm (1.8/1.0 ϕ 7.2 in).
Current consumption ≦ 1 mA.
Weight: 300 g (10 oz) net, 780 g (1.5 lb) shipping.

Included Accessories: SA 41 stand adapter.

tional accessory, the H-22 tie bar, is capable of holding one *or* two CK 6713 heads—thus making dual feed or redundant backup of lavaliers convenient.

The C 535 EB

This condenser vocalist microphone is designed for professional use in recording studios and on stage. The C 535 EB is a pressure gradient receiver with cardioid polar response.

The microphone head is made of satin charcoal shock-absorbing stainless steel wire mesh having a built-in pop filter, with the transducer elastically suspended to minimize handling noise.

Light reflection is reduced by a satin charcoal chromium-plated housing. For quick service, the entire head can be detached and replaced if necessary. See Table 12-6.

Bass Cut and Bass Rolloff. A built-in switch provides for pre-attenuation and filter (bass out and bass rolloff). The switch, built in the microphone shaft, is a four position type, and is used to produce these characteristics:

- ☐ Full sensitivity, full frequency response
- ☐ Full sensitivity, cut-off below 100 Hz, with about 12 dB/octave
- ☐ −14 dB sensitivity, full frequency response
- ☐ −14 dB sensitivity, rolloff with about 6 dB/octave below 500 Hz

Figure 12-4 shows the switch positions.

The C422 (Large Diaphragm)

The C422 is a studio condenser microphone that has been specially designed for sound studio and radio broadcasting. The microphone head holds two twin diaphragm condenser capsules elastically suspended to protect against handling noise.

The rugged wire mesh cap protects the capsules from mechanical damage. It is differently colored at the two opposing grille sides (bright = front grille side; dark = rear grille side) thereby allowing the relative position of the two systems to be easily checked. The microphone head as a whole may be rotated by 45° about the shaft in counter-clockwise direction, position "0°" and "45°" being lockable. This will allow quick and exact change-over from 0° (for MS-stereophony) to 45° (for XY-stereophony), even when the microphone is rigidly mounted. The upper system may be rotated by 270° against the lower one. A scale on the housing adjustment ring and arrow-shaped mark on the upper system will allow the base angle to be exactly adjusted. Any desired readjustment can be easily effected by means of the exact angle marking. The C422 is characterized by another special feature: since in sound studio work and radio broadcasts it is often necessary to recognize the respective positions of the two systems even from a greater distance, two light emitting diodes with a particularly narrow light emitting angle have been incorporated both in the upper rotatable and in the lower system.

The LED's are energized by a conventional 9V battery (IEC 6F22) or 9V accumulator cell which is continuously charged during operation via phantom power supply. The LED's are switched on/off by a separate switch. Enclosed within the microphone shaft there are separate FET-preamplifiers for each channel which are characterized by particularly high input impedance, extremely low internal noise, and high crosstalk attenuation. The output level of both channels can be simultaneously lowered by 10 or 20 dB.

The C422 is connected to a S42E remote control unit which allows any one of 9 polar patterns to be selected for each channel. Due to noiseless selection, polar pattern changeover is possible even during recording. See Table 12-7.

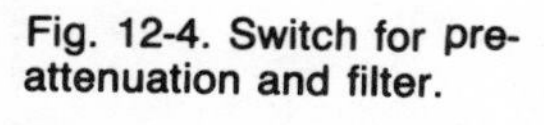
Fig. 12-4. Switch for pre-attenuation and filter.

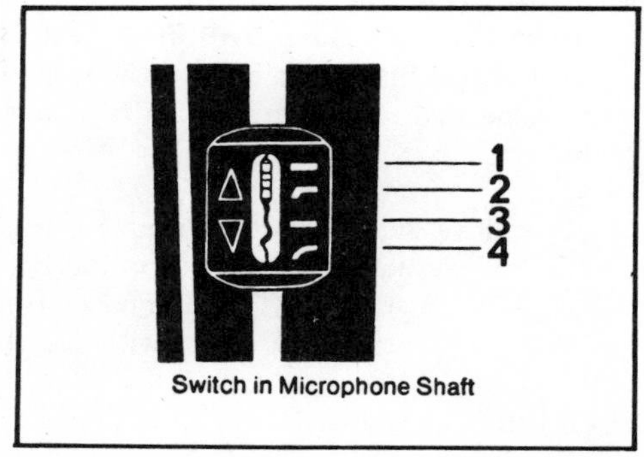

Table 12-7. Technical Data for the C422.

Mode of Operation: 2 twin diaphragm systems, pressure gradient receivers

Frequency Range: 20-20kHz

Sensitivity at 1000Hz: 6mV/Pa, −64.5dBV on no-load operation.

Electrical Impedance: 200 ohms ± 20%, balanced, floating

Normal Load Impedance: ≥ 500 ohms

Equivalent Noise Level: ≤ 22dB (filter CCITT-C, DIN45405)

Weighted Noise Level: Approx. 1.6uV RMS (filter CCITT-C, DIN45405)

Crosstalk Attenuation:
≥ 70dB (20 to 10kHz)
≥ 40dB (20 to 15kHz)
(measured without capsules)

Magnetic Field Interference Factor:
(at 50Hz) 5uV/5uT

Max. Sound Pressure Level: (for 100 Hz) and 500 ohms load impedance, harmonic distortion k = 0.5% 92 Pa ≙ 133 dB SPL

Admissible Climatic Conditions: Temperature Range (−20°C/−4° to 60°C/148°F)
Relative Humidity (99% at 20°C/−4°F)

Supply Voltage: 9-52V according to DIN45596 via S42E with phantom power supply

Current Consumption:
5mA max. for misc. (each ch.)
Approx. 50mA for LED's

Dimensions: 33/42 mm (1.29/1.65 in) ϕ × 235 mm (9.25 in) in length

Connector Type: 12 pica miniature DIN plug

Finish: Satin-black, chromium-plated all-metal housing.

Weight: Approx. 430g (15.18 oz.)

Included Accessories: W42 foam windscreen, H15/9 elastic suspension, MK42/20 cable (20m), S42E remote control unit, individual frequency curves, case

Recommended Accessories:
SA42 stand adapter, H42 suspension ring, N62E power supply unit (2 ch), N66E power supply unit (65 ch)

Frequency and Polar Response Curves. The frequency response characteristics of the C422 are shown as omni, cardioid and figure-eight in Fig. 12-5.

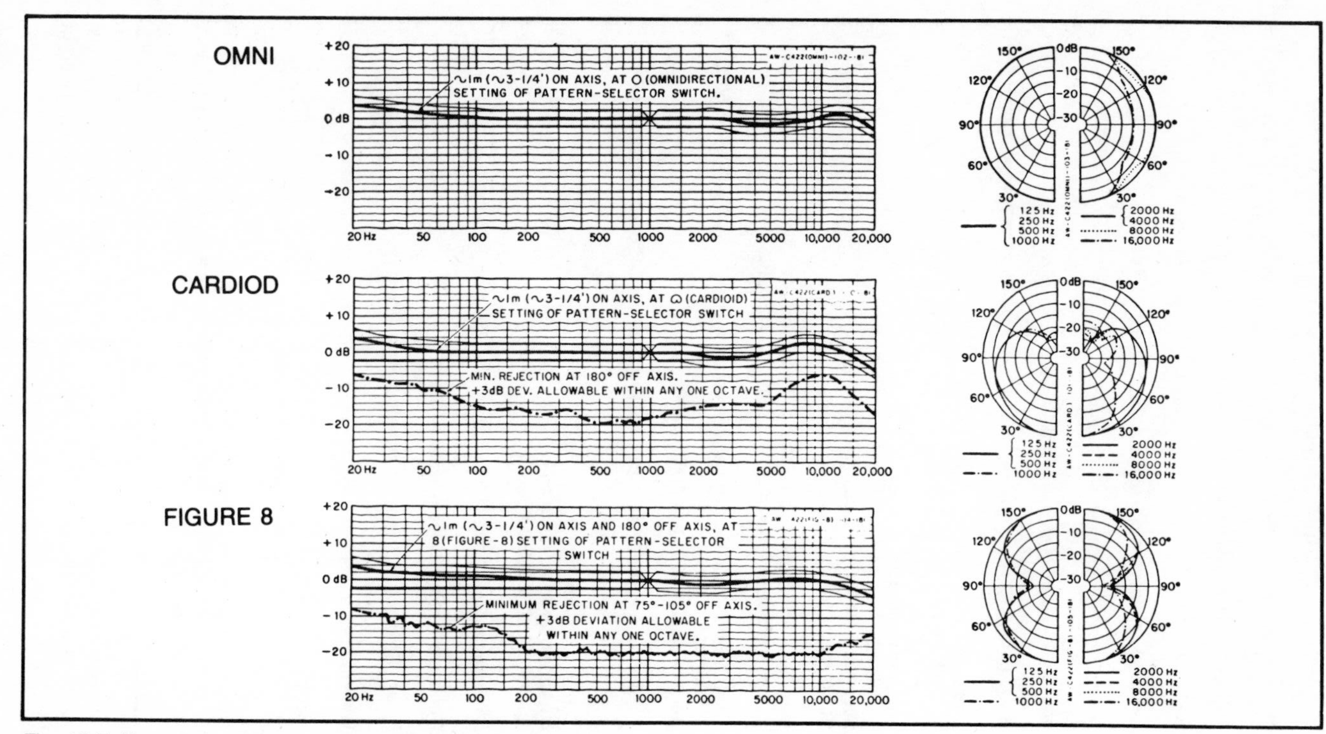

Fig. 12-5. Frequency and polar response curves of the C422.

The C34 (Small Diaphragm)

The C34 is a stereo condenser microphone specially designed for sound studio, radio broadcast and mobile use. The microphone head holds 4 condenser capsules elastically suspended to protect them from handling noise.

A rugged wire mesh cap protects the capsules from mechanical damage. The upper system can be rotated by 180°, whereas the lower one is rigidly connected with the microphone shaft. This allows different base angles to be easily adjusted for stereo transmissions. The wire mesh cap is differently colored at the two opposing grille sides (bright = front grille side, dark = rear grille side) thereby allowing the relative position of the two systems to be easily checked. An additional scale on the rigid housing ring and an arrow-shaped mark on the rotatable housing ring allows the base angle to be adjusted exactly and readily readjusted, if desired. Enclosed within the microphone shaft there are separate FET-preamplifiers for each channel which are characterized by particularly high input impedance, extremely low internal noise, and high crosstalk attenuation. The satin-black finish of the microphone assists in preventing disturbing light reflections in film and TV-studio work. Technical data for the C34 are supplied in Table 12-8.

The C34 is connected to an S42E remote control unit which allows any one of 9 polar patterns to be selected for each channel. Due to noiseless selection, polar pattern changeover is possible even during recording.

Frequency and Polar Response Curves. The frequency response characteristics of the C34 are shown as omni, cardioid and figure-eight in Fig. 12-6.

The D-109

The AKG D-109 is an omnidirectional microphone electroacoustically designed for optimum performance when worn lavalier-mounted. A compact and light-weight instrument, the D-109 combines accurate speech reproduction with unobtrusive appearance. It is therefore ideal for television, videotape, and film productions requiring an inconspicuous on-camera microphone. The D-109 is also recommended for "hand-free" use in churches, schools, lecture halls, and conference rooms.

Unlike a conventional microphone (which exaggerates low frequencies and muffles high frequencies when "adapted" for lavalier use), the D-109 has a specially contoured frequency response that enables it to provide naturally balanced speech reproduction when placed against the user's chest. Response is intentionally rolled off below 300 Hz to reduce tonal coloration caused by chest-cavity resonance. Furthermore, with the ring-shaped lavalier clip raised fully above the level of the microphone grille, response is boosted between 2 kHz and 6 kHz to compensate both for the lack of high-frequency propagation below the chin and for the filtering effect of clothing that may be used to conceal the microphone. (The peak at 6 kHz may be reduced as required simply by lowering the lavalier clip, as shown

Table 12-8. Technical Data for the C34.

Mode of Operation: 4 condenser capsules, pressure gradient receivers.

Frequency Range: 20-20 kHz

Sensitivity at 1000 Hz: 0.45 mV/μbar ≙ 4.5 mV/Pa, −62.5 dBv on no-load operation

Electrical Impedance: 200 ohms ±20 %, balanced, floating

Normal Load Impedance: ≥ 500 ohms

Equivalent Noise Level: Approx. 26 dB (filter C1 TT-C, DIN45405)

Weighted Noise Level: Approx. 1.8u V RMS (filter CCITT-C, DIN45405)

Crosstalk Attenuation:
≥ 70 dB (20 to 10 kHz)
≥ 40 dB (20 to 15 kHz)
(measured without capsules)

Magnetic Field Interference Factor: (at 50 Hz) 3.5 mV/5uT

Max. Sound Pressure Level: (for 1000 Hz and 500 ohms load impedance, harmonic distortion k = 0.5%) 80 Pa ≙ 132 dB's PL

Admissible Climatic Conditions:
Temperature range (−20°C/−4°N to 60°C/148°F)
Relative Humidity (99% at 20°C/−4°F)

Supply Voltage: 9-52 V according to DIN45596 via remote control unit S42E

Current Consumption: 5 mA maximum for each channel

Dimensions: 33/26.5 mm (1.29/1.04 in) ϕ × 196 mm (7.92 in) in length

Connector Type: 12-pin miniature DIN-plug

Finish: Satin-black, chromium-plated all-metal housing.

Weight: 280g (9.88 oz.)

Included Accessories: W34 foam windscreen, H15/6 elastic suspension, individual frequency curves, MK42/20 cable (20m,),
S42E remote control unit, case

Recommended Accessories: SA30 stand adaptor, N62E power supply unit (2 ch), N66E power supply unit (6 ch)

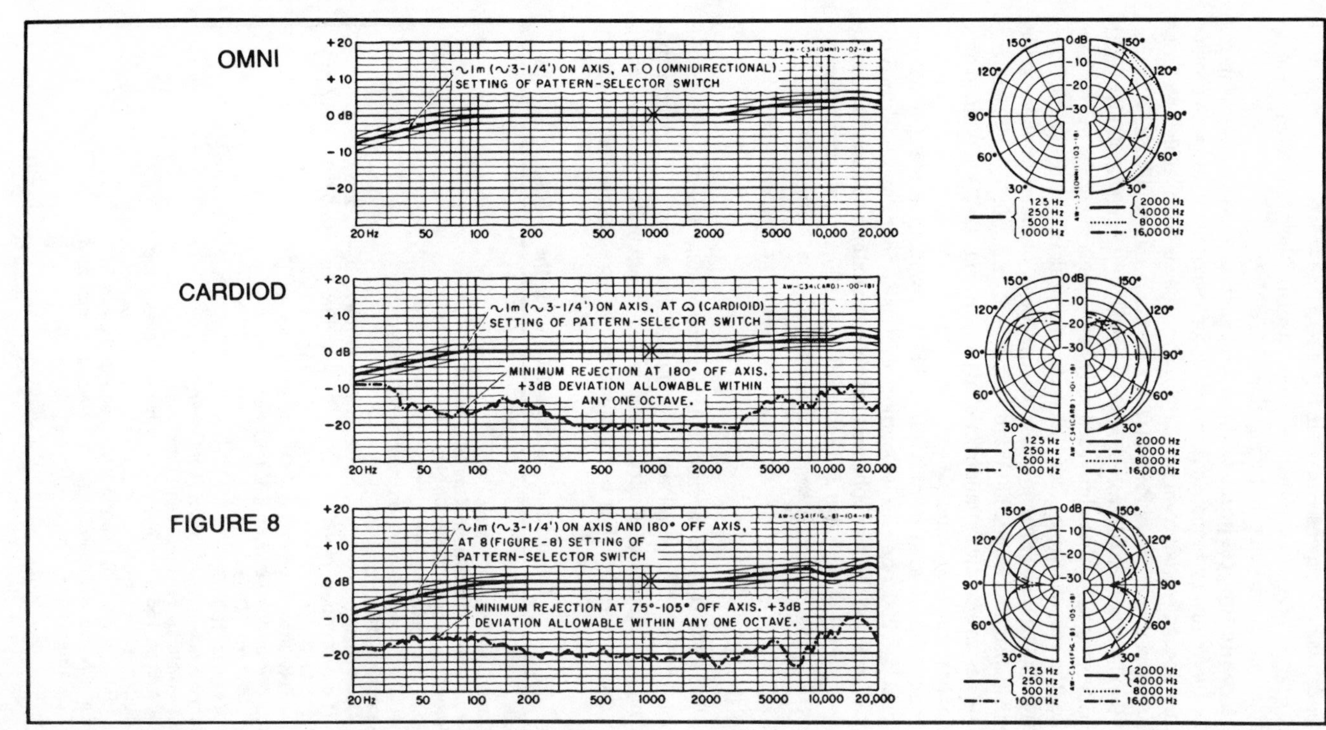

Fig. 12-6. Frequency and polar response curves of the C34.

in the frequency-response curves in Fig. 12-7.) The D-109's wire-mesh grille and rugged brass case effectively encapsulate the transducer element against metal particles and dust. See Table 12-9.

Frequency Response. The microphone is a dynamic moving-coil type. Its frequency response is specially contoured, as follows, to provide natural speech reproduction when the microphone is worn lavalier-mounted against the user's chest: Response is essentially flat at 300 Hz, 1000 Hz, and 9000 Hz; with 10 dB rolloff attenuation at 70 Hz and 12,000 Hz. With the ring-shaped lavalier clip raised fully below the level of the microphone grille, the response rises from 2000 Hz to a peak of +5 dB at 6000 Hz; with the lavalier clip lowered fully avove the level of the microphone grille, the response falls from 2000 Hz to a dip of −2 dB at 6000 Hz. The microphone has an effective omnidirectional pattern (Fig. 12-7) maintained over the entire frequency range.

Impedance and Sensitivity. The microphone has a nominal impedance of 200 ohms. The output level is −58 dBm (re: 1 mW/10 dynes/cm^2), and the microphone is capable of handling a maximum sound-pressure level of 630 μbar (130 dB SPL) at 1000 Hz with distortion not exceeding 1%. The EIA sensitivity rating (G_m) is −151 dBm.

The D-190E

The AKG D-190E cardioid dynamic microphone was designed for the serious recordist or performer who, though limited by his budget, is unwilling to compromise quality. A rugged, versatile unit, it is an excellent general-purpose speech or music microphone as well as an ideal hand-held

Table 12-9. Technical Data for the D-109.

Transducer Type: Dynamic
Directional Characteristic: Omnidirectional
Frequency range: 70-12,000 Hz
Nominal Impedance: 200 ohms
Recommended Load Impedance: ≥ 500 ohms
Sensitivity at 1 kHz:

Open circuit:	0.11 mV/μb; −79.2 dB V
Maximum power level:	−58 dBm (re: 1 mW/10 dynes/cm^2)
EIA Gm:	−151 dBm
Tolerance:	+3, −1 dB

Sound Pressure Level for 1% THD:
40 Hz: 130 dB
1000 Hz: 130 dB
Hum Sensitivity: −103 dBm (1 mG field)
Case Material: Nickel-plated brass
Net Weight: 155 g (≈ 5-½ oz) w/cable and lavalier clip
Included Accessories:
9 m (≈ 29-½ ft) integral 2-cond shielded cable
Lavalier clip w/adj tie clasp, adj/removable neck cord
Foam-lined vinyl case

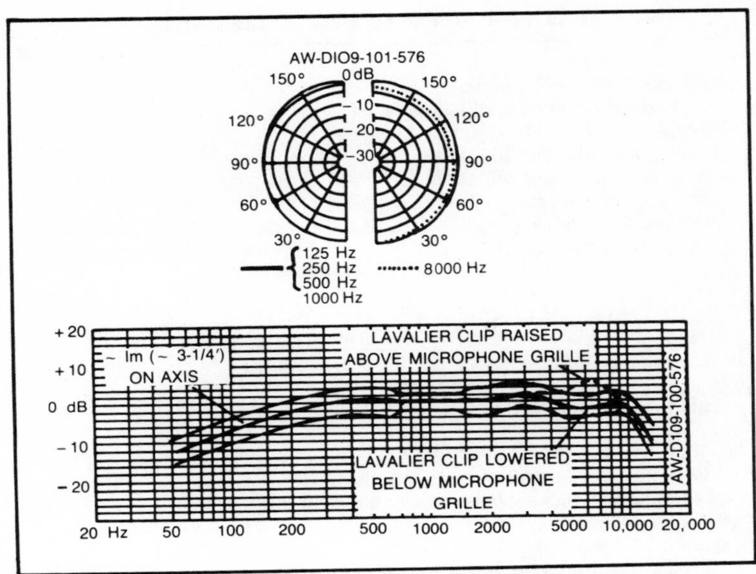

Fig. 12-7. Frequency response and polar pattern of the D-109.

vocalist's microphone for contemporary-music recording sessions and concerts.

The microphone's smooth frequency response and uniform cardioid directional pattern produce an open, effortless sound and relative immunity to feedback even under acoustically unfavorable conditions. Bass response is intentionally tailored to compensate for proximity effect when the microphone is used within close working distances, and to reduce pickup of low-frequency room rumble, floor vibrations, and acoustic-feedback components. To add brilliance and permit pickup of instruments and voices from greater distances, the on-axis response is slightly emphasized at 4 kHz.

The transducer element is elastically suspended within the housing, a feature which greatly reduces sensitivity to handling noise, mechanical shock, and spurious vibrations. An integral sintered-bronze windscreen reduces the effects of breath "pop" and wind noise. It also protects the transducer from dust, metal particles, and moisture. See Table 12-10.

Frequency Response. The microphone is a dynamic pressure-gradient type with a frequency range of 30-15,000 Hz (Fig. 12-8). Its on-axis response is slightly emphasized at 4000 Hz for added brilliance and improved pickup of distant instruments and voices. It has a cardioid directional pattern. The front-to-rear discrimination exceeds 16 dB at 1000 Hz at a sound-incidence angle of 180 degrees, and an effective cardioid pattern is maintained over the entire frequency range, as shown in Fig. 12-8.

Impedance and Sensitivity. The microphone has a nominal impedance of 200 ohms. The output level is −52 dBm (re: 1 mW/10

Table 12-10. Technical Data for the D-190E.

Transducer Type: Dynamic	
Directional Characteristic: Cardioid	
Frequency Range: 30-15,000 Hz	
Nominal Impedance: 200 ohms	
Recommended Load Impedance: $\geq$ 500 ohms	
Sensitivity at 1 kHz:	
Open circuit:	0.23 mV/μb; −72.8 dB V
Maximum power level:	−51 dBm (re: 1 mW/10 dynes/cm^2)
EIA Gm:	−144.5 dBm
High-Z output w/MC-25T, −25TS (optional):	−63 dB V at 1 μb
Tolerance:	+0, −1.5 dB
Sound Pressure Level for 1% THD:	
40 Hz: 125 dB	
1000 Hz: 130 dB	
Hum Sensitivity: −102 dBm (1 mG field)	
Case Material: Nickel-plated brass; sintered bronze	
Net Weight: 185 g ($\approx$ 6-1/2 oz)	
Included Accessories:	
SA-11 stand adapter with 5/8-in.-27 thread	
Foam-lined vinyl case	
Optional Accessories:	
SA-11/1 metal-base stand adapter with 5/8-in. -27 thread	
SA-23/2 snap-out stand adapter with 5/8-in. -27 thread	
H-24 shock mount	
GN-7E, GN-14E, GN-20E Modular Flexible Gooseneck Kits	
KM-221C flange adapter	for use with gooseneck
KM-237 clamp adapter	for use with gooseneck
KM-238 clamp adapter	for use with gooseneck
ST-5 table stand	for use with gooseneck
ST-305 anti-shock table stand	for use with gooseneck
W-8 foam windscreen	
MC-series heavy-duty cable assemblies	

dynes/cm^2) and the microphone is capable of handling a maximum sound-pressure level of 630 μbar (130 dB SPL) at 1000 Hz with distortion not exceeding 1%. The EIA sensitivity rating (G_m) is −144.5 dBm.

The D-224E

The D-224E is the premier, studio-quality member of AKG's family of two-way cardioid dynamic microphones. It represents the most sophisticated application of AKG's exclusive two-way design principle. The D-224E is suggested for musical applications in recording, scoring, and broadcast studios—especially those requiring accurate reproduction of instrumental soloists, chamber or jazz ensembles, and large choruses or orchestras. Because of its characteristics, the D-224E is also dais or lectern microphone in sound-reinforcement applications.

This microphone employs *two*, coaxially mounted, dynamic transducers: one designed for optimum performance at high frequencies and placed closer to the front grille, the other for optimum performance at low frequencies and positioned behind the first. Each transducer incorporates a hum-bucking compensating winding to cancel the effects of stray magnetic fields. Both transducers are coupled to a 500-Hz inductive-capacitive crossover network that is electroacoustically phase corrected. (This is essentially the same design technique used in modern two-way speaker systems, but applied in reverse.)

As a result, the D-224E exhibits several characteristics that make it superior to conventional cardioid dynamic microphones for its intended applications: (1) smooth and wide-range on-axis frequency response—rivaling that of a condenser microphone at frequencies up to and beyond 20 kHz; (2) a predominantly frequency-independent directional pattern—producing more linear frequency response at the sides of the microphones and more constant discrimination at the rear of the microphone; (3) an absence of proximity effect at working distances down to 15 cm ($\approx$ 6 in.); (4) low harmonic distortion at high sound-pressure levels.

In all applications—recording, broadcasting, and sound reinforcement—these qualities contribute to more natural, uniform and uncolored tonal qual-

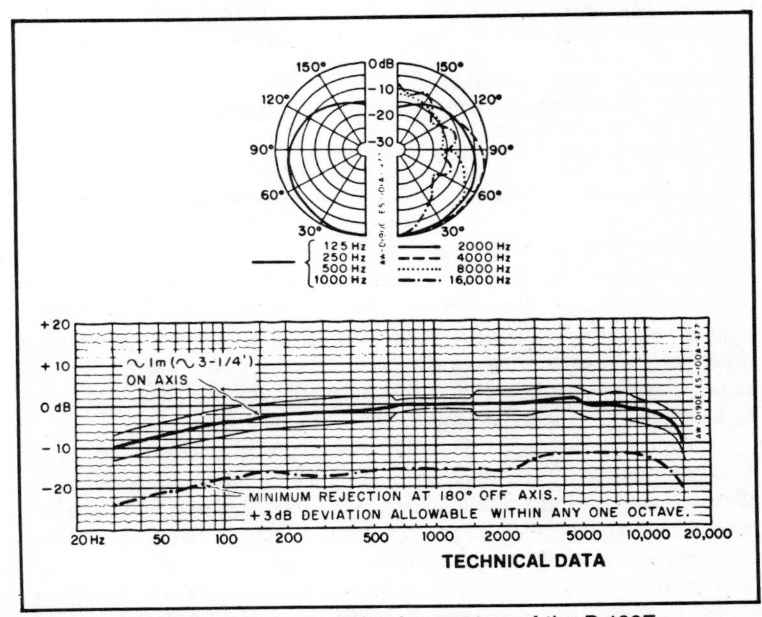

Fig. 12-8. Frequency response and polar patterns of the D-190E.

Table 12-11. Technical Data for the D-224E.

Transducer Type: Dynamic, two-way system
Directional Characteristic: Cardioid
Frequency Range: 20 Hz-beyond 20,000 Hz (filter switch at "0")
Crossover Frequency: 500 Hz
Nominal Impedance: 200 ohms
Recommended Load Impedance: $\geq$ 500 ohms
Sensitivity at 1 kHz:

Open circuit:	0.13 mV/μb; −77.7 dBV
Maximum power level:	−56.5 dBm (re: 1 mW/10 dynes/cm^2)
EIA Gm:	−149.5 dBm
Tolerance:	+0, −1 dB

Sound Pressure Level for 0.5% THD:
40 Hz: 128 dB
1000 Hz: 128 dB
Hum Sensitivity: −142 dBm (1 mG field)
Case Material: Nickel-plated brass
Net Weight: 285 g ($\approx$ 10 oz)
Included Accessories:
SA-40 stand adapter with 5/8-in.-27 thread
W-2 foam windscreen (for front of microphone)
W-2A foam windscreen (for rear of microphone)
Foam-lined vinyl case
Optional Accessories:
SA-26 clothespon stand adapter with 5/8-in. −27 thread.
SA-70/3 boom-suspension adapter for use with H-70 below.
H-9 clamp for surface-mounting or hanging H-10 below.
H-10 stereo bar for stand-mounting 2 microphones.
H-70 boom-suspension shock mount for use with Sa-70/3.
W-22 wire-mesh windscreen.
KM-series floor and boom stands, stand accessories.
ST-series table stands.
MC-series heavy-duty cable assemblies.

ity, regardless of the relative position or distance of performers and instruments within a semicircle around the front and sides of the microphone. Stereo separation is improved and greater isolation (lower leakage) is achieved in multiple-microphone installations. In sound-reinforcement applications, these same qualities also permit greater freedom in microphone and speaker placement, more effective and predictable suppression of acoustic feedback, and higher overall system gain.

A built-in three-position low-filter switch provides an additional 7 dB or 12 dB rolloff at 50 Hz, as required. This feature is useful in speech applications and in acoustically unfavorable environments with excessive low-frequency ambient noise, reverberation, or feedback. See Table 12-11.

Frequency Response. The microphone incorporates a low-frequency filter network with a three-position selector switch to shape frequency-response characteristics at 1m ($\approx$3 1/4 ft) on axis as follows: (1) the "0" position of the switch produces an unmodified frequency range of 20 Hz to beyond 20,000 Hz with 3 dB rolloff attenuation at 50 Hz; (2) the " – 7" dB" position of the switch introduces an additional 7 dB rolloff at 50 Hz for a total attenuation of 10 dB at that frequency; (3) the " – 12 dB" position of the switch introduces an additional 12 dB rolloff at 50 Hz for a total attenuation of 15 dB at that frequency.

The microphone has a predominantly frequency-independent cardioid directional pattern (Fig. 12-9) throughout most of its frequency range. The off-axis frequency response is linear from 125-8000 Hz at a sound-incidence angle of 90 degrees. The typical front-to rear discrimination remains a constant 20 dB from 125-8000 Hz at a sound-incidence angle of 180 degrees.

Impedance and Sensitivity. The D-224E microphone has a nominal impedance of 200 ohms. The output level is – 56.5 dBm (re: 1 mW/10 dynes/cm^2), and the microphone is capable of handling a maximum sound-pressure level of 500 μbar (128 dB SPL) at 1000 Hz with distortion not exceeding 0.5 %. The EIA sensitivity rating (G_m) is – 149.5 dBm.

The D-310

With its variable bass-versus-distance contour (marked bass emphasis or "proximity effect" when used close up; progressively diminishing bass response when used farther away) and smooth presence-rise contour (for added crispness and "punch"), the D-310 cardioid dynamic microphone is a *highly* creative tool—one that offers its users flexible personal control

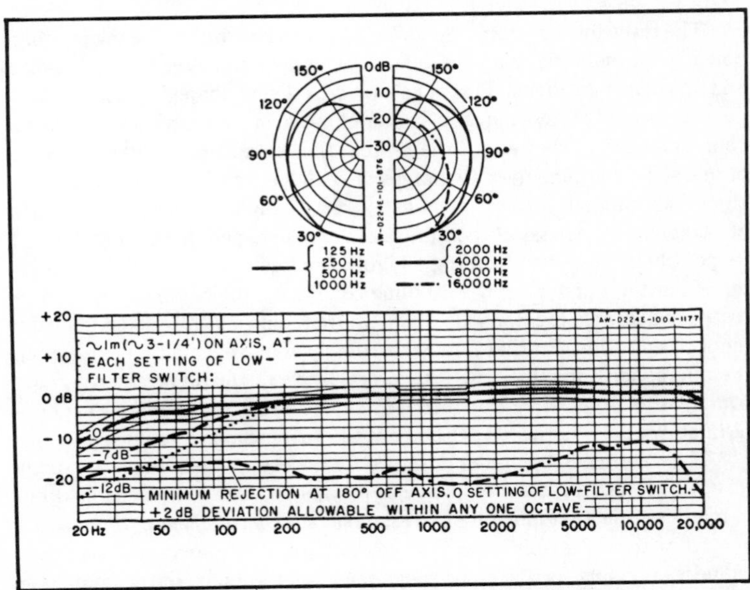

Fig. 12-9. Frequency response and polar patterns of the D-224E.

over the tonal "shading" of voices or instruments to suit a variety of locales, or musical moods and styles.

In addition, the D-310's cardioid directional pattern is *unusually* uniform with respect to frequency. This uniformity—extremely important in broadcasting and utterly indispensable for sound reinforcement—produces negligible coloration of desired sounds at the sides of the microphone while providing exceptional (and highly predictable) suppression of unwanted acoustic feedback and background noise towards the rear of the microphone.

The microphone's design and construction begin with a *three-layer* windscreen/pop-filter assembly. This assembly consists of a shock absorbing stainless-steelwire-mesh outer layer, a blast-diffusing fabric middle layer and phosphor-bronze wire-mesh inner layer (sandwiched together into a removable liner) plus a threaded retaining ring that secures the entire unit to the microphone housing. So effective is this combination in reducing the effects of wind noise, breath pop and similar acoustic interference, that an external windscreen or pop filter is seldom, if ever, required.

As tough as the windscreen/pop filter is alone, it is *further* reinforced against impact damage by a special two-piece safety basket assembly. Note that the safety basket is contoured to support the inside of the windscreen/pop filter *and* to surround the front of the transducer system—thus also isolating all *internal* parts from head-on impact damage. The safety basket consists of a resilient dome-shaped ribbed cage mated with a reinforced open-frame casting. The casting, in turn, is fitted with a fine-wire-mesh screen that coincides with side ports in the microphone housing. (Table 12-12.)

The transducer system "floats" in *all* directions within the microphone housing for isolation from the effects of impact damage, handling noise and spurious vibrations. This is achieved by a ring-shaped elastomer suspension *around* the system at its center of mass, in conjunction with a brass counterweight at the *rear* of the system that establishes neutral balance of mass. To combine the advantages of both hard *and* soft suspension designs, this ring suspension has a dense, relatively hard body and a series of compliant, *progressively* compressible dome-shaped projections around its periphery—in effect forming a highly damped low-pass mechanical filter. Therefore, under normal handling conditions, the compliant domes effectively decouple the transducer system from mechanically *and* motionally induced vibration. Further, under extremely abusive conditions (dropping the microphone or subjecting it to lateral impact), the domes increasingly compress, in proportion to applied g-force, in a progressive *braking* action—ultimately allowing the harder body of the suspension to act as a bumper.

The transducer, computer designed for widest possible frequency range (Fig. 12-10) and transient response, is encapsulated in a replaceable drop-in module. Injection-molded of stress-resistant thermoplastic, the module is fitted with two readily accessible solder lugs. These serve as convenient connection points for the color-coded leads that are part of the connector assembly within the microphone housing.

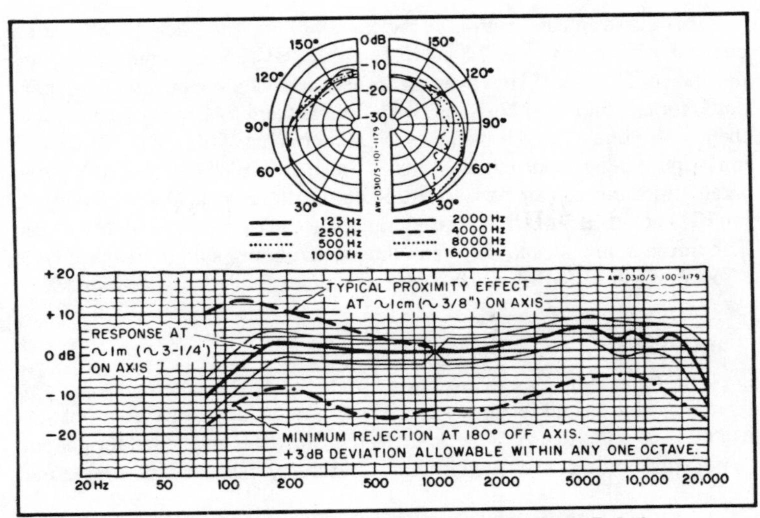

Fig. 12-10. Frequency response and polar patterns of the D-310.

The microphone housing itself is a lightweight, impact resistant zinc-alloy casting—balanced and finished in glare-free matte nickel. For maximum wear resistance, the finish is *electroplated* (not merely painted). In conjunction with the windscreen/pop filter and the safety basket, this housing encapsulates all internal parts against metal particles, dust and moisture spray.

Should field service ever become necessary, the windscreen/pop filter can be unscrewed, the two-piece safety basket lifted out, and the transducer module removed for replacement by unsoldering the two color-coded leads, removing the module's ring suspension (to free the leads), and withdrawing the module from the microphone housing. Spare assemblies and parts are available, and the "in-phase" lugs of all replacement transducer modules are coded for ease of polarity identification. See Table 12-12.

Frequency Response. This microphone has a frequency range of 80-18,000 Hz, accompanied by the following on-axis characteristics: (1) an 18-dB variable bass-versus-distance contour at 100 Hz that ranges from −6 dB response rolloff at a working distance of 1 m (≈3 1/4 ft) to +12 dB typical proximity-effect boost at a working distance of 1 cm (≈3/8 in.); (2) a fixed presence-rise contour from 1500 Hz to 16,000 Hz that boosts response +6 dB at 5000 Hz.

The microphone has a highly uniform cardioid directional pattern (Fig. 12-10) with respect to frequency as follows: (1) at a sound-incidence angle of 90 degrees, typical off-axis frequency response does not vary more than 5-6 dB from 125-16,000 Hz; (2) at a sound-incidence angle of 180 degrees, minimum front-to-rear discrimination exceeds 14 dB at 1000 Hz. An effective cardioid pattern is maintained over the entire frequency range.

Impedance and Sensitivity. At 1000 Hz, the microphone has an open-circuit sensitivity of 1.3 mV/Pa (−57.7 dBV) and a nominal (true) impedance of 270 ohms. However, in accordance with various industry standards for calculating other sensitivity figures from this open-circuit value, the microphone's rated impedance is categorized as follows: (1) for a maximum power-level output of −57.5 dBm (re: 1 mW/10 dynes/cm^2), the rated impedance is categorized as 250 ohms; (2) for an EIA sensitivity rating (G_m) of −149.5 dBm, the rated impedance is categorized as 150 ohms. The microphone is capable of handling a maximum sound-pressure level of 50 Pa (128 dB SPL) at 1000 Hz with distortion not exceeding 1%.

The D-558B

The AKG D-558B was designed for close-talk use wherever high ambient noise levels and feedback-prone environment acoustics are problems. Mounted on a flexible gooseneck shaft, the D-558B is suggested for general radio communications, paging/public address, talkback/intercom, and

Table 12-12. Technical Data for the D-310.

Transducer Type: Dynamic
Directional Characteristic: Cardioid
Frequency Range: 80-18,000 Hz
Nominal Impedance at 1 kHz: 270 ohms
Rated-Impedance Category: 250 ohms
Recommended Load Impedance: ≥ 600 ohms
Sensitivity at 1 kHz:

Open circuit:	1.3 mV/Pa; −57.7 dBV*
Maximum power level:	−57.5 dBm (re: 1 mW/10 dynes/cm^2)
EIA Gm:	−149.5 dBm
High-Z output w/MCH-25T, −25TS (optional):	−47.5 dBV at 1 Pa*
Tolerance:	+2, −1.5 dB

Sound Pressure Level for 1% THD:
40 Hz, 1000 Hz, 5000 Hz: 128 dB
Hum Sensitivity (1 mG field): −142 dBm
Case Material: Nickel-plated die-cast zinc alloy
Net Weight: 255 g (≈ 9 oz).
Included Accessories:
SA-31 flex. snap-in stand adapter with 5/8-in. −27 thread
Foam-lined vinyl case
Optional Accessories:
SA-26 clothespin stand adapter with 5/8-in. −27 thread
PF-10 foam pop filter (red, blue, yellow, off-white, gray)
GN-7E, GN-14E, GN-20E modular flexible-gooseneck kits
KM-series floor and boom stands, stand accessories
ST-series table stands
MC-series heavy-duty microphone cable assemblies
*1 Pa (Pascal) = 10 μb = 10 dynes/cm^2 ≈ 94 dB SPL

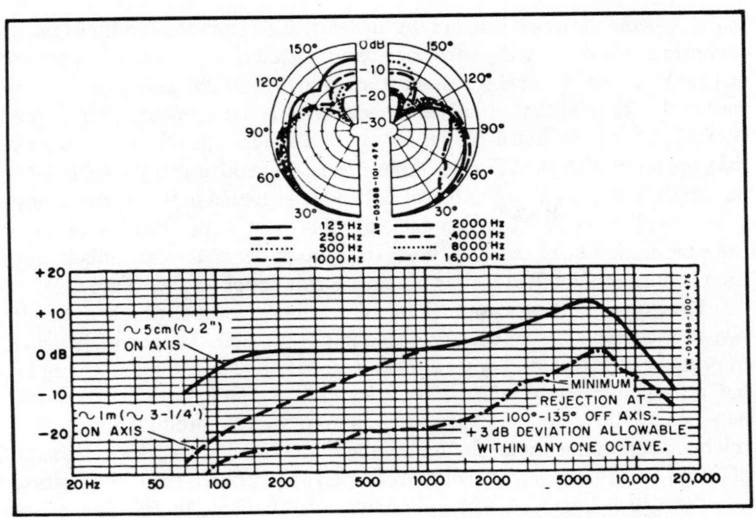

Fig. 12-11. Frequency response and polar patterns of the D-558B.

specialized broadcast applications. (Because of its noise-cancelling characteristics, the D-558B is excellent as a newsdesk microphone in the clatter of the newsroom or—used with the W-3 windscreen—as a sports announcer's microphone in crowded, noisy stadiums and arenas.) See Fig. 12-11. The D-558B achieves its high noise rejection and relative immunity to acoustic feedback by combining a dynamic pressure-gradient transducer with a front- and side-ported housing in a differential design technique. As a result, the microphone clearly reproduces speech originating within its recommended working distance (approximately 5 cm or 2 in.), but greatly attenuates low-frequency on-axis noise and feedback components originating at a distance of 1 m (≈3 1/4 ft) or more. Higher-frequency noise and feedback components are attenuated by the microphone's tight hypercardioid directional pattern at these frequencies. For improved speech intelligibility in narrowband communications channels, on-axis response is intentionally emphasized between 1 kHz and 6 kHz. This rising-response characteristic also contributes to better noise penetration in paging applications where the loudspeakers are likely to be in extremely noisy areas.

The integral gooseneck shaft provides adjustable—and noise-free—means for positioning the microphone to suit both the personal convenience of the user and the acoustical requirements of the working environment. A brass case and wire-mesh grille encapsulate the microphone's transducer element against metal particles and dust. The D-558B operates satisfactorily over a wide range of temperatures and withstands moderately high humidity levels (Fig. 12-12.)

A low-impedance balanced-output unit, the D-558B is supplied complete with a 1.15 m (≈3 3/4 ft) non-detachable 2-conductor shielded cable having stripped and tinned leads at its free end. Also included is a kit of basic mounting hardware. This hardware can be used in various combinations to install the unit several ways—custom-mounted on virtually any flat surface (either flush with the surface or recessed), mounted on any microphone stand having a standard 5/8-in. -27 male thread, or used in conjunction with the optional AKG ST-series table stands. (Several other optional mounting accessories—listed in Technical Data—are available.) Depending on how the microphone is installed, the shielded cable may be routed either through the slot on the *side* of the shaft boss, or through the *bottom* of the boss and then through the hollow-center mounting bolt supplied. See Table 12-13.

Frequency Response. The microphone incorporates a dynamic pressure-gradient transducer enclosed in a front- and side-ported housing to produce a differential noise-cancelling characteristic. The microphone has a frequency range of 70-15,000 Hz, with an on-axis rising response between 1000 Hz and 6000 Hz for improved speech intelligibility in narrowband communications channels. When used within a working distance of 5 cm ($\approx$2 in.), the microphone's rejection of an on-axis noise originating at a distance of 1 m ($\approx$3 1/4 ft) or more exceeds 16 dB at 100 Hz. The microphone has a hypercardioid directional pattern. The off-axis discrimination

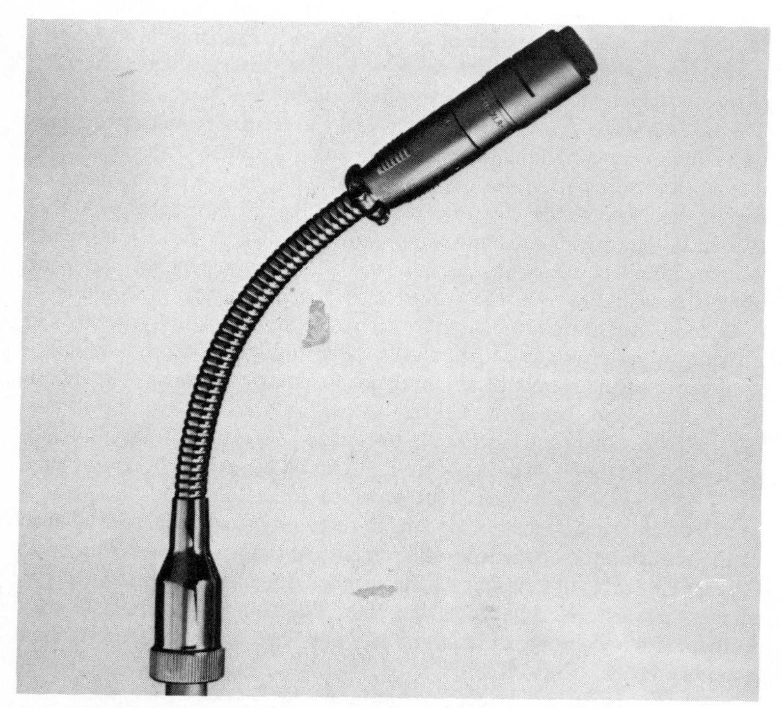

Fig. 12-12. Flexible shaft supplies variety of mic positions.

Table 12-13. Technical Data for the D-558B.

Transducer Type: Dynamic
Directional Characteristic: Hypercardioid
Frequency Range: 70-15,000 Hz
Nominal Impedance: 200 ohms
Recommended Load Impedance: $\geq$ 500 ohms
Sensitivity at 1 kHz:

Open circuit:	0.072 mV/μb; −82.9 dBV
Maximum power level:	−62 dBm (re: 1 mW/10 dynes/cm^2)
EIA Gm:	−154.5 dBm
Tolerance:	±2.5 dB

Sound Pressure Level for 0.5% THD:
1000 Hz: 128 dB
Hum Sensitivity: −127 dBm (1 mG field)
Temperature Range: −10° C ($\approx$ +14°F) to +60° C ($\approx$ +140° F)
Maximum Relative Humidity: 80%
Microphone-Case, Shaft-Boss Material: Nickel-plated brass
Flexible-Gooseneck Shaft Material: Nickel-plated steel
Net Weight: 325 g ($\approx$ 11-1/2 oz) w/cable and mounting hardware

Included Accessories:
1.15 m ($\approx$3-3/4 ft) non-detachable 2-conductor shielded cable
Mounting hardware
Optional Accessories:
KM-221C flange adapter
KM-237 clamp adapter
KM-238 clamp adapter
ST-5 table stand
ST-305 anti-shock table stand
W-20 foam windscreen

exceeds 20 dB at 1000 Hz at a sound-incidence angle of 100-135 degrees, and an effective hypercardioid pattern is maintained over the entire frequency range.

Impedance and Sensitivity. The microphone has a nominal impedance of 200 ohms. The output level is −62 dBm (re: 1 mW/10 dynes/cm^2), and the microphone is capable of handling a maximum sound-pressure level of 500 μbar (128 dB SPL) at 1000 Hz with distortion not exceeding 0.5%. The EIA sensitivity rating (G_m) is −154.5 dBm.

MICROPHONE ACCESSORIES

There are a number of accessories available for use with microphones—a variety of cables, windscreens, suspensions, booms, table stands, stand adapters, flexible shafts, and floor stands, to mention just a few.

The right accessory can be extremely convenient. A flexible shaft, or gooseneck can change microphone positioning from a nuisance to an easily handled recording situation. The best way to become aware of the accessories that are available is to consult the manufacturer's catalog or to visit your dealer's showroom and discuss your recording problem with your dealer salesman.

MICROPHONE CARE

To protect your microphones from dirt and damage, store and transport them in their original packages or in a specialized microphone case.

To clean the integrated windscreen, unscrew the microphone head and wash the windscreen in a soap solution.

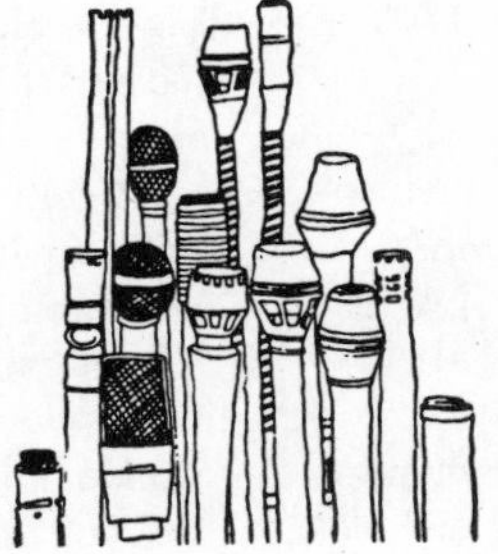

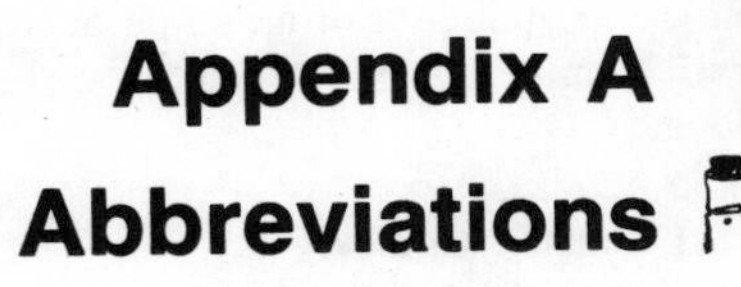

Appendix A
Abbreviations

ac	Alternating current. Also used as an abbreviation for alternating voltage.
a/d	Analog to digital converter.
af	Audio frequency.
alc	Automatic level control.
arll	Audible relative loudness level. A technique for the measurement of phono rumble.
aux	auxiliary.
bit	a binary digit.
ccitt-c	A C-weighting filter in accordance with CCITT, International Telephone and Radio Authority.
comb	A combination, often of a mic and associated equipment or accessories.
cps	Cycle per second. A measurement of the frequency of a wave, now replaced by the Hertz. Thus, 60 cps = 60 Hz.
d/a	Digital to analog converter.
dB	Decibel.
db SPL	dB sound pressure level, related to the SPL of 20 μPa or 2 $\times 10^{-4}$ μb.
dBm	Output level related to 0.775 volt at 60 ohms.
dBV	Output level related to 1 volt.
dc	Direct current. Also used to indicate a direct voltage.
DIN	Deutsche Industrie Norm (German Industry Standards).
emf	Electromotive force. (Voltage).
ENL	Equivalent noise level.

eq	Equalization. A technique for altering bass, midrange or treble tone level.
fet	Field effect transistor.
gnd	Ground. A common connection.
Hz	Hertz. The cycle per second.
IC	Integrated circuit.
J	Jack. A female receptacle for accommodating a plug.
k	Multiplication by 1,000. One kiloHertz (1 kHz) is 1,000 Hz.
LED	Light emitting diode.
LSI	Large scale integrated circuit.
M	Meg. Prefix used to represent 1,000,000. 1 megohm is 1,000,000 ohms.
mA	Milliampere. A thousandth of an ampere.
μA	Microampere. A millionth of an ampere.
μF	Microfarad. A millionth of a farad.
mic	Microphone.
mono	Monophonic.
mV	Millivolt. A thousandth of a volt.
μV	Microvolt. A millionth of a volt.
mW	Milliwatt. A thousandth of a watt.
μW	Microwatt. A millionth of a watt.
NAB	National Association of Broadcasters.
neg	Negative.
omni	Omnidirectional.
PC	Printed circuit.
pos	Positive.
preamp	Preamplifier.
pwr	Power.
pxe	Piezoelectric transducer.
PZM	Pressure zone microphone.
rc	Remote control.
rf	Radio frequency.
sig	Signal.
S/N	Signal-to-noise ratio.
spec	Specifications.
SPL	Sound pressure level.
stereo	Stereophonic.
THD	Total harmonic distortion.
TIM	Transient intermodulation distortion.
VU	Volume unit.
xformer	Transformer.
XLR	A 3 or 5 contact studio connector according to IEC 268-14B (International Electric Standard).
xtal	Crystal.
W	Watt. Basic unit of electrical power.
Z	Impedance.

Appendix B
Microphone Manufacturers

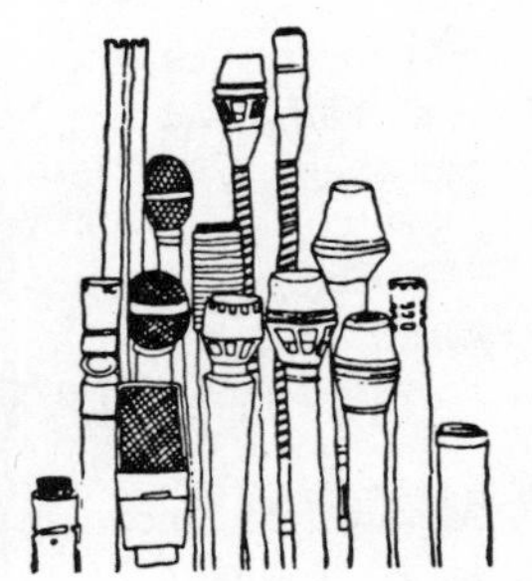

There are two kinds of manufacturers of microphones. The first is the manufacturer of a complete line, making a specialty of the design and development of microphones. Such manufacturers have their own factories, including research and development divisions. The second is the manufacturer who may supply a microphone intended specifically for use with his products.

The following includes as many names as possible. No name has been intentionally omitted and if that has happened it is an accidental oversight.

Aiwa America, Inc.
350 Oxford Drive
Moonachie, NJ 07074

AKG Acoustics, Inc.
77 Selleck St.
Stamford, CT 06902

Astatic Corp.
Harbor and Jackson Sts.
Conneaut, OH 44030

Audio Technica U.S. Inc.
1221 Commerce Drive
Stow, Ohio 44224

Beyer Corp.
Burns Audiotronics, Inc.
5-05 Burns Ave.
Hicksville, NY 11801

C-Tape Developments, Inc.
P.O. Box 1069
Palatine, IL 60078

Countryman Associates, Inc.
417 Stanford Ave.
Redwood City, CA 94063

Crown International, Inc.
1718 Mishawaka Rd.
Elkhart, IN 46517

Electro-Voice, Inc.
600 Cecil St.
Buchanan, MI 49107

GC Electronics, Inc.
400 South Wyman St.
Rockford, IL 61101

HM Electronics, Inc.
6151 Fairmount Ave.
San Diego, CA 92120

JVC Company of America
41 Slater Dr.
Elmwood Park, NJ 07407

Marantz Co., Inc.
20525 Nordhoff St.
Chatsworth, CA 91311

Nakamichi U.S.A. Corp.
1101 Colorado Ave.
Santa Monica, CA 90401

Neumann*

Numark Electronics Corp.
503 Raritan Center, Box 493,
Edison, NJ 08818

Pioneer Electronics (USA) Inc.
1925 E. Dominguez St.
Long Beach, CA 90810

Quasar Co.
9401 W. Grand Ave.
Franklin Park, IL 60131

Realistic Co.
P.O. Box 2625
Fort Worth, Texas 76113

Sansui Electronics Corp.
1250 Valley Brook Ave.
Lyndhurst, NJ 07071

Schoeps Schalltechnik*

Sennheiser Electronic Corp.
10 West 37th St.
New York, NY 10018

Shure Brothers, Inc.
222 Hartrey Ave.
Evanston, IL 60204

Signet Co.
4701 Hudson Dr.
Stow, OH 44224

Sony Corp. of America,
Sony Drive
Park Ridge, NJ 07656

Soundcraftsmen, Inc.
2200 S. Ritchey St.
Santa Anna, CA 92705

TEAC Corp. of America,
7733 Telegraph Rd.
Montebello, CA 90640

Technics,
One Panasonic Way,
Secaucus, NJ 07094

Toshiba America, Inc.
82 Totowa Rd.
Wayne, NJ 07470

Turner Div. of Conrac Corp.
909 - 17th St., N.E.
Cedar Rapids, IA 52402

Yamaha International Corp.
P.O. Box 6600,
Buena Park, CA 90620

---

*address not available

Appendix C

Microphone Buying Guide

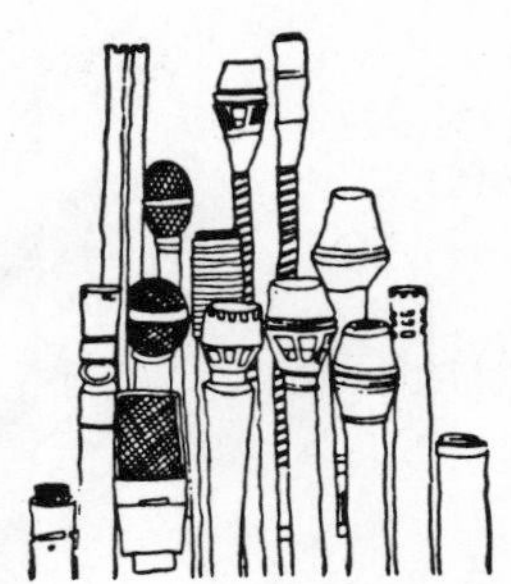

The kind of sound you will record will depend on the kind of microphone you use, the number of microphones, their positioning and the acoustics of the recording room. It will also depend on the materials of the microphone, how the microphone was made and the kind and amount of quality control. You should not buy a microphone solely by weight, shape, size, color, finish, or general appearance.

The chart on the following pages lists the various manufacturers of microphones, model numbers, microphone types, directional patterns, impedance in ohms, frequency response, the type of connector on the microphone, the type of connector on the cable, cable length (when supplied by the manufacturer), the case material, the weight and the cost.

Of all the data supplied in the tables there are two variables: the model number and the price. Microphone models do change. Older models are removed from the market, newer ones take their place. As for price, manufacturers may suggest or recommend retail selling prices but dealers have the right to sell microphones at any price they wish.

The advantages of this listing is that it gives you an immediate comparative analysis so when you do your shopping for a microphone you are aware of what is available and you have on hand a fair estimate of the cost.

Recordists, whether professional or dedicated hobbyists, gradually acquire a collection of microphones. However, if it is the intention to increase the number of mics, possibly a better route to follow would be the prior acquisition of a mixer. The advantage is that the mixer will supply the required combination of connectors and it may even have a built-in transformer. A quality transformer isn't a low-cost item. Using a mixer will not only minimize or eliminate the need for a complicated cabling arrangement, but will save on transformer costs as well.

Table C-1. Microphone Buying Guide.

MANUFACTURER	Model	Directional Pattern Plus Channels: Stereo = x2, Quad = x4, M-S = M-S	Operating Principle	Case Material	Most Common Use	Actual Impedance, 1 kHz, Ohms; Balanced = B	Operating Range, Hz to kHz	Open-Circuit Sensitivity, 1 kHz, −dB, re: 1 Volt/1 μbar	Mike Connector Type	Cable Length, Feet	Connector Type on Equipment End of Cable	Weight, Ounces	Wind Screen = W, On/Off Switch = S, Switched Low or Proximity Filter = F	Price, $	Notes
AIWA	CM-Z7	Sel.	Electret Condenser	Alum.	Vocal/ Record	Lo-Z	60-14	Var.	Cannon XLR-3		Phone, Mini	6	WSF	150.00	
	DM-D6	Uni	Dynamic	Zinc	Vocal/ Record	600	80-12	55	Cannon XLR-3		Phone, Mini	8.8	S	50.00	
	DM-D30	Uni	Dynamic	Alum.	Vocal/ Record	600	100-10	75		10	Phone, Mini	6	S	20.00	
	CM-70	Card. x2	Condenser	Plastic	Hall/ Record	250	100-12	65			Mini	0.8	S	40.00	Tie clip.
	CM-60	Omni	Condenser	Plastic	Voice/ Record	250	100-12	48		2.6	Mini	0.5	SF	29.50	As above.
	CM-30A	Card. x2	Condenser	Alum.	Hall/ Record	1k	80-15	54		1.4	Mini	2.2	WS	39.00	
	CM-Z3	Var.	Condenser	Alum.	Voice/ Record	250	100-13	Var.		4	Mini	1.4	WS	39.00	
AKG ACOUSTICS	C-34	Multi x2	Condenser	Brass	Studio	200B	20-20	47	Multi	66	XLR	9¾	WF	1649.00	Remote control with 9-pattern selector.
	C-422	Multi x2	Condenser	Brass	Studio	200B	20-20	44.5	Multi	66	XLR	15½	WF	2500.00	As above.
	C414EB/P48	Multi	Condenser	Brass	Studio	200B	20-20	41	XLR			12	WF	795.00	Four patterns.
	C-460B	Multi	Condenser	Brass	Multi	200B	20-20	40	XLR			4	WF	464.00	Modular system.
	C-535EB	Card.	Condenser	Brass	Vocal	200B	20-20	Var.	XLR			10	F	295.00	
	C-568EB	Lobe	Condenser	Brass	Record	200B	50-20	42	XLR			6¼	WF	295.00	Shotgun.
	C-567E	Omni	Condenser	Brass	Speech Instr.	200B	20-20	45	XLR	4	XLR	3½	W	235.00	Lavalier.
	The Tube	Multi	Condenser	Brass	Studio	200B	20-20	60	12-Pin Tuchel	30	XLR	24	WSF	1700.00	Remote control with 9-pattern slector.
	D-12E	Card.	Dynamic	Zinc	Instr.	200B	30-15	53	XLR			1½		295.00	
	D-40	Card. x2	Dynamic	Zinc	Record	600	80-15	59	Attached	15	Phone	7		99.00	
	D-58E	Card.	Dynamic	Zinc	Instr.	200B	75-15	63	XLR			1½		99.00	Noise cancelling.

	D-80	Card.	Dynamic	Brass	Vocal	200B	60-15	56	XLR	15	Phone	7¼	S	90.00	With XLR plug, $95.00.
	D-109	Omni	Dynamic	Zinc	Speech	200B	70-15	59		30		5½		99.00	
	D-125E	Card.	Dynamic	Zinc	General	200B	90-18	54	XLR			8		110.00	
	D-130E	Omni	Dynamic	Zinc	General	200B	50-15	56	XLR			8		105.00	
	D-190E	Card.	Dynamic	Brass	Record	200B	30-16	56	XLR			6½	S	115.00	
	D-202E1	Card.	Dynamic		Record	200B	30-15	53	XLR			10	F	350.00	Two-way system.
	D-222EB	Card.	Dynamic	Zinc	Record	200B	20-17	56	XLR			9	F	290.00	As above.
	D-224EB	Card.	Dynamic	Zinc	Record	200B	20-20	58	XLR			10	F	500.00	As above.
	D-310	Card.	Dynamic	Zinc	Vocal	200B	80-18	58	XLR			8½	S	135.00	
	D-320B	Hyper Card.	Dynamic	Zinc	Vocal	200B	80-18	57	XLR			10½	F	170.00	
	D-330BT	Hyper Card.	Dynamic	Zinc	Vocal	200B	50-20	58	XLR			12	F	210.00	EQ switching.
	D-900	Lobe	Dynamic	Zinc	Sound Reinf.	200B	60-15	50	XLR			18	WF	310.00	Shotgun.
	D-1200	Card.	Dynamic	Brass	Vocal	200B	40-17	73	XLR			8½	F	140.00	
ASTATIC	985cn	Card.	Dynamic	Zinc	Vocal	250B	40-15	75	A3F	18	A3M	9.5	WS	119.00	
	980cn	Card.	Electret Condenser	Zinc	Record	600B	40-20	68	A3F	18	A3M	6.5	WS	192.00	
	975cn	Card.	Dynamic	Zinc	Vocal	500B	60-14	74	A3F	18	A3M	8.8	WS	98.00	
	970cn	Omni	Dynamic	Zinc	Brdcst. Record	200B	50-15	82	A3F	18	A3M	7.5	WS	128.80	
	965cn	Card.	Dynamic	Zinc	Vocal	230B	50-15	74	A3F	18	A3M	9.5	WS	123.40	
	960cn	Card.	Dynamic	Zinc	Vocal/ Record	250B	40-16	75	A3F	18	A3M	9.5	WS	151.00	
	827	Card.	Electret Condenser	Black Chrome	†	600B	80-18	72	A3F			6.5	W	66.70	†Lectern; 13- or 17-inch gooseneck.
	840s	Card.	Dynamic	Zinc	Sound Reinf.	200B	50-12	82	None	30	None	1.3	S	93.50	Lavalier.
	1070	Card.	Dynamic	Plastic	Conf.	250B	100-15	75	None	21	None	14.5		199.00	Spring shock mount, hum filter.
	BL14	Card.	Dynamic	Zinc Alloy	Instr.		40-15	53		20		21	SF	92.00	
	BL24	Omni	Dynamic	Alum.	Acous.	200	50-16	82		30	A3M	5.5	S	86.00	Lavalier.
	BL34	Card.	Dynamic		Instr.		60-14							49.95	
	BL44	Card.	Dynamic	Zinc Alloy	Vocal	230	40-15	74	A3F	25	A3M	10.5	SF	110.00	
	BL54	Omni	Dynamic	Zinc Alloy	Acous.	200	50-15	82	A3F	25	A3M	7.5	SF	113.00	
	BL64	Card.	Dynamic	Zinc Alloy	Instr.	230	40-15	74	A3F	25	A3F	9	SF	115.00	
	BL74	Card.	Dynamic	Zinc Alloy	Vocal	230	50-16	74	A3F	25	A3M	9.5	S	127.00	
	BL94	Card.	Dynamic	Zinc Alloy	Vocal	250	40-16	75	A3F	25	A3M	9.5	SF	152.00	

MANUFACTURER	Model	Directional Pattern Plus Channels: Stereo = x2, Quad = x4, M-S = M-S	Operating Principle	Case Material	Most Common Use	Actual Impedance, 1 kHz, Ohms: Balanced = B	Operating Range, Hz to kHz	Open-Circuit Sensitivity, 1 kHz, −dB, re: 1 Volt/1 μbar	Mike Connector Type	Cable Length, Feet	Connector Type on Equipment End of Cable	Weight, Ounces	Wind Screen = W, On-Off Switch = S, Switched Low or Proximity Filter = F	Price, $	Notes
	BLJT30	Omni	Crystal	Zinc Alloy	†	500	30-10	49	Mini	20		8.5		78.75	†Harmonica.
	BLJT30P	Omni	Crystal	Zinc Alloy	†		30-10	49			A3M	8.5		58.10	
	BLJT30PC	Omni	Crystal	Zinc Alloy	†		30-10	49	A3F	20	A3M	8.5		88.75	
	BLJT30PS	Omni	Crystal	Zinc Alloy	†		30-10	49	A3F	20	A3M	8.5		98.75	
AUDIO-TECHNICA	AT9000	Omni	Electret Condenser	Alum.	Vocal	1.5k	60-10	63†	Attached	10	Mini	1.65	S	13.95	†0 dB = 1 mW per 10 dynes/cm^2.
	AT9100	Uni	Dynamic	Plastic	Vocal	600	60-15	63†	Attached	10	Mini	6.7	S	24.95	
	AT9200	Uni	Electret Condenser	Plastic	Music	1.5k	60-17	48†	Attached	10	Mini	7.1	S	34.95	
	AT9300	Uni/Omni	Electret Condenser	Alum.	Video	1k	40-10	61†	Attached	6	Mini	2.75	SW	79.95	
	AT9400	Uni x2	Electret Condenser	Plastic	Music	1.5k	60-17	53†	Attached	10	Mini	7.1	S	49.95	
	AT9500	Omni	Electret Condenser	Alum.	Vocal	2k	50-16	53†	Attached	10	Mini	0.18	S	29.95	
	AT9600	Uni	Dynamic	Alum.	Vocal/Music	600	60-16	62†	Attached	13	Mini	6.7	SF	49.95	
	AT9700	Uni	Electret Condenser	Alum.	Vocal/Music	600	50-17	52†	Attached	16	Mini	5	SF	59.95	
	AT9800	Uni	Dynamic	Alum.	Music	250	50-18	58†	A3M	16	Mini	10	SF	99.95	
	AT9900	Uni	Electret Condenser	Alum.	Music	600	4-20	56†	A3M	16	Mini	6.5	SF	129.95	
BEYER DYNAMIC	M-69	Hyper Card.	Dynamic	Brass	Vocal	200B	50-16	52	3-Pin Male XLR			11.4		165.00	
	M-69S	Hyper Card.	Dynamic	Brass	Vocal	200B	50-16	52	3-Pin Male XLR			11.4	F	200.00	

M-88	Hyper Card.	Dynamic	Brass	Vocal	200B	30-20	52	3-Pin Male XLR			11.4		320.00	
M-201	Hyper Card.	Dynamic	Brass	Instr.	200B	40-18	57	3-Pin Male XLR			7.9	W	190.00	
M-160	Hyper Card.	Double Ribbon	Brass	Instr.	200B	40-18	59	3-Pin Male XLR			5.6		360.00	
M-260	Hyper Card.	Ribbon	Brass	Instr.	200B	50-18	57	3-Pin Male XLR			10.7		200.00	
M-260S	Hyper Card.	Ribbon	Brass	Instr.	200B	50-18	57	3-Pin Male XLR			10.7	S	210.00	
M-101	Omni	Dynamic	Brass	Vocal	200B	40-20	57	3-Pin Male XLR			5.7	W	220.00	
M-130	Figure 8	Ribbon	Brass	Instr.	200B	40-18	59	3-Pin Male XLR			5.4		440.00	
M-111	Omni	Dynamic	Zinc Alloy	Vocal	200B	60-15	62	3-Pin Male XLR			2.7		230.00	Lavalier.
M200	Card.	Dynamic	Alum.	Vocal	600B	50-15	56.6	3-Pin Male XLR			5		100.00	
M200S	Card.	Dynamic	Alum.	Vocal	600B	50-15	56.6	3-Pin Male XLR			5	S	110.00	
M300	Card.	Dynamic	Alum.	Vocal	250B	50-15	58.5	3-Pin Male XLR			8.6		125.00	
M300S	Card.	Dynamic	Alum.	Vocal	250B	50-15	58.5	3-Pin Male XLR			8.6	S	135.00	
M400	Super Card.	Dynamic	Brass	Vocal	200B	40-16	53	3-Pin Male XLR			9.2		160.00	
M400S	Super Card.	Dynamic	Brass	Vocal	200B	40-16	53	3-Pin Male XLR			9.2	S	170.00	
M500	Hyper Card.	Ribbon	Alum	Vocal	200B	40-18	57	3-Pin Male XLR			9		240.00	
M500S	Hyper Card.	Ribbon	Alum	Vocal	200B	40-18	57	3-Pin Male XLR			9	S	250.00	
M600	Hyper Card.	Dynamic	Brass	Vocal	250B	40-16	57	3-Pin Male XLR			8.75		270.00	
M600S	Hyper Card.	Dynamic	Brass	Vocal	250B	40-16	57	3-Pin Male XLR			8.75	SF	280.00	
M260.80	Hyper Card.	Ribbon	Brass	Vocal	200B	100-18	57	3-Pin Male XLR			8.2	F	210.00	
M411	Card.	Dynamic	Brass	Vocal	200B	200-12	56	Tuchel			5.4		130.00	
M412	Card.	Dynamic	Rubber	Vocal	200B	200-12	56	Tuchel			5.4		135.00	
M64	Card.	Dynamic	Brass	Vocal	200B	100-12	59	DIN†			4		100.00	†For gooseneck mounting.
M420	Hyper Card.	Dynamic	Brass	Vocal	200B	100-12	57	3-Pin Male XLR			5.3		155.00	

MANUFACTURER	Model	Directional Pattern Plus Channels Stereo = x2, Quad = x4, M-S = M-S	Operating Principle	Case Material	Most Common Use	Actual Impedance, 1 kHz, Ohms: Balanced = B	Operating Range, Hz to kHz	Open-Circuit Sensitivity, 1 kHz, −dB, re: 1 Volt 1 μbar	Mike Connector Type	Cable Length, Feet	Connector Type on Equipment End of Cable	Weight, Ounces	Wind Screen = W, On/Off Switch = S, Switched Low or Proximity Filter = F	Price, $	Notes
BEYER DYNAMIC	M422	Super Card.	Dynamic	Brass	Vocal	200B	100-12	59	3-Pin Male XLR			2.5		75.00	
	MC734	Card.	Condenser	Alum.	Vocal	150B	20-18	43.8	3-Pin Male XLR				F	830.00	
	MC736	Card./Lobe	Condenser	Alum.	Vocal	150B	40-20	28.2	3-Pin Male XLR			8.6	F	725.00	Short shotgun.
	MC737	Lobe	Condenser	Alum.	Vocal	150B	40-20	28.2	3-Pin Male XLR			15.7	F	825.00	Long shotgun.
	CK701	Omni	Condenser	Alum.	Vocal	200B	40-20	41	3-Pin Male XLR			0.6	F	210.00	
	CK702	Omni	Condenser	Alum.	Vocal	200B	40-20	41	3-Pin Male XLR			0.6	WF	250.00	Elastic suspension of capsule.
	CK703	Card.	Condenser	Alum.	Vocal	200B	40-20	39	3-Pin Male XLR			1	F	260.00	
	CK704	Card.	Condenser	Alum.	Vocal	200B	40-20	39	3-Pin Male XLR			1	WF	300.00	As above.
	CK706	Card./Lobe	Condenser	Alum.	Vocal	200B	40-20	39	3-Pin Male XLR			3	F	400.00	
	CK707	Lobe	Condenser	Alum.	Vocal	200B	40-20	39	3-Pin Male XLR			8.7	F	500.00	
	CK708	Figure 8	Condenser	Alum.	Vocal	200B	40-20	39	3-Pin Male XLR			4.4	F	460.00	
	MCE-5.11	Omni	Electret Condenser	Brass	Vocal	200B	20-20	36	3-Pin Male XLR	10	3-Pin Male XLR	0.25	W	275.00	Lavalier.
	MPC-50	Omni	Boundary Layer	Wood	Vocal	200B	20-20	33	3-Pin Male XLR			18		530.00	
CROWN INTERNATIONAL	PZM-30GP	PZM Hemi.	Electret Condenser	Alum.	Piano/ General Orch.	150B	20-20	70	Swcft. A3M			6½	W	359.00	
	PZM-31S	PZM Hemi.	Electret Condenser	Alum.		150B	20-20	72	Swcft. A3M			6½	W	359.00	

	PZM-6LP	PZM Hemi.	Electret Condenser	Alum.	Conf.	150B	20-20	70	Swcft. A3M	15	Swcft. A3M	5	W	359.00	
	PZM-6S	PZM Hemi.	Electret Condenser	Alum.	Orch.	150B	20-20	72	Swcft. A3M	15	Swcft. A3M	5	W	359.00	
	PZM-20RMG	PZM Hemi.	Electret Condenser	Alum.	Conf.	150B	20-20	70	Swcft. A3M			6½		299.00	
	PZM2.5	PZM Hemi.	Electret Condenser	Alum.	†	150B	20-12	64	Swcft. A3M			61		359.00	†Stage floor, lectern.
	PZM-3LVR	PZM Hemi.	Electret Condenser	Plastic		150B	20-15	70		10	Swcft. TA4F	½		329.00	Redundant lavalier.
	PZM-3LV	PZM Hemi.	Electret Condenser	Plastic		150B	20-15	70		15	Swcft. TA4F	½		239.00	Lavalier.
	PZM-2LV	PZM Hemi.	Electret Condenser	Alum.	Stage Props	150B	20-20	70		15	Swcft. A3M	1		269.00	
	PZM-12SP	PZM Hemi.	Electret Condenser	Nylon	General	150B	20-20	70	Swcft. A3M			2	W	259.00	
	PZM-180	PZM Hemi.	Electret Condenser	Nylon	General	150B	20-15	70	Swcft. A3M			2	W	160.00	
	Sound Grabber	PZM Hemi.	Electret Condenser	Nylon	Conf.	1.6k	40-15	55		10	Mini	2	W	99.00	
ELECTRO-VOICE	RE20	Card.	Dynamic	Steel	Music/Voice	50B 150B 250	45-18	57†	A3F	15	None	26	WF	484.50	†0 dB = 1 mW per 10 dynes/cm^2.
	RE18	Super Card.	Dynamic	Steel	Music/Voice	150B	80-15	57†	A3F	15	None	8	W	279.25	
	RE16	Super Card.	Dynamic	Steel	Voice	150B	80-15	56†	A3F	15	None	8	WF	269.50	
	RE15	Super Card.	Dynamic	Steel	Voice	150B	80-15	56†	A3F	15	None	6	WF	256.75	
	RE11	Super Card.	Dynamic	Steel	Voice	150B	90-13	56†	A3F	15	None	8	WF	179.50	
	RE10	Super Card.	Dynamic	Steel	Voice	150B	90-13	56†	A3F	15	None	6	WF	166.50	
	DS35	Card.	Dynamic	Steel	Voice/Music	150B	60-17	60†	A3F	15	None	9.2	WF	159.00	Bass-boost proximity effect.
	PL77AA	Card.	Condenser	Zinc & Alum.	Voice/Music	150B	60-17	60†	A3F	15	None	12	W	183.75	Phantom and/or battery power, pop filter.
	CS15P	Card.	Condenser	Steel	Voice/Music	150B	40-18	45†	A3F	15	None	8	W	263.00	
	CO15P	Omni	Condenser	Steel	Music	150B	20-20	45†	A3F	15	None	7.5	W	283.25	Phantom powered.
	RE55	Omni	Dynamic	Steel	Music	150B	40-20	57†	A3F	15	None	8.5	W	259.00	
	DO54	Omni	Dynamic	Steel	Music	150B	50-18	58†	A3F	15	None	6.5	W	150.00	
	DO56	Omni	Dynamic	Steel & Alum.	Voice	150B	80-18	61†	A3F	15	None	6.5	W	125.00	Integral shock mount.
	DO56L	Omni	Dynamic	Steel & Alum.	Voice	150B	80-18	61†	A3F	15	None	5.5	W	141.00	As above, long handle.
	RE50	Omni	Dynamic	Alum.	Voice	150B	80-15	55†	A3F	15	None	9.5	W	156.00	Integral shock mount.
	635A	Omni	Dynamic	Steel	Voice	150B	80-13	55†	A3F	15	None	6	W	95.50	
	DL42	Super Card.	Dynamic	Alum. &	Voice	150B	50-12	50†	A3F	1	A3M	27	W	525.00	Shotgun, shock mount.

MANUFACTURER	Model	Directional Pattern Plus Channels: Stereo = x2, Quad = x4, M-S = M-S	Operating Principle	Case Material	Most Common Use	Actual Impedance, 1 kHz, Ohms; Balanced = B	Operating Range, Hz to kHz	Open-Circuit Sensitivity, 1 kHz, −dB, re: 1 Volt/1 μbar	Mike Connector Type	Cable Length, Feet	Connector Type on Equipment End of Cable	Weight, Ounces	Wind Screen = W, On/Off Switch = S, Switched Low or Proximity Filter = F	Price, $	Notes
	667A	Card.	Dynamic	Steel Alum.	Voice	50B/150B/250B		51†	A3F	2	A3M	24	WF	525.00	Boom mount, selectable patterns.
	CO94	Omni	Condenser	Brass & Alum.	Voice	150B	80-15	45†	A3F	15	A3M	0.7	W	231.75	Lavalier.
	CO90	Omni	Condenser	Brass & Alum.	Voice	150B	40-15	57†	A3F	15	None	0.7	W	145.00	Lavalier, battery powered.
	CO90P	Omni	Condenser	Brass & Alum.	Voice	150B	40-15	57†	A3F	15	None	0.7	W	176.50	Lavalier, phantom powered.
	CO90E	Omni	Condenser	Brass & Alum.	Voice	150B	40-15	57†	A3F	15	None	0.7	W	98.00	Lavalier, for wireless.
	RE85	Omni	Dynamic	Steel	Voice	150B	90-10	61†	A3F	15	None	8	W	145.50	
	649B	Omni	Dynamic	Alum.	Voice	150B	80-10	61†	A3F	15	None	1.1	W	132.00	
	PL80	Super Card.	Dynamic	Zinc & Alum.	Voice	150B	60-17	56†	A3F	0	None	12.5	W	216.00	
	PL91A	Card.	Dynamic	Zinc	Voice	150B	60-15	59.5†	A3F	0	None	8	W	132.00	
	PL95A	Card.	Dynamic	Steel	Voice	150B	60-17	60†	A3F	0	None	9.2	W	180.00	
	PL76B	Card.	Condenser	Zinc & Alum.	Voice	150B	50-20	55†	A3F	0	None	12	WS	177.00	Battery powered.
	PL77B	Card.	Condenser	Zinc & Alum.	Voice	150B	50-20	50†	A3F	0	None	12	WF	210.00	Battery or phantom powered.
	PL5	Omni	Dynamic	Steel	Music	150B	80-13	55†	A3F	0	None	6	W	110.00	
	PL6	Super Card.	Dynamic	Zinc	Music	150B	90-13	56†	A3F	0	None	10.5	W	119.00	
	PL9	Omni	Dynamic	Steel	Music	150B	50-18	58†	A3F	0	None	6.5	W	169.00	
	PL11	Super Card.	Dynamic	Steel	Music	150B	90-13	56†	A3F	0	None	6	W	204.00	
	PL20	Card.	Dynamic	Steel	Voice/Music	50B/150B/250B	45-18	57†	A3F	0	None	26	WF	570.00	
	681	Card.	Dynamic	Steel	Voice/Music	150B/Hi-Z	60-14	59.5†	A3F	15	None	8	W	135.00	
	644	Super Card.	Dynamic	Zinc & Brass	Voice/Music	150B/Hi-Z	40-12	53†	QC-4M	15	None	41	W	244.00	Shotgun.

	RE30	Omni	Condenser	†	Voice	200B	40-15	54†	A3F	15	None	11.7	WS	400.00	†Various; special ENG/EFP with limiter.
	RE34	Card.	Condenser	†	Voice	200B	40-15	54†	A3F	15	None	11.8	WS	400.00	As above.
GC ELECTRONICS	30-2373	Uni	Dynamic	Alum.		30k	50-17	58	2-Pin Screw	16.5	Phone		S	39.95	Lavalier strap.
	30-2374	Uni	Dynamic	Alum.		500/50k	80-15	72/52	4-Pin Screw	20	Phone		S	27.95	
	30-2376	Uni	Dynamic	Alum.		500	100-13	85	2-Pin Screw	15	Phone		S	37.95	
	30-2372	Uni	Dynamic	Alum.		200	60-15	75	3-Pin	20	Phone		S	70.00	
	30-2378	Uni	Electret Condenser	Alum.		600	30-16	68	Auached	20	None		WS	30.95	Built-in preamp.
	30-2382	Uni x2	Electret Condenser	Alum.		600	50-16	68	Attached	9.9	None		S	43.00	
	30-2398	Cmni	Electret Condenser	Alum.		600	50-16	65	Attached	20	Phone		WS	23.95	As above.
	30-2388	Omni	Dynamic	Alum.		250/50k	100-10	78/60	4-Pin Screw	15	Phone		S	38.00	
	30-2300	Omni	Dynamic	Plastic		200	100-10	70	Attached	4	Micro, Mini		S	6.75	
	30-2302	Omni	Dynamic	Plastic		30k	50-13	60	Attached	4.5	Micro, Mini		S	10.35	
	30-2308	Omni x2	Dynamic	Plastic		500	100-10	74	Attached	4.3	Phone		WS	21.40	
	30-2384	Omni	Electret Condenser	Alum.		1k	50-16	63	Attached	13.2	Mini			20.95	Lapel style.
	30-2383	Omni	Dynamic	Alum.		30k	70-12	57	Attached	16.5	Mini		S	16.95	Lavalier.
JVC	M-201	Card. x2	Condenser	Alum.	Hall	600	40-18		(2) Phones	6			WS	59.95	
MARANTZ	EC-1	Omni	Electret Condenser			2k	60-13	164		10	Mini	3.5		18.00	
	EC-3	Card.	Electret Condenser			1.5k	50-15	164		10	Mini	8.8		28.00	
	EC-5	Card.	Electret Condenser			2.2k	40-15	164		10	Mini	4.1		42.00	
	EC-7	Card.	Electret Condenser			250B	40-16	164		10	Phone	10.3		64.00	
	EC-9P	Card.	Electret Condenser			250B	30-17	177	XLR	10		13.8		110.00	
	EC-12B	Omni	Electret Condenser			250B	100-15	164		10	Mini	2.3		54.00	
	EC-15P	Omni	Electret Condenser			250B	70-16	164		15	XLR	1		100.00	
	EC-33S	Card. x 2	Electret Condenser			1k	50-15	145		10	Mini	6.2		66.00	

MANUFACTURER	Model	Directional Pattern Plus Channels: Stereo = x2, Quad = x4, M-S = M-S	Operating Principle	Case Material	Most Common Use	Actual Impedance, 1 kHz, Ohms; Balanced = B	Operating Range, Hz to kHz	Open-Circuit Sensitivity, 1 kHz, −dB, re: 1 Volt/1 μbar	Mike Connector Type	Cable Length, Feet	Connector Type on Equipment End of Cable	Weight, Ounces	Wind Screen = W, On-Off Switch = S, Switched Low or Proximity Filter = F	Price, $	Notes
NAKAMICHI	DM-1000	Card.	Dynamic	Alum.	General	250B	30-18	76		16½	Phone		WSF	300.00	
	DM-500	Card.	Dynamic	Alum.	General	200B	50-15	73		16½	Phone		W	100.00	
	CM-300	Card. or Omni	Electret Condenser	Alum.	General	200B	30-18	76		16½	Phone		WSF	170.00	
	CM-100	Card.	Electret Condenser	Alum.	General	200B	30-18	76		16½	Phone		WSF	110.00	
NEUMANN	KM 83	Omni	Condenser	Alum.		200	40-20	7†	A3M	0	A3F	3	W	349.00	†mV/Pa (1 Pa = 94 dB SPL).
	KM 84	Card.	Condenser	Alum.		200	40-20	10†	A3M	0	A3F	3	W	349.00	
	KM 85	Card.	Condenser	Alum.		200	40-20	9†	A3M	0	A3F	3	W	349.00	
	U 89	Sel.	Condenser	Alum.	Studio	150	40-18	8†	A3M	25	A3F	14	WF	1048.00	
	TLM 170	Sel.	Condenser	Alum.	Studio	150	40-18	8†	A3M	25	A3F	22	WF	1148.00	Transformerless.
	USM 69fet	Sel. x2	Condenser	Alum.	Stereo Music	150	40-16	10†	A5M	33	A3F	16		3305.00	
NUMARK	UD885	Card.	Dynamic	Alum.		500	60-12			10	Phone		WS	26.95	With mini adaptor.
	UD9100	Card.	Dynamic	Alum.		600	50-12			10	Phone		WS	29.95	
	UD9100S	Card. x2	Dynamic	Alum.		600	50-12			10	Phone		WS	60.00	
	UD9200	Card.	Dynamic	Alum.		600	50-12		XLR	10	Phone		WS	37.25	
	UD9500	Card.	Dynamic	Alum.		600/50k	60-12			10	Phone		W	33.50	
	UD925	Card.	Dynamic	Alum.		600	60-15		XLR	10	Phone		WS	58.00	
	STD272	Omni	Electret Condenser	Alum.		600	30-16			15	Phone		S	69.95	Calibrated measurement standard.
	UC935	Card.	Electret Condenser	Alum.		600	30-16		XLR	10	Phone		WS	69.95	
	TC995	Omni	Electret Condenser	Alum.		800	50-16			10	Phone			49.95	Lavalier.
	UD940	Card.	Dynamic	Alum.		250B	45-16		XLR	10	Phone		WS	90.00	
PIONEER	DM-61		Dynamic		Vocal	600	80-12	75	Cannon	16.4	Phone	8.6	S	129.95	

	DM-51		Dynamic		Vocal	600	80-14	72	Cannon	16.4	Phone	5.4	S	99.95	
	DM-21		Dynamic		Vocal	500	100-15	75		16.4	Phone	5.6	S	29.95	
QUASAR	KT585SE													5.95	
REALISTIC	33-919	Card. x 2	Condenser	Alum.	Vocal	600	30-15		Attached	10	Phone		S	39.95	
	33-1066	Card. x 2	Condenser	Plastic	Vocal	600	50-18		Attached	8	Mini		WS	29.95	
	33-1065	Card. x 2	Condenser	Plastic	Vocal	600	50-15		Attached	6.5	Mini		WS	19.95	
	33-984	Card.	Dynamic	Alum.	Vocal	600	80-15		Attached	16	Phone		S	49.95	
	33-1070	Omni	Dynamic	Alum.	Vocal	600	40-17		Attached	16	Phone			39.95	
	33-1071	Card.	Dynamic	Alum.	Vocal	600	50-15		Attached	12	Phone		S	29.95	
	33-992	Card.	Dynamic	Alum.	Vocal	600	80-12		Attached	6	Phone		S	24.95	
	33-985	Omni	Dynamic	Alum.	Vocal	600	50-13		Attached	6	Phone		S	19.95	
	33-986	Uni	Dynamic	Plastic	Vocal	600	80-15		Attached	9.8	Phone		S	14.95	
	33-1089	Omni	Condenser	Plastic	Vocal	1k	20-18			5	Mini		S	12.95	
	33-1090	Omni	Condenser	Alum.	Vocal	600	20-18		Attached	18	Phone		WS	39.95	
	33-1056	Omni	Condenser	Alum.	Vocal	600	30-12		Attached	10	Phone		S	19.95	Tie clip.
	33-1052	Omni	Condenser	Alum.	Vocal	600	50-15			5	Mini		S	12.95	As above.
	33-1062	Card.	Condenser	Alum	Vocal	600	80-12		Attached	18	Mini		WS	49.95	For video cameras.
	33-1050	Omni	Condenser	Alum.	Vocal	600	20-13		Attached	9	Phone		WS	17.95	
	33-990	Omni	Dynamic	Alum.	Vocal	10k	150-10		Attached	6				15.95	Lavalier with neck cord and mike stand.
SANSUI	D-M7	Uni	Ribbon		Vocal/ Music	600B				20				80.00	
	D-M5	Uni	Ribbon		Vocal/ Music	500B				17				35.00	
	D-M3	Dynamic	Ribbon		Vocal/ Music	500B				10				20.00	
	E-M5	x2	Ribbon		Vocal	1000B				7				37.00	Lavalier.
SCHOEPS SCHALLTECHNIK	CMC 32U	Omni	Condenser	Nickel/ Brass	Orch.	20B	20-20		XLR-3M			3		640.00	
	CMC 34U	Card.	Condenser	Nickel/ Brass	Orch.	20B	40-20		XLR-3M			3		640.00	
	CMC 35U	Card./Omni	Condenser	Nickel/ Brass	Orch.	20B	20-20		XLR-3M			3		835.00	
	CMC 36U	Card./Omni/ Bi	Condenser	Nickel/ Brass	Orch.	20B	20-20		XLR-3M			3		985.00	
	CMC 38U	Figure 8	Condenser	Nickel/ Brass	Orch.	20B	40-16		XLR-3M			3		780.00	
	CMC 341U	Hyper Card.	Condenser	Nickel/ Brass	Film/TV	20B	40-20		XLR-3M			3		730.00	
	BLM-33U	Hemi.	Condenser	Alum.	Orch:	20B	20-20		XLR-3M			23		790.00	
	CMH 34U	Card.	Condenser	†	Vocal	20B	60-20		XLR-3M			7	WF	835.00	†Black anodized brass.

MANUFACTURER	Model	Directional Pattern Plus Channels: Stereo = x2, Quad = x4, M-S = M-S	Operating Principle	Case Material	Most Common Use	Actual Impedance, 1 kHz, Ohms; Balanced = B	Operating Range, Hz to kHz	Open-Circuit Sensitivity, 1 kHz, −dB, re: 1 Volt/1 μbar	Mike Connector Type	Cable Length, Feet	Connector Type on Equipment End of Cable	Weight, Ounces	Wind Screen = W, On/Off Switch = S, Switched Low or Proximity Filter = F	Price, $	Notes
	CMTS 301U	Card./Omni/ Bi x2	Condenser	Nickel Brass	Orch.	20B	40-16		XLR-5			12½		2125.00	
SENNHEISER	MD 200	Omni	†	PVC	General	600	60-13		Phone	10		3.7		33.00	†Pressure transducer.
	MD 400	Card.	†	PVC	Record	600	60-13		Phone	10		3.7		41.00	
	MD 402 U	Super Card.	Dynamic	Metal	Record	200B	80-12.5	151	XLR	15		5.4		85.00	
	MD 402 K	Super Card.	Dynamic	Metal	Record	200	80-12.5	151	Phone	10		6.7		80.00	
	MD 421	Card.	Dynamic	Plastic	Voice	200	30-17	146	3-Pin XLR			18		332.00	
SHURE	SM11-CN	Omni	Dynamic	Alum.	Voice	150B	50-15	85	None	4	3-Pin Male	0.3†		98.00	†Without cable; lavalier.
	SM57LC	Card.	Dynamic	Alum.	Instr.	150B	40-15	75.5	3-Pin Male			10		127.75	
	SM58LC	Card.	Dynamic	Alum.	Vocal	150B	50-15	75.5	3-Pin Male			10.5		164.75	
	SM81LC	Card.	Condenser	Steel	Instr./ Vocal	150B	20-20	65	3-Pin Male			8	WF	336.75	
	SM83CN	Omni	Condenser	Brass	Voice	150B	80-20	69	None	10	3-Pin Mini	1.6	WF	210.00	Lavalier.
	SM85LC	Card.	Condenser	Alum.	Vocal	150B	50-15	74	3-Pin Male			6.3		251.25	
	SM87LC	Super Card.	Condenser	Alum.	Vocal	150B	50-18	74	3-Pin Male			6.3		329.00	
	SM80LC	Omni	Condenser	Steel	Instr./ Vocal	150B	20-20	65	3-Pin Male			8	WF	336.75	
	515SAC	Card.	Dynamic	Zinc	Voice/ Music	Hi-Z	80-13	59	None	15	Phone	18	S	56.75	
	545L	Card.	Dynamic	Alum.	Voice	150	50-15	77.5	None	20	None	12½	S	100.50	As above.
	545SDLC	Card.	Dynamic	Zinc	Voice Instr.	150	50-15	78	3-Pin Male			9	S	109.50	
	565SDLC	Card.	Dynamic	Zinc	Vocal	150	50-15	76	3-Pin Male			10½	S	121.50	
	588SBLC	Card.	Dynamic	Zinc	Voice Vocal Music	150	80-13	82	3-Pin Male			12	S	64.25	
	586SBLC	Card	Dynamic	Zinc	Voice Music	150	50-13	56	3-Pin Male				S	94.25	

	579SB 578	Omni Omni	Dynamic Dynamic	Zinc Steel	Voice Voice/ Music	150 150	50-14 50-15	78 80	3-Pin Male None	20 15	None	16	S	99.50 113.75	
SIGNET	RK-201	Card.	Electret Condenser	Alum.	Music	600	45-17.5	64	Attached	16½	Phone	6½	WS		
SONY	F-V30T	Card.	Dynamic	Alum.	Vocal	200	80-12	74	Attached	16	Mini/ Phone	6.2		29.95	
	F-V50T	Card.	Dynamic	Alum.	Vocal/ Music	200	80-15	75	Attached	16	Mini/ Phone	9.4		44.95	
	F-V6ET	Omni	Dynamic	Alum.	Music	200	100-12	73	Attached	16	Mini/ Phone	4.9		54.95	Variable echo.
	F-99T	Card. x2	Dynamic	Alum.	Music	200	80-12	65	Attached	5	Mini/ Phone x2	6.4		39.95	
	ECM-220T	Card.	Electret Condenser	Alum.	Instr.	200	50-14	65	Attached	16	Mini/ Phone	8.3		49.95	
	ECM-23FM	Card.	Electret Condenser	Alum.	Studio	250B	20-20	74	Attached	16	Phone	6.7	W†	115.00	
	ECM-929LT	M-S	Electret Condenser	Alum.	Stereo/ Music	200	70-15	67	Attached	6	Mini/ Phone	3.8		85.00	
	ECM-939LT	M-S	Electret Condenser	Alum.	Stereo/ Music	200	70-15	57	Attached	6	Mini/ Phone	2.6	W	115.00	
	F-V200	Uni	Dynamic	Alum.	Vocal	600B	70-15	74	XLR-3		None	9.2	WF	150.00	
SOUND-CRAFTSMEN	SAM II	Omni	Electret Condenser	Alum.	RTA	600	20-18	65†	Phone	15		3		69.00	†dBm.
SWINTEK	Mark 50A-dbs		†	Alum.			50-12		A3F			¼		580.00	†Wireless lavalier transmitter in high VHF band between 130 and 250 MHz.
TECHNICS	RP-V340 RP-V370	Card. Card.	Dynamic Dynamic	Alum. Alum.	Vocal Music/ General		100-10 40-12					9 12	W W	26.00 40.00	With adaptor. As above.
	RP-3500E	Card.	Electret Condenser	Alum.	Music		50-12						W	60.00	With desk tripod.
	RP-3215E	Card. x 2	Electret Condenser	Alum.	Music		50-10						W	60.00	
	RP-3545E	Card.	Electret Condenser	Alum.	Vocal		40-14						W	70.00	With adaptor.

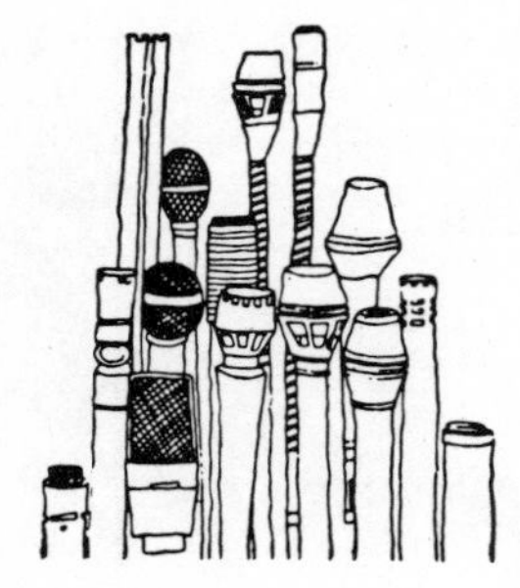

Glossary

absorption—The acceptance of sound energy by some material and its conversion to heat energy by that material.

absorption coefficient—Some materials absorb more sound energy than others. Absorption coefficient is a measure of that ability and is also dependent on frequency. Absorption coefficients range from 0% to 100%, but are supplied as decimals rather than percentages.

A-B time-intensity stereo—Left/right spaced microphone pair.

acoustic power output—The total sound power output of a device such as a speaker.

acoustics—The study of sound waves and their behavior indoors and outdoors. Acoustics also include a study of various materials and their effect on sound.

additive noise—The increase of noise level by 3 dB for each track laid down on tape.

ambient noise—The amount of *background noise* present during recording or playback.

amp—Abbreviation for *amplifier*. Unit can be pre-amp (preamplifier) or power amp (power amplifier). Amp is also an abbreviation for *ampere*, the basic unit of current.

amplifier—An electronic device for increasing the strength of an electrical signal applied to its input. The electrical signal is ac and is, or should be, an electrical replica of the sound supplied to the microphone. There are two basic types of audio amplifiers: the *preamp*, or voltage amplifier, and the *power* amp, an amplifier capable of delivering relatively large amounts of signal current to a speaker.

amplitude—The strength of an electrical wave of voltage or current or sound. The instantaneous amplitude is the value of a wave at a particular moment; peak amplitude is the maximum value.

anechoic chamber—Room with highly sound absorbent surfaces with all possible reverberant sound totally absorbed.

antinode—A region of sound reinforcement.

artificial reverb—A method of producing reverberation by vibrating springs, a reverberant chamber, by an endless magnetic tape or by analog or digital methods.

asperity—Same as modulation noise. Caused by non-homogenous particles on a magnetic tape as it moves past a head gap.

attenuation—The process of weakening a sound, voltage or current.

attenuator—It is possible to supply a preamp with an excessive amount of signal voltage from a microphone, a condition known as input overload. Some preamps have greater signal input handling capabilities than others. Overload can be controlled by inserting an attenuator, also known as a *loss pad*, or simply as a *pad*, between the microphone and the preamp. The attenuator reduces the amount of microphone signal input to the preamp.

axis—An imaginary line perpendicular to the head of a microphone. Signals that are directed at the head of the microphone are *on axis*. Those that reach the microphone at some angles are *off axis*.

balanced cable—Pair of wires surrounded by metallic shield braid used for connecting the output of a microphone to a preamplifier or to a mixer/preamp.

bar—Unit of sound pressure over a certain area. Corresponds to the dyne per square centimeter (dyne/cm^2).

bass reflex—Ported arrangement in a speaker in which sound energy at the rear of the speaker is permitted to exit from the front in phase with the sound exiting at the front of the speaker.

behind-the-signal noise—Also known as modulation noise or asperity. Caused by non-homogeneous magnetic particles of magnetic tape as it slides past a head gap.

bidirectional microphone—Also known as *figure-eight* microphone. Unit has receptivity pattern in the front and rear of the microphone with areas of sound rejection on the sides of the microphone. This characteristic is used for rather special recording cases, where rejection of 90° off-axis sounds is important.

bimorph—Dual piezoelectric crystalline structure. Pair of crystal plates cemented together.

black sound—Sound so weak in strength that it is completely inaudible.

boost—An increase in response at some section of a frequency-response curve. Bass boost is an increase in response to low-frequency tones. Treble boost is an increase in response to high-frequency tones. Boost can occur anywhere along the frequency-responsecurve of a component.

bottoms—Loudspeaker cabinet module with woofers. A woofer is a speaker for bass-range response.

cable—See *balanced cable* and *unbalanced cable.*

capsule—The elements of the acoustic transducer system in a microphone.

carbon microphone—Microphone containing a diaphragm resting against one or two buttons containing carbon granules. Sound pressure on the granules makes the button function as a variable resistor. Current is supplied to the button from a dc source, such as a battery. The resistance variation of carbon granules causes this current to change in step with sound reaching the diaphragm.

cardioid—A type of microphone that picks up sound better from front than back. A heart-shaped response pattern. This is the type of microphone characteristic most commonly used in both recording and live performances, where ambient noise should be suppressed.

ceramic microphone—Microphone that uses barium titanate as its transducing element. Substance has piezoelectric properties, producing a voltage when it is twisted, pressed, or squeezed. Sound exerts pressure on this element, producing the signal output voltage.

close micing—The positioning of a microphone very close to the lips of a vocalist or to the sound from an instrument so as to minimize extraneous noise pickup, the sounds of other instruments, or to produce desired audio effects.

condenser microphone—Microphone which depends on the ability of facing conducting surfaces to store an electric charge. Except in cases of electret microphones, condenser microphones require a polarizing voltage, a dc potential ranging from about 50 to 200 volts. The diaphragm is usually flat and can be a metallic substance or metallicized plastic. Separation between the conducting surfaces is about 1 mil (0.001 inch). The signal output is taken from the conducting surfaces and brought directly to an amplifier, generally a part of the microphone.

The condenser microphone can contain one or two diaphragms and a base plate. Variations in the distance between the diaphragm (or diaphragms) and the base plate produces changes in capacitance, resulting in a corresponding variation in output voltage.

crystal microphone—Microphone using Rochelle salt as the transducer. Rochelle salt has piezoelectric properties, producing a voltage when it is subjected to some sort of strain.

cycle—A single complete waveform of sound, voltage or current. The second is the basic unit of time associated with a cycle. 60 cycles usually means 60 cycles per second. In this example one complete cycle requires 1/60 second.

decibel—Abbreviated as dB, this is a unit indicating a ratio between two voltages, currents, or powers. Two to three decibels is usually the smallest change in sound level which can be detected by the human ear. An increase of 6 dB means the sound pressure level has doubled.

A sound pressure level (SPL) of 0 dB is the threshold of human hearing. It is the lowest SPL of which you can be aware.

diaphragm—Moving element of a microphone. It may be flat, as in the case of condenser microphones, or sometimes convex, as in dynamic microphones.

differential microphone—Also known as a *noise-cancelling microphone*, especially designed for speech communications in high-level noise environments.

diffraction—When sound encounters an object in its path it will bend around it, an effect that takes place more easily with low frequencies. Diffraction is sometimes called scattering.

diffused sound—A sound field in which the intensity of the sound is about the same in all parts of the room.

directional characteristics (polar response)—Indicates the microphone's sensitivity to sound pressure from every angle of incidence.

directivity—Area or areas of response of a microphone.

double-element microphone—Also called a two-way or coaxial microphone, it uses two microphone elements instead of one, with each designed to cover a particular band of sound frequencies.

dry sound—Primary sound without reverberation. Sound arriving directly from a source, without reflections.

dynamic range—The sonic level between a very soft and a very loud sound.

dynamic microphone—A microphone which contains a coil of wire positioned between the poles of a permanent magnet, with the coil fastened to a diaphragm. A moving-coil microphone.

dyne—The *dyne per square centimeter* is a unit of sound pressure.

echo—A reverberant sound which can be distinguished aurally from a direct sound. A sound whose time displacement from a dry sound enables the ear to recognize it as a distinct, separate sound.

echo chamber—A room completely free of any sound absorbing substances. A room having optimum sound reflectivity.

eigentones—Room resonances.

electret microphone—A condenser microphone using a capacitor-like device, an *electret*, which is precharged by the manufacturer. This eliminates the need for a polarizing voltage, a requirement of ordinary condenser microphones.

electrical noise—Noise produced as a result of the movement of electrons through a conductor, such as wire, or through some part such as a transistor.

element—See *diaphragm.*

equal loudness contours—Also known as Fletcher-Munson characteristics, these are a set of curves of various levels of loudness throughout the audio range from 20 Hz to 20 kHz. These curves indicate how much additional sound power is needed at various frequencies to get sounds of equal loudness. These curves show that our ears are not as good for low and high sound frequencies as they are for frequencies between 3 kHz and 4 kHz.

far field—A sound field sufficiently removed from a sound source so that the sound waves have become parallel.

feedback—Feedback consists of an electrical or audible signal fed back from some output to an input. The sound produced by a speaker is an output, that going into a microphone is an input. When sound from a speaker is fed back at a high enough level, in phase, to a microphone, the result can be ringing or howling.

figure-eight—Bilateral frequency response. A microphone response pattern resembling the digit 8.

Fletcher-Munson curves—Equivalent loudness curves. These show that more audio power is needed in bass and treble than in midrange to get sounds of equal loudness for the human ear.

frequency—Frequency refers to the number of vibrations of sound per second or the number of complete cycles per second of an electrical wave. The number of vibrations per second constitutes the *pitch* of a tone. Frequency is measured in cycles per second (cps) or in Hertz, abbreviated as Hz. The Hertz corresponds to the cycle per second. —Hz is 60 Hertz or 60 cps.

frequency limit—The upper and lower limits of a band of frequencies. Generally taken to mean a 3 dB decrease in signal level.

frequency response—The way in which an electronic device, such as a microphone, amplifier, or speaker responds to signals having a varying frequency. A flat frequency response indicates that the device or component handles signals of all frequencies in the same way, without favoritism. Some components boost certain frequencies or do not respond well to all frequencies.

front-to-back discrimination—Comparison of sound picked up by the front of a microphone compared to that picked up at the rear of that microphone. In the case of a cardioid, the front-to-back discrimination can be as much as 20 dB, indicating that the rear of the microphone is 20 dB less sensitive to pickup of sound than the front of the microphone. This refers to results obtained in an anechoic chamber only.

fuzz box—An amplifier, ordinarily used with an instrument such as an electric guitar, for the production of sounds other than those supplied by the guitar. A distortion-type amplifier.

gain—*Amplification*. An increase in strength or amplitude of a signal. The ratio of output signal level to input signal level.

gain control—Variable element for adjusting the amount of gain, most often a variable resistor. A component for controlling the signal level of a device.

gobo—A heavy frame or frames used to surround a performer to minimize sound pickup from other musicians.

Haas effect—Same as precedence effect. The time it takes for us to become aware of the location of a sound source after it has moved.
harmonics—Waves that are multiples in frequency of a fundamental wave. A second harmonic has twice the frequency of its fundamental, a third harmonic has three times the frequency of its fundamental, and so on. Harmonics are also known as partials or overtones.
heads—Preamplifiers or mixers. A device that receives signals from microphones.
Hertz—A measure of frequency. Abbreviated as Hz. Cycles per second or cps.
hiss—Noise of random sound distributed through the mid- and high-frequency audio spectrum.
H-pad—An attenuator pad inserted in the line between the output of a microphone and input to an amplifier. So called because of its circuit resemblance to the letter H. A network composed of fixed resistors designed to introduce a predetermined reduction in signal strength, such as 10 dB, 20 dB, etc. H-pads are designed to not interfere with impedance matching between the microphone and amplifier.
hum—A cyclical type of noise whose frequency is either 60 Hz or 120 Hz.
hum bucking coil—A coil used in a microphone to counteract the effects of hum induced voltages, decreasing hum pickup in the order of 25 dB.
hypercardioid—A cardioid-type of microphone which has a narrow segment of response toward the rear, but with 90° off-axis response more attenuated than that of the cardioid.

impact noise—A sound vibration produced by one object striking another. Hand clapping is an impact noise; so are footsteps.
impedance—The opposition of a component, such as a microphone, speaker, or input or output of an amplifier, to the flow of current. Microphones are designated as either low- or high-impedance types. Impedance is measured in ohms and is symbolized by the letter Z. Low impedance or low Z is in a range of approximately 50 to 250 ohms; high impedance or high Z usually means an impedance of 10,000 ohms or more.
input overload distortion—Distortion produced by supplying too strong a signal from the output of a microphone to a preamp. Manufacturers of amplifiers indicate in their specification (spec) sheets the maximum amount of signal the preamplifier can accept before overloading. An excessively strong signal going into a preamp cannot be controlled or reduced by the gain control on the preamp, since such controls are located circuitwise following the input amplifier.
inverse square law—A law which describes how *microphone* output (or *loudspeaker* output) decreases as the listener moves away from the *micro-*

phone (or *loudspeaker*). The law states that the *sound pressure* will drop 6 dB every time the distance is doubled. Strictly speaking, this law applies only to a situation where no reverberation is present, as in a large field, out-of-doors, Indoors, the *sound pressure* drop with distance eventually ceases. However, the law still holds true for any normal *microphone working-distance*, and is also true for distances from *loudspeakers* up to about 10 feet. The precise distance is dependent on room acoustics, increasing for "dead" rooms and decreasing for highly reverberant rooms.

Kelvin temperature scale—Degress Celsius plus 273.18.

lavalier microphone—Microphone designed to be worn around the neck or attached via a clip to a shirt or dress.

line-matching transformer—Also known as an *impedance-matching transformer,* or just an *impedance transformer*. It is used to change the impedance of a device from *high* to *low*, or vice versa. Line-matching transformers can be built into the microphone or may be external to it. They can be used to change the output impedance of a microphone from low to high or high to low.

loudspeaker—Transducer, used for changing electrical energy to sound energy.

loudness—The subjective impression of the strength of a sound. It is the amount of sound perceived by the ears. Loudness is measured in *sones.* One sone is the subjective loudness of a 1 kHz note whose SPL is 40 dB above a standard reference level.

masking effect—The ability of high level sounds to cover or hide low level sounds, possibly hiss or other noise.

mechanical shock sensitivity—The measure of a microphone's reaction to handling or bumping. A microphone with high shock sensitivity will readily transmit the effects of cable friction, ring clankings, and finger thumpings to the sound system. Microphones vary widely in their shock sensitivity.

mel—The subjective unit of pitch. (See *pitch*). One thousand (1000) mels is the subjective perception of a 1 kHz tone having an SPL of 40 dB above a reference level of 0 dB, or the threshold of hearing.

microphone—A transducer for changing sound energy to electrical energy.

microphone noise—Noise produced by a microphone independently of all other sounds reaching it. A self-generated voltage.

microvolts per micro tesla—Unit of hum sensitivity. 5 μT $\geqq$ 50 mG = related hum field μV/μT.

millibar—Thousandth of a bar, a unit of sound pressure.

millivolts per microbar—10 mV/Pa = output voltage per rated sound pressure of 1 microbar (1 μb $\geqq$ 0.1 Pa) mV/μb.

millivolts per Pascal—mV/N/m^2 = output voltage per rated sound pressure of 1 Pa (1Pa ≧ μb) mV/Pa.

mixer—An electronic component that will accept a number of different signal inputs from microphones, combining them into a single electrical output. Since each of the input signals from the various microphones is adjustable by a control, or a group of controls, on the mixer, the relationship of a single sound to all the other sounds can be adjusted.

modulation noise—Also known as asperity or behind-the-signal noise. Caused by non-homogenous magnetic particles and by vibrations across the tape as it slides past a head gap.

monophonic sound system (mono)—A system in which all sound is carried by one channel or path. The system may use a number of microphones, one or more pre- and power amps, and possibly more than one speaker. However, at some point, all the inputs from the various microphones are combined into a single channel. If two or more speakers are used, the sound is divided between them, but all the speakers reproduce the same sound.

moving-coil microphone—See *dynamic microphone.*

noise cancelling microphone—See *differential microphone.*

node—Region of sound cancellation in a room.

noise—Random sound containing frequencies not harmonically related.

octave—The frequency range between one frequency and another having twice its value. Thus, the range between 125 Hz and 250 Hz is an octave; so is the range between 250 Hz and 500 Hz. Each is often referred to as an octave range. An octave band is sometimes identified by its center frequency.

ohm—Basic unit of resistance or impedance. Multiples are the kilohm or a thousand ohms and the megohm, a million ohms.

omni—Abbreviation for an omnidirectional microphone.

omnidirectional—Uniformly responsive to an entire sound field at a particular frequency or a band of frequencies. A microphone capable of responding to sound from all directions. These microphones should only be used when the ambience is an essential part of the recording or where reverberation or acoustical feedback is no problem.

overtones—See *harmonics.*

pad—See *attenuator.*

partials—See *harmonics.*

phasing or phase—The time relationships, usually of a pair of waves. Two waves, starting at the same time, increasing and decreasing at the same time, and terminating at the same time, are *in phase*. The two waves need not have the same amplitudes. Waves that do not start or stop at the same time are *out of phase*. Waves can be partially or completely out of phase.

phon—A unit used for measuring the loudness of a pure tone.

piezoelectric—A type of transducer. A substance, usually crystalline, that produces a voltage when subjected to stress, such as bending or twisting.

pink noise—A noise that has a 3 dB/octave slope from 20 Hz to 20 kHz.

pitch—The fundamental or basic tone of a sound determined by its frequency.

polar pattern—Areas of response of a microphone graphed in polar (circular) form. The main patterns are omnidirectional, bidirectional or figure-eight, and unidirectional (cardioid, hypercardioid, and supercardioid).

polarizing potential—Voltage (EMF) for a condenser microphone for establishing a charge on the plates of the microphone. The dc voltage required for a condenser microphone.

pop filter—Shield positioned above the microphone diaphragm or around the exterior of the microphone to reduce the pressure level of certain vocal sounds which can cause a popping effect. The shield is acoustically transparent and does not interfere with the movement of sound toward the microphone diaphragm.

pop sensitivity—Vocal sounds vary, depending on how they are produced by the lips and tongue. Some, such as words starting with the letters P, F, or T, can result in a popping sound. *Popping* is a short-lived "boominess" in the bass end of the sound spectrum. Not all microphones have the same amount of pop sensitivity.

potential—Voltage or electromotive force (EMF).

preamp—Voltage amplifier connected to the signal output of a microphone. Usually solid-state, incorporating transistors for the amplification of the signal. Can be an individual unit or incorporated in a mixer.

precedence effect—Same as Haas effect. The time it takes for us to become aware of the location of a sound source after it has moved.

pressure gradient—Difference in acoustic pressure. Difference in sound pressure between the front and back of a microphone element.

pressure gradient receiver—In a unidirectional microphone both sides of the transducer diaphragm have to be exposed to sound pressure. Either the sum or the difference of the sound pressures (the pressure gradient) will actually move the diaphragm and in turn will produce an output voltage related to the sound pressure and angle of incidence.

proximity effect—The increase in low-frequency response in most cardioid and bidirectional microphones when the microphone is used close to a spherical wave sound source.

pure tone—A fundamental tone only, without harmonics. The tone produced by a tuning fork.

quadraphonic sound system—Sometimes called four-channel sound. At least two microphones are used to record left front and right front sound; two more microphones to pick up left rear and right rear sound.

reverberant sound—Sound that has been reflected from a surface.

reverberation time—The time it takes for reflected sound to decrease by 60 dB.

ribbon microphone—Microphone with a metallic ribbon suspended between the poles of a magnet. The movement of the ribbon in the magnetic field provided by the permanent magnet produces a voltage. A microphone functioning on the principle of electromagnetic induction, producing an EMF corresponding to amplitude changes of an incident sound wave. A velocity microphone.

rms—Abbreviation for root mean squared. The square root of the mean or average value of the squares of the instantaneous values taken over a complete cycle.

rolloff—A decrease in frequency response at the bass or low-frequency end of a response curve or at the treble or high-frequency end, or at both. Indicated in decibels per octave.

sabin—A unit of sound absorption obtained by multiplying the absorption coefficient of a material by its area.

scattering—See *diffraction*.

sensitivity—Response of a microphone to sound pressure. The output voltage of microphone given in dB referenced to 1 mW at an SPL of 10 dynes/cm^2 at a specific frequency or in mV/Pa (millivolts per Pascal).

shock mounts—Devices for subduing or eliminating the transfer of mechanical vibrations to the microphone diaphragm. Shock mounts can be external to the microphone or built into it.

shock sensitivity—Reaction of a microphone to handling. The movement of a microphone cable, jarring of a microphone housing, or accidentally hitting it against a hard object can result in microphone sound output. The higher the shock sensitivity of a microphone, the more readily it converts such mechanical shocks to sound.

sibilance—Hissing sound produced when a speaker uses words having S or Z. Sometimes caused by performer speaking directly into a microphone. Correct this condition by talking across the microphone instead of into it, or by using a windscreen.

sone—Subjective loudness of a 1 kHz tone at an SPL of 40 dB above the reference level.

sound pressure level—Abbreviated as SPL, it is the deviation above and below normal atmospheric pressure. Atmospheric pressure exists in the presence or absence of sound. Through vibration, voices or musical instruments cause variations in atmospheric pressure. When these variations reach our ears we interpret them as sound.

sound system—System for the reproduction of sound, using one or more microphones. A sound system generally includes, in addition to microphones, devices such as attenuators (also known as pads), preamplifiers, mixers, combined pre-amp/mixers, power amplifiers and speakers. Public sound systems can include other sound sources such as AM/FM broadcasts, phono records or recorded tape.

stage monitor—A *loudspeaker* so placed that it directs *sound* to the performer, since many good *sound systems* set up to adequately cover the audience do a poor job of letting the performer hear himself.

standing waves—Sound waves in a room that are in or out of phase. Regions in a room in which there are dead sound spots or very loud sound spots.

stereophonic sound system (stereo)—Sound system with two channels or sound paths. Channels are identified with respect to dry sound. Sounds produced from center to left of stage are called left channel; sounds from center to right are called right channel.

supercardioid microphone—A modified cardioid microphone with two lobes, front and rear, but with a rear lobe much smaller than the front. A microphone with an elongated front lobe.

thermal noise—Noise produced by temperature or by a temperature change.

threshold of hearing—Sound level at which sound becomes perceptible to the human ear.

tone compensation—A switchable filter used to change the performance characteristics of a microphone, emphasizing bass, midrange or treble.

tops—Loudspeaker cabinet modules with midrange or tweeters. Midrange is a speaker that reproduces midrange frequencies; a tweeter is a speaker that reproduces the treble range.

tranducer—Device for the conversion of one form of energy to another. All microphones are transducers. So are speakers, motors, and batteries.

transient—Wave having very short or no sustain time. Wave that starts and stops rapidly. Wave having very short lifetime compared to other waves.

transformer—See *line-matching transformer.*

unbalanced cable—Cable consisting of central conductor (wire) surrounded by shield braid.

unidirectional (cardioid, hypercardioid)—A microphone sensitive to sounds coming from the front, relatively insensitive to sounds from the rear. The shape of the polar pattern resembles a heart or cardioid. This is the type of microphone characteristic most commonly used in both recording and live performances, where ambient noise should be suppressed.

velocity microphone—See *ribbon microphone.*

weighted noise—Noise voltages sent through a weighting filter.

wet sound—Reverberant sound.

white noise—Noise that has a constant amplitude from 20 Hz to 20 kHz.

white sound—Any sound which is audible.

windscreen—Shield for protection against movement of wind or rapid transfer of the microphone from one place to another. Windscreens generally cover the microphone only partially. The screen only has to cover the end of an omni and is placed over any rear sound holes in the body for directional microphones. Only shotgun and slotted-in-body microphones need screens covering the entire microphone. Such types are rare. Head only screen services the majority of microphone designs.

working distance—Distance from the sound source to the diaphragm of the microphone. Sound pressure input to a microphone decreases by 6 dB when the distance from the performer to the microphone is doubled.

X-Y intensity stereo—Microphone pair mounted on a bar with heads almost touching.

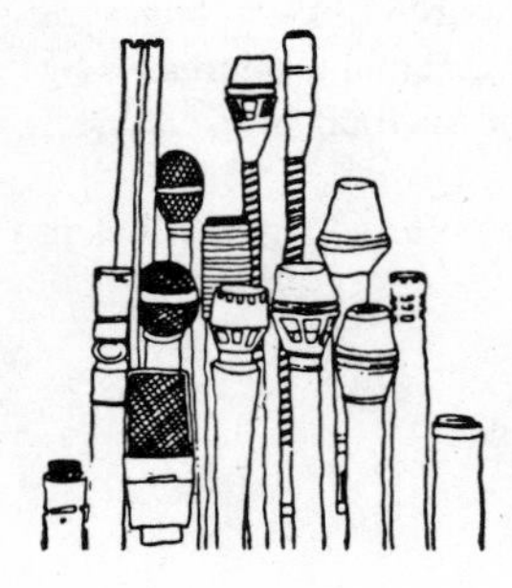

Index